# Applications of Solids Processing Unit Operations 2018



Topical at the 8th World Congress on Particle Technology

Orlando, Florida, USA
22 - 26 April 2018

ISBN: 978-1-5108-6969-1

**Printed from e-media with permission by:**

Curran Associates, Inc.
57 Morehouse Lane
Red Hook, NY 12571

**Some format issues inherent in the e-media version may also appear in this print version.**



Printed by Curran Associates, Inc. (2018)



AIChE
120 Wall Street, FL 23
New York, NY 10005-4020

Phone: (800) 242-4363
Fax: (203) 775-5177

www.aiche.org

**Additional copies of this publication are available from:**

Curran Associates, Inc.
57 Morehouse Lane
Red Hook, NY 12571 USA
Phone: 845-758-0400
Fax: 845-758-2633
Email: curran@proceedings.com
Web: www.proceedings.com

# TABLE OF CONTENTS

**Some abstracts in this book refer to images that were not made available by the publisher**

# 508388 An Empirical Comparison of Two Different Cyclone Designs in the Usage of a Third Stage Separator


**Abstract:**


## Background and Motivation

Actually the Third Stage Separator (TSS) in a FCC-process is mainly constructed as a standard reverse flow cyclone which can be designed by approved calculation models, as for example by using the model of *Barth/Muschelknautz* [1]. The FCC-process efficiency mainly depends on the cyclone performance i.e. on its separation efficiency and its pressure loss. The conical shape of the lower separation body of a reverse flow cyclone leads to high tangential velocities and therefore to high erosion which strongly affects its life time. A second cyclone type, which is not yet well investigated, is the uniflow cyclone. Its essential advantages compared to the reverse flow cyclone are its compact construction and its easy integration into pipes. A further advantage is its low pressure loss and therefore its low energy consumption. *Postma et al.* investigated the usage of uniflow cyclones (swirl tubes) in the FCC-Process, which requires high process reliability [2]. A study on the applicability of uniflow cyclones as a TSS was carried out in 2010 at PSRI laboratories and showed acceptable performance data for the usage within FCC-process engineering [3].

## Standard Reverse Flow Cyclones vs. Uniflow Cyclones

The simplicity of design and the lack of moving parts in cyclone separators are key to their ubiquitous presence in process industry. In some applications, where high gas flow rates must be processed, the physical size of a conventional reverse flow cyclonic unit can become very large. The uniflow cyclone or axial cyclone, is a interesting alternate design, which is able to deliver comparable efficiencies in a very compact geometry. The name is derived from the flow field pattern generated inside the cyclone. Unlike reverse flow cyclone, the gas enters and leaves the cyclone in the same direction (see Figure 1).

Fig. 1: Depiction of a uniflow cyclone and its flow field pattern.

Compared to a reverse flow cyclone design, the uniflow cyclones can be built approximately two-third smaller in volume at comparable efficiencies. Wide spread adoption of this design has been limited by the absence of a universally accepted and validated design approach. The existing literature mainly focuses on special applications, as shown in the research work of *Gauthier et al.* or *Zhang et al.* [4,5]. The highly turbulent particle flow patterns offer a complex parameter system which makes the dimensional analysis rather complicated. For instance, numerous design variations of the inlet swirl generator (Figure 1) are possible, and each approach will result in remarkably different flow field within the cyclone. Therefore, an empirical approach is most apt for studying uniflow cyclones.

To facilitate the understanding of its operation, the design structure of a uniflow cyclone can be divided into four major sections in the direction of the flow, namely Swirl Vane Inlet, Separation Chamber, Vortex Finder and Particle Outlet.

## State of the art

Even though the working principle of this cyclone type is well-known, there is hardly any literature which investigates the design, the estimated efficiency or pressure drop. To close this gap the MCI is driving systematic research activities since 2008 to improve the separation efficiency of uniflow cyclones by optimizing the geometric design [6,7,8,9]. As outcome of these activities, there are several applications established in process and automotive industry, as for example in air-pre-cleaning systems for internal combustion engines used in off-road vehicles (e.g. building and agricultural engines).

**Experimental**

To compare the two cyclone systems on a fair basis, the executed test series show data for efficiency and pressure drop and the erosion potential at comparable operating conditions. The experimental testing is performed on a standard reverse flow cyclone and a uniflow cyclone with various operation configurations differing from each other regarding the vortex finder diameter, the volume flow rate and pressure recovering systems. A test stand which allows comparing effectiveness and optical accessibility for phenomenological observations is set up at MCI laboratories, by using selected FCC-Powder which typically reaches the Third Stage Separator.

References:

[1] *VDI Heat Atlas,* 11th Edition. VDI-Book. Berlin [u.a.]: Springer Vieweg, 2013.

[2] Postma, R.S., Hofmann, A.C., Dries, H.W.A., Williams, C.P., The Use of Swirl Tubes for Dedusting, World Congress on Particle Technology 3, Brighton, UK, 1998.

[3] Kraxner, M., Muschelknautz, U., Karri, S.B.R., Cocco, R., Knowlton, T.M., Applicability of a Uniflow Cyclone as a Third Stage Separator in the FCC-Process, AIChE - American Institute of Chemical Engineers - Annual Meeting, Minneapolis / MN, USA, 2011.

[4] Gauthier, T.A., Briens, C.I., Bergougnou, .A., Galtier, P., Uniflow cyclone efficiency study, Powder Technology, Nr.62: 217-225, 1990.

[5] Zhang, Y., Wang, X., Riskowski, G.L., Christianson, L.L., Ford, S.E. Particle separation efficiency of a uniflow deduster with different types of dust, Transactions of ASHRAE, 1999.

[6] Muschelknautz, U., Pattis, P., Reinalter, M., Kraxner, M., Design Criteria of Uniflow Cyclones for the Separation of Solid Particles from Gases, CFB10 â€“ 10th International Conference on Circulating Fluidized Beds and Fluidized Bed Technology, Sunriver / OR, USA, 2011.

[7] Kraxner, M., U. Muschelknautz, S. Wechner, S. Ackermann, V. Greif, J. Bolda, Influence of the Inlet Vane Geometry on the Uniflow Cyclones Performance, AIChE-Annual Meeting, Pittsburgh / PA, USA, 2012.

[8] Kraxner, M., Skarke, B., Kofler, T., Pillei, M., Pressure Drop in Uniflow Cyclones: Investigation on an Empirical Calculation Model, CFB-11, Beijing, China, 2014.

[9] Kraxner, M., Empirical Evaluation of Design Criteriaâ€™s for Uniflow Cyclones in Multicycloneboxes, Doctor Thesis, TU MÃ¼nchen, Germany, 2013.

# 506435 Numerical Study of Hygroscopic Particle Removal Using a Cyclone Separator with Saturated Water Vapor

**Abstract:**


**Introduction:** Cyclone separator is one of the most widely applied equipments for separating particles from the air. However, it is difficult to separate particles with diameters smaller than 2.5 micron using the cyclone separator. In light of the particle-vapor interaction, it is possible to increase the particle diameter via forced condensation, especially when the particle contains soluble component. This study has examined the possibility of using a cyclone separator to collect the soluble particle, i.e., sodium chloride (NaCl) as test example, numerically. Specifically, a steam of saturated water vapor with elevated pressure (0.7 MPa) is introduced vertical to the air-particle inlet to increase the particle size by hygroscopic growth, i.e., forming a NaCl-water droplet due to absorbing water vapor. Comparisons of trajectories between the particles with and without hygroscopic growth are presented.

**Methods:** A cyclone separator model and its mesh, i.e., tetrahedral mesh with prism layers, were created (see Fig 1.). The velocities of the air and the water vapor were 10 m/s and 1 m/s, respectively. The NaCl particle density was 2165 kg/m$^3$. The Euler-Lagrange method was applied to predict the airflow and particle transport. If The NaCl-water droplet collided with the cyclone surface, deposition was assumed to occur when considering the particle-vapor interaction. For comparison, the particle separation without hygroscopic growth was also simulated to examine the separation efficiency. Simulations were carried out by ANSYS Fluent with user-defined functions based on validated particle-vapor interaction model (J. Aerosol Sci. 2017, 105: 108-127).

**Results and Discussion:** The trajectories of the 2.5 micron NaCl particles are shown in Fig. 2 when including the effect of particle-vapor interaction. The particle / droplet size increases rapidly when traveling through the air with high relative humidity. Thus, these large droplets would deposit on the cyclone surface due to inertial impaction, which leads to a high separation efficiency. The solution could be collected at the bottom of the cyclone separator with extra water trickled from the top of the cyclone, if the deposited droplets and condensed water cannot form a liquid film on the cyclone surface. Fig. 3 illustrates the trajectories of the 2.5 micron NaCl particles without hygroscopic growth.

Fig.1 Geometry and mesh for the cyclone separator

Fig. 2 2.5 micron particle trajectories with hygroscopic growth

Fig. 3 2.5 micron particle trajectories without hygroscopic growth

**Conclusions:** The numerical study validated the feasibility of separating small soluble particle (<2.5 micron) using cyclone separator by introducing the saturated water vapor with elevated pressure. This method could be useful for the design of the fine particle removal device with simple structure and easy-maintenance.



**Acknowledgements:** The authors gratefully acknowledge the financial support of the National Natural Science Foundation of China (grant No. 51606041), National Natural Science Foundation of Jiangsu Prov. (grant No. BK20160688). The Fundamental Research Funds for the Central Universities are also acknowledged.

# DESIGN CRITERIA FOR MULTI CYCLONES IN A LIMITED SPACE

*Ulrich Muschelknautz, MK Engineering, Heidelberg, Germany*

## Abstract

Multi cyclones for gas-solids separation consist of many parallel cyclone cells within a common housing having a common solids hopper. They are preferably used for high-efficiency separation of a gas-solids flow. Principally increasing the number of cyclone cells and decreasing their size improves the efficiency of a multi cyclone without changing its pressure drop, provided a uniform distribution of gas and solids feed into each single cyclone cell can be achieved and bypass flows through the solids discharge openings from one cyclone cell to the other can be avoided.

In many cases multi cyclones use standard reverse flow cyclone cells with an axial inlet, also called swirl tubes. Analogously multi cyclones can be made from uniflow cyclone cells. Deviating from swirl tubes uniflow cyclones have gas and particles passing through them in only one direction exiting at the same end of the device which makes them much more compact than swirl tubes. In recent years comprehensive experimental and theoretical studies of uniflow cyclones have strongly improved their understanding and led to approved design criteria and calculation methods similar to those which have already been achieved for standard reverse flow cyclones. In the present investigation well proven calculation programs have been applied to compare multiclones based in swirl tube cells with those based on uniflow cyclone cells regarding their efficiency to separate a specified gas-solids feed at a given pressure drop. The investigation has been performed by varying the available space for the dedusting device, the number of cyclone cells, the cyclone pressure drop, the particle size distribution of the feed and the particle density. The results indicate that, under the considered operation conditions, a uniflow multiclone is more efficient than its swirl tube counterpart when the space is limited onto about the volume needed by the uniflow multiclone for its optimum performance and the available pressure drop is low.

**KEYWORDS**

Multi cyclone, multiclone, uniflow cyclone, standard cyclone, swirl tube, axial cyclone.

## Introduction

Cyclone separators are widely used in industry for separating solid particles from gases. They are applied preferably for reasons of high operational safety due to their simple and robust construction, usability for product recovery and in many applications comparable low invest and operation costs. In addition, they can be operated under extreme conditions including large gas volume flow rates as well as high solids concentrations, temperatures and pressures, representing a comparative advantage over other gas-solids separation technologies. The minimum particle size that can be collected by a cyclone generally decreases with decreasing cyclone size. This effect is used in multi cyclones, i.e. systems of many parallel cyclone cells within a common housing having a common solids hopper for solids discharge. Multi cyclones are generally used to increase the separation efficiency beyond the level which is achievable by a single cyclone. Principally increasing the number of parallel cyclone cells and decreasing at the same time their size improves the efficiency of a multi cyclone without affecting its pressure drop, provided a uniform distribution of the gas and the solids feed into each single cyclone cell can be achieved and bypass flows through the solids discharge openings from one cyclone cell to the other can be avoided.

In many cases multi cyclones use standard reverse flow cyclone cells with an axial inlet, also called swirl tubes, which are shown in Fig. 1 left (In the strict sense of the word swirl tubes denote that type of a cyclone with a pure cylindrical shape (1)). Analogously multi cyclones can be made from uniflow cyclone cells, Fig. 1 right. Often those devices are used in space limited applications, e.g. as third stage separators in fluid catalytic cracking (FCC) processes (2). In both cases axial inlets with swirl vane inserts for vortex generation are used. The main difference between both cyclone types is that in swirl tubes the gas flow reverses its direction and pure gas and collected particles are leaving the device at opposite sides, whereas uniflow cyclones have gas and particles passing through them in only one direction exiting at the same end of the device. Due to those design differences unifow cyclones can be built much more compact and can easily be integrated into pipe lines.

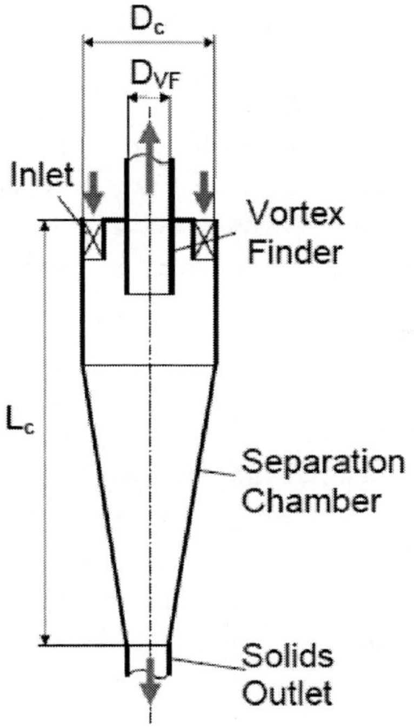


Fig. 1. Schematic illustration of a reverse flow cyclone with axial inlet (left), often applied in multi cyclones, and a uniflow cyclone with axial inlet and vertical downward orientation.

Even though the principle of uniflow cyclones has been known for a long time, not much literature has been published on how to design and to calculate them. So far the investigation of uniflow cyclones has been restricted to special applications such as a short-contact time reactor for high solids loadings (3). To fill this gap systematic experimental studies have been performed recently (4-16). A major challenge to make uniflow cyclones accessible to general industrial practice is an accurate, reliable and fast design method. For calculating the performance data of conventional reverse flow cyclones, approved analytical models with high precision exist which are often preferred over methods based on numerical simulation with long computing times. With this background, a new analytical approach for calculating the separation efficiency and the pressure drop of uniflow cyclones for practical design work has been developed (17, 18). This approach is based on an equilibrium orbit concept similar to that applied in the Barth-Muschelknautz model for reverse flow cyclones, which has been proven to work successfully in a wide range of industrial applications. More detailed information about this model can be found e.g. in (1, 19, 20).

Applying the calculation models for standard cyclones and for uniflow cyclones which both are based on the same physical concepts indicated that uniflow cyclones need much less space for separating a specified solids gas mixture than standard reverse flow cyclones but are generally also slightly less efficient, at least in their present state design (21,22).

Often space for a dedusting unit in an industrial plant is limited. If saving of space is of high priority or if there is even only a limited space available for purifying a given gas volume flow the question arises whether under that restriction of *limited space* a multicyclone made from uniflow cyclone cells may be more favorable than its swirl tube counterpart. This question is addressed by applying well proven calculation programs for standard cyclones according to (19,20) and for uniflow cyclones as described in (17,18).

## Outline of the Calculation Method for Uniflow Cyclones

The model for standard cyclones assumes a limited loading capacity of the gas stream splitting the separation procedure in two steps. At any solids loading $\mu$ in excess of this critical loading $\mu_{lim}$, the solids are immediately separated from the gas at the inlet to the cyclone. The solids remaining in the gas are separated in the cyclone barrel and in the inner vortex below the gas outlet tube with a second, inner separation efficiency $\eta_i$. The total separation efficiency of the cyclone is described by the equation

$$\eta = (1 - \mu_{\text{lim}} / \mu) + (\mu_{\text{lim}} / \mu)\eta_i \tag{1}$$

Analogous to the model for standard cyclones, it is assumed that both separation processes also occur in uniflow cyclones, Fig. 3. The first separation takes place inside the swirl vane inserts for swirl generation and subsequently in the separation chamber due to exceeding the limited load ratio $\mu_{lim}$ of the uniflow cyclone. If the load ratio at the inlet $\mu$ exceeds the limited load ratio, $\mu_{lim}$ , the excess mass fraction will be removed immediately after the gas jet enters the cyclone, and only a small fraction that is restricted by $\mu_{lim}$ will undergo the centrifugal separation process in the inner vortex of the cyclones, Fig. 3.

Deviating from standard cyclones a third separation process has to be taken into account in uniflow cyclones: In contrast to standard cyclones the solids discharge in uniflow cyclones is located closely to the gas exit. Furthermore, a considerable part of the gas flow passes through the ring chamber between the cyclone wall and the vortex finder pipe before it leaves the device through the gas exit pipe. Due to those characteristics particles carried into the ring chamber can be re-entrained back into the gas outlet. The re-entrainment rate depends on the cyclone geometry and on the operation data. The efficiency of the gas particle separation within the ring chamber is called $\eta_{RC}$. Finally, the collection efficiency of the bunker $\eta_b$ has to be taken into account, as in standard cyclones. Thus, the total separation efficiency as well as the fractional separation efficiency of the cyclone are functions of those four single separation efficiencies.

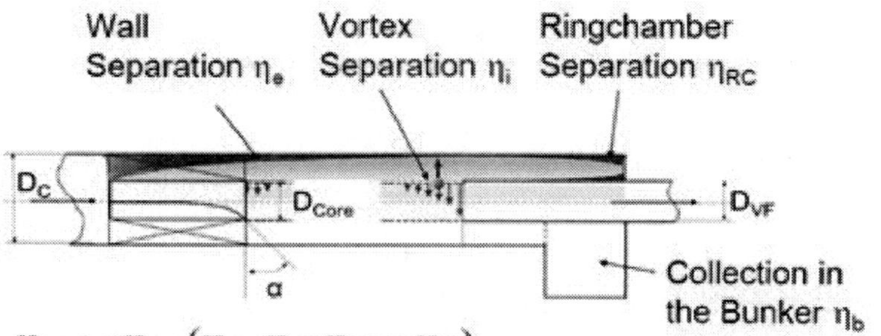


$$\eta_{tot} = \eta_{tot}(\eta_e, \eta_i, \eta_{RC}, \eta_b)$$

$$\eta_{F,tot}(d) = \eta_{F,tot}(d, \eta_e, \eta_i, \eta_{RC}, \eta_b)$$

Fig. 2. Overall structure of the separation process in a uniflow cyclone.

The calculation of the *separation in the inner vortex* $\eta_i$ has been described in detail elsewhere (17,18). For low loadings $\mu$ close to 0.001 this separation mechanism is dominant and its efficiency is expected to be close to the total separation efficiency. With increasing solids loadings the wall separation becomes more and more important and already below a solids loading of 0.01 the wall separation can become the dominant separation mechanism depending mainly on the cyclone size, the tangential velocity at the inlet and on the size distribution and density of the particles.

In uniflow cyclones with an axial inlet the gas flow is set in rotation within a very short time. The entering particles, driven by the drag force of the gas flow, follow this rotational motion, but with a certain delay which is determined by the particle relaxation time. At the end of the separation chamber it may occur that some particles will still not have the circumferential velocity of the gas. This applies in particular to coarser particles with a relaxation time similar to the residence time of the gas within the device. Thus, for small uniflow cyclones and not too fine particles that slip between the circumferential gas velocity and the particle velocity within the separation chamber can be significant and has to be taken into account when calculating the separation efficiency.

Analogous to the calculation model for reverse flow cyclones the pressure drop of uniflow cyclones is calculated as the difference of the total pressures between a position in front of the cyclone inlet and a position m far beyond the opening of the gas outlet where the swirl strength of the vortex flow in the gas outlet has decreased to approximately zero due to wall friction. As in standard reverse flow cyclones the total pressure drop is divided into three parts, the pressure drop in the inlet, the pressure drop in the separation chamber between the mean entrance radius and the position i at the vortex finder radius $r_{VF}$, and the pressure drop between the position i and the measurement position m in the gas outlet tube. A detailed description of the three pressure drop parts is given in (18).

In the present study the performance data of the considered multi cyclones are determined by calculating the performance data of single cyclone cells. This means a uniform distribution of the gas solids feed onto each cyclone cell has been assumed and bypass flows between the solids discharge openings of neighboring cyclone cells have been excluded.

## Results

To compare both multicyclone types a typical industrial application has been considered and investigated in a systematic way: 30 kg/h fine particles shall be removed from 14,000 $m^3/h$ exhaust gas of a reactor at a specified cyclone pressure drop. Regarding the pressure drop two cases are studied, 1000 Pa (0.145 Psi) and 3000 Pa (0.290 Psi). The gas density is $\rho_g = 1.2$ kg/$m^3$ and the gas viscosity is $2 \cdot 10^{-5}$ Pa s. Regarding the particle feed only fine powders are taken into consideration in order to make a meaningful comparison. Note, that the difference between both cyclone types becomes most clear when considering the collection of fine particles. With increasing particle size of the feed the efficiencies of both cyclone types increase and converge (20). Four fine powders are considered characterized by two different particle densities $\rho_s$ = 1490 kg/$m^3$ and $\rho_s$ = 2700 kg/$m^3$ and two different particle size distributions $Q_3(d)$ which are assumed to be describable by RRSB-functions. The RRSB-parameters are (1) $d_{min}$ = 0.1 μm, $d_{max}$ = 30 μm, d' = 19 μm, n = 0.8, $d_{50}$ = 12.2 μm and (2) $d_{min}$ = 0.2 μm, $d_{max}$ = 200 μm, d' = 25.5 μm, n = 0.95, $d_{50}$ = 17.5 μm.

A multi cyclone variant often applied to solve such a problem is Variant ST16-L shown in Fig. 3, left. Its elementary cyclone cells are identical swirl tubes operated in parallel and properly designed for the given gas volume flow. The swirl tube design with an axial inlet has the advantage (compared to the standard cyclone design with an tangential inlet) that those cells a) are more compact (diameter and height) and b) can be packed very densely with very narrow distances between neighboring cells. The characteristic of a swirl tube multi cyclone is that the pure gas and the collected solids leave the device at opposite sides. The pure gas pipelines cross the crude gas inlet duct before they are led into a common clean gas discharge channel located on the roof of the device. The collected solids fall into the common hopper from where they are discharged e.g. by a rotary valve. The walls of the hopper have to be inclined steep enough to prevent caking of particles on the walls. This condition requires a certain height of the hopper and creates thus a large volume of the construction used only for discharging collected particles with no other function else.

A more compact variant for cleaning the same gas-solids feed at the same pressure drop is shown in Fig. 4. This variant, made of uniflow cyclone cells, has both, gas and collected particles, exiting at the bottom of the device, saving considerably height. In addition, due to the small diameter of the uniflow cyclone cells they need less base area than the swirl tube based variant ST16-L. In this construction the pure gas pipelines cross the particle hopper before entering a common pure gas discharge plenum. Caking of particles at the pipelines within the hopper can be avoided by appropriate claddings. The distance between neighboring uniflow cyclone cells has been chosen larger than in the case of the swirl tube multiclone (see Fig. 3, left and right) in order to minimize bypass flows from one uniflow cyclone cell into the other.

The calculation results indicate that both multicyclone variants (ST16-L and UC16) have similar efficiencies, if the allowed pressure drop is 1000 Pa (see Fig. 5). However the uniflow variant needs considerably less space and construction material than the swirl tube cyclone variant, see Fig. 3 and Fig. 4. To see whether a smaller swirl tube variant might compensate that disadvantage without losing too much efficiency swirl tube variant ST16-S with the same base area as the uniflow cyclone variant has been investigated in addition (Fig. 3, center). Note, that for this purpose the smallest possible diameter of the swirl tube cells has been chosen which just allows for complying with the required low pressure drop of 1000 Pa. This variant, however, provides significantly lower efficiencies than the uniflow cyclone device, see Fig. 5.

If a higher pressure drop of 3000 Pa is allowed the results shift in favor of the swirl tube multiclone: In that case its efficiency is in all considered cases (varied particle sizes and particle densities) higher than that of the uniflow multiclone, see Fig. 6. This is true also for the small swirl tube variant ST16-S having the same base area as the uniflow multiclone.

Increasing the number of cyclone cells from 16 to 30 (for purifying the same gas-solids feed) and performing an analogous investigation for those more complex variants indicate similar results as for the 16-cell variants, but with all efficiencies increased compared due to the higher cell number (Figs. 7-10).

There seems to be no principal difference between a multiclone based on swirl tube cells and a multiclone based on uniflow cyclone cells when it comes to erosion in the entrance zone. Both cyclone types have the same concept of swirl generation which is realized by swirl vane inserts. Critical erosion zones are in both cases the blades of the swirl vane inserts as well as the cyclone wall in the entrance region. Due to the smaller cyclone diameter erosion in the uniflow cyclone cells may be more pronounced than in their swirl tube counterpart if the entrance velocity is in both devices the same. However, slightly reducing the entrance velocity in the uniflow cyclones can make that difference disappear.

A uniform distribution of the gas and particles on all cyclone cells is in both multiclone types challenging and can principally be solved with the same concepts in both cases, e.g. by declining the separation wall between the common upstream and downstream plenums as is shown in Fig. 3.

The pressure drop of both multiclone types can be reduced very efficiently and by about the same amount by installing additional swirl vane inserts into the gas outlet pipes of each cyclone cell (23). Those curved blades transform the rotational flow entering the gas outlet pipe into a pure axial flow and recover in this way rotational energy of the gas flow back into static pressure. Without this measure that energy is lost due to dissipation. This measure has been successfully applied in numerous industrial cyclones in a wide range of industrial applications, see e.g. (24).

## Multiclones with 16 Cells

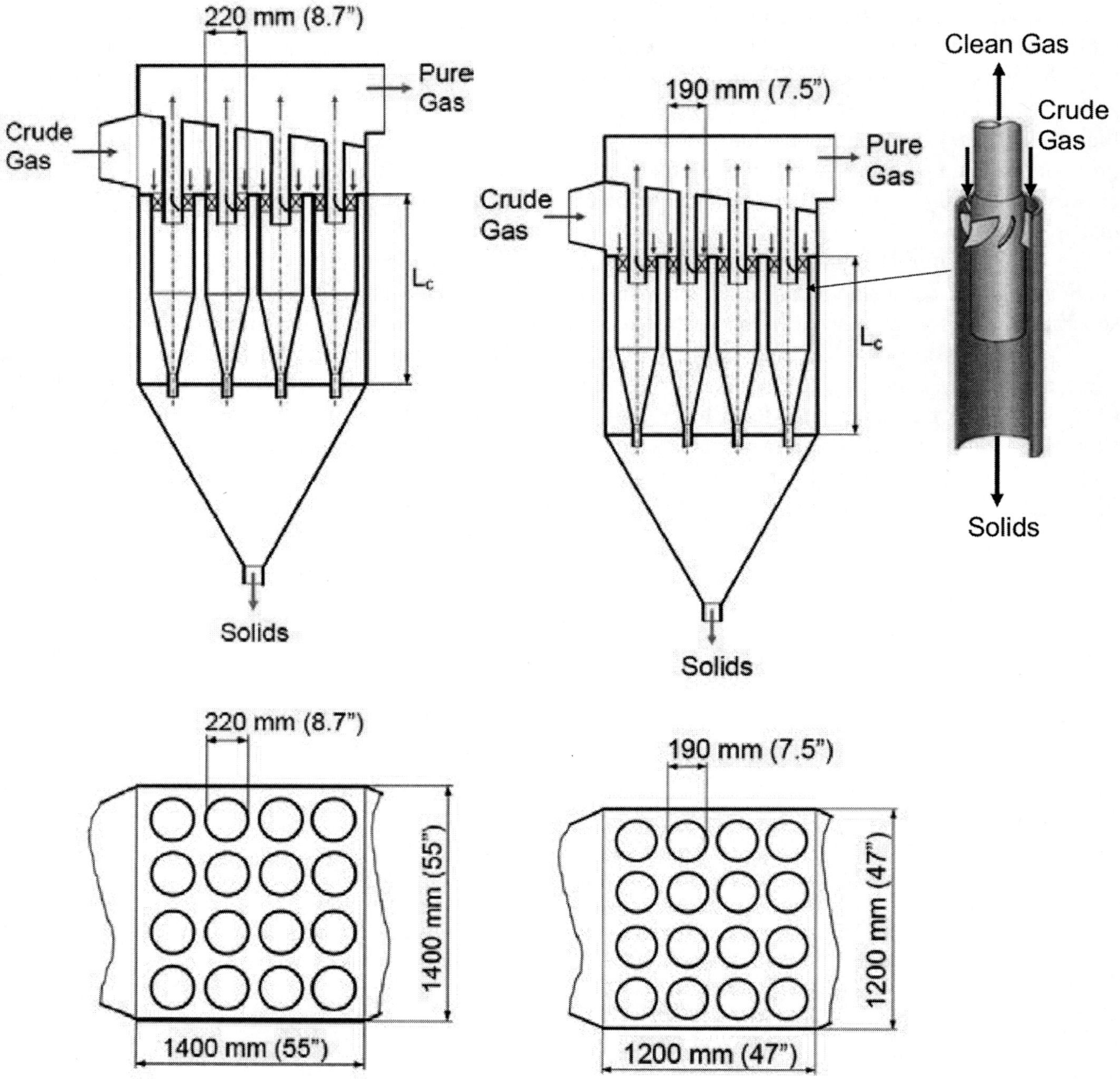


Fig. 3. Multicyclone-Variants ST16-L (left) and ST16-S (right) with 16 Swirl Tubes.

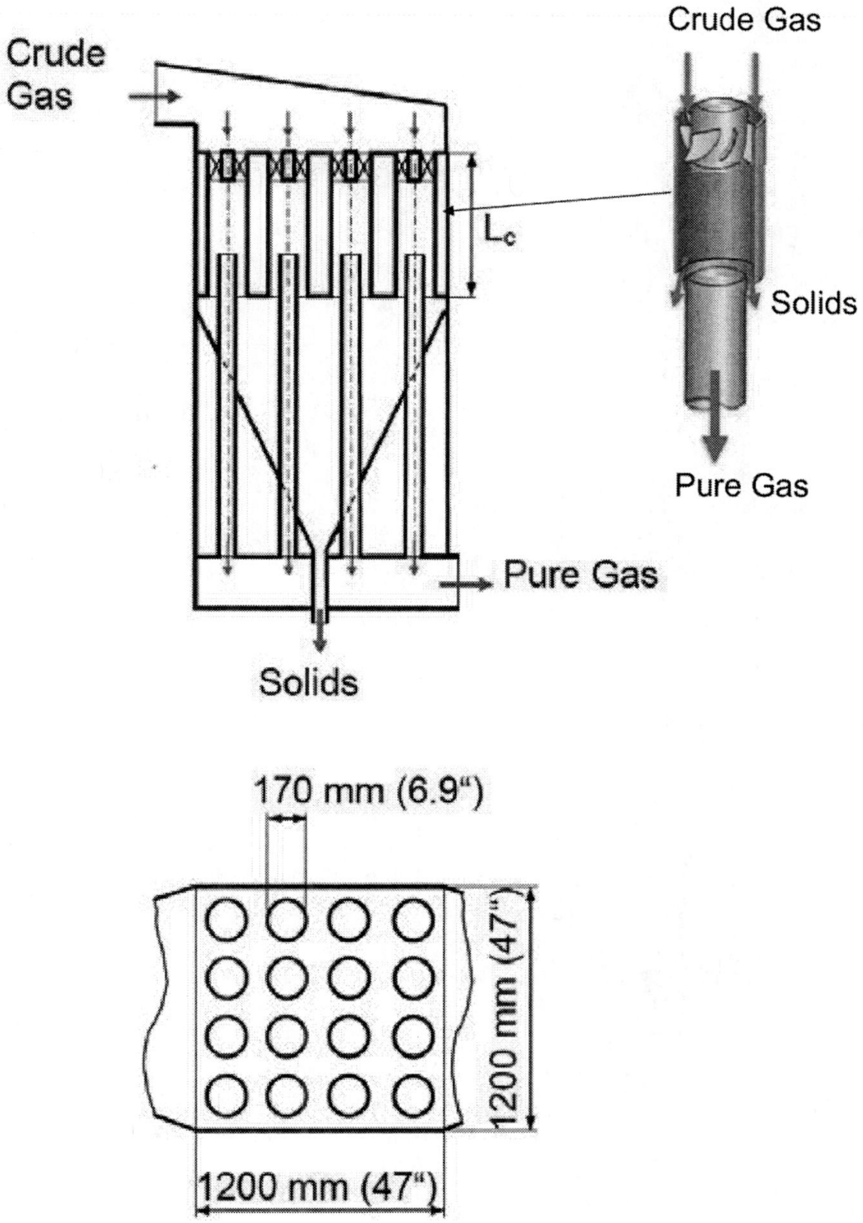


Fig. 4. Multicyclone-Variant UC16 with 16 Uniflow Cyclones.

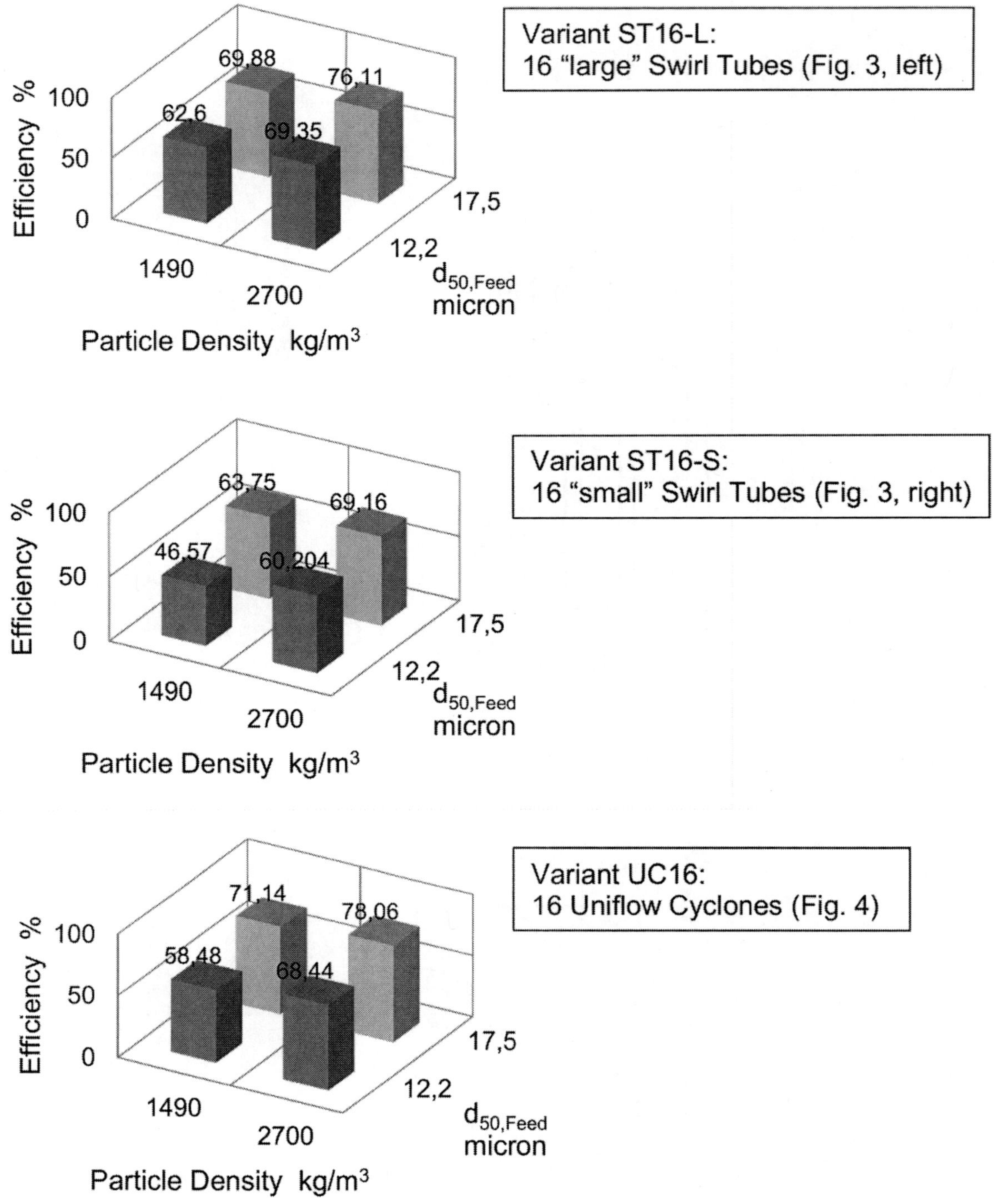


Fig. 5. Calculated Separation Efficiencies of Multicyclone-Variants ST16-L (top), ST16-S (center) and UC16 (bottom) with $\Delta p = 1000$ Pa.

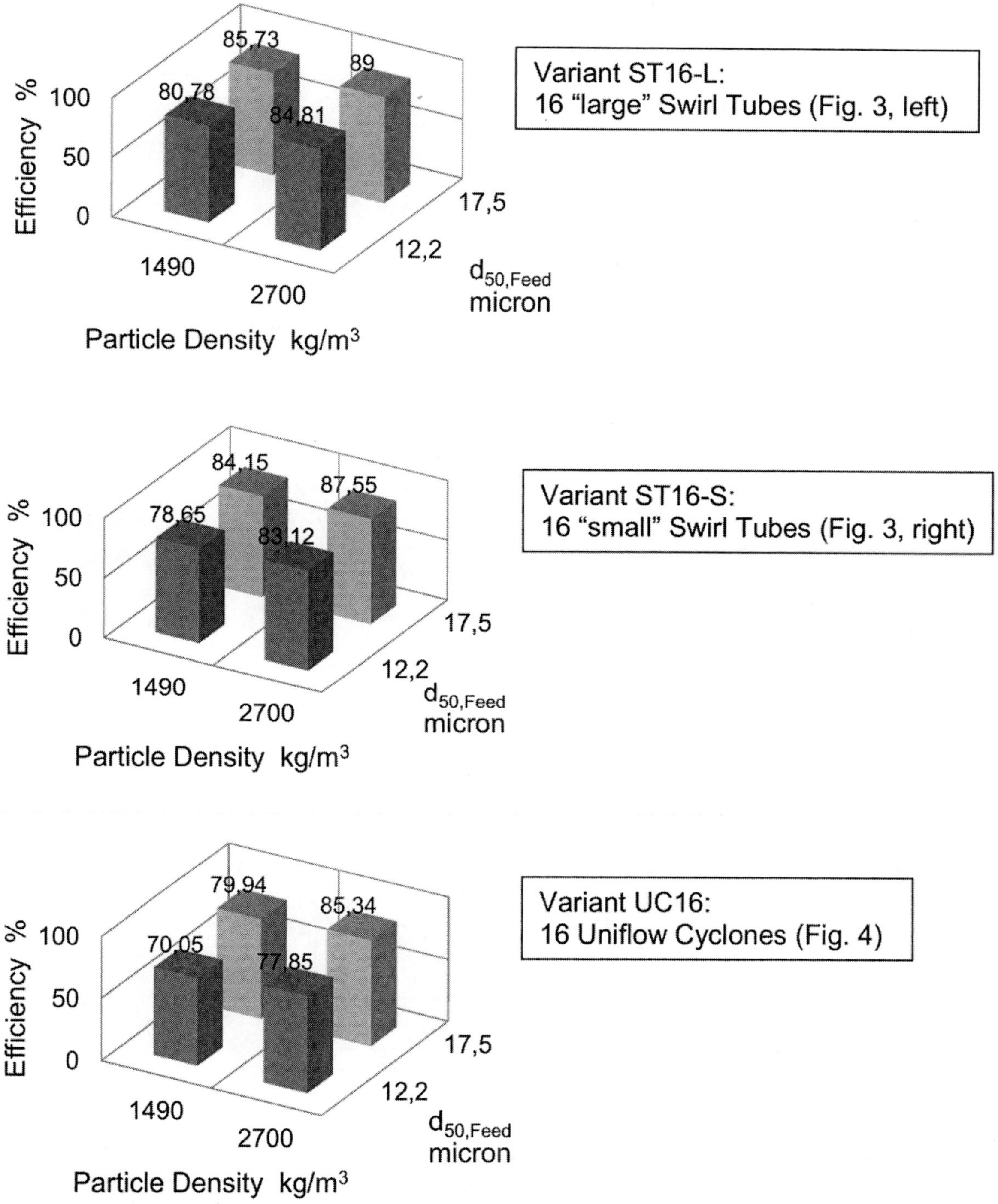


Fig. 6. Calculated Separation Efficiencies of Multicyclone-Variants ST16-L (top), ST16-S (center) and UC16 (bottom) with $\Delta p = 3000$ Pa.

## Multiclones with 30 Cells

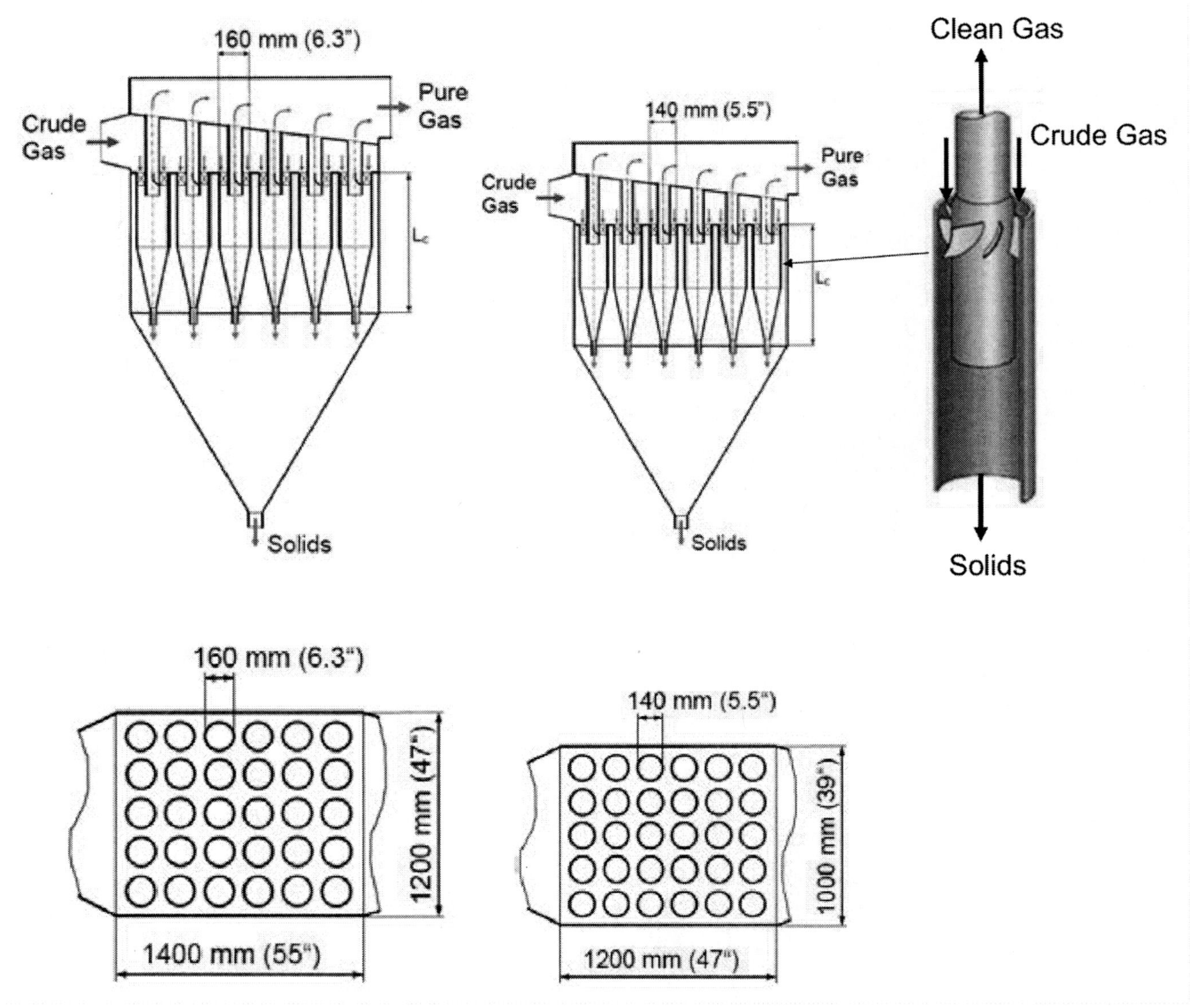


Fig. 7. Multicyclone-Variants ST30-L (left) and ST30-S (right) with 30 Swirl Tubes.

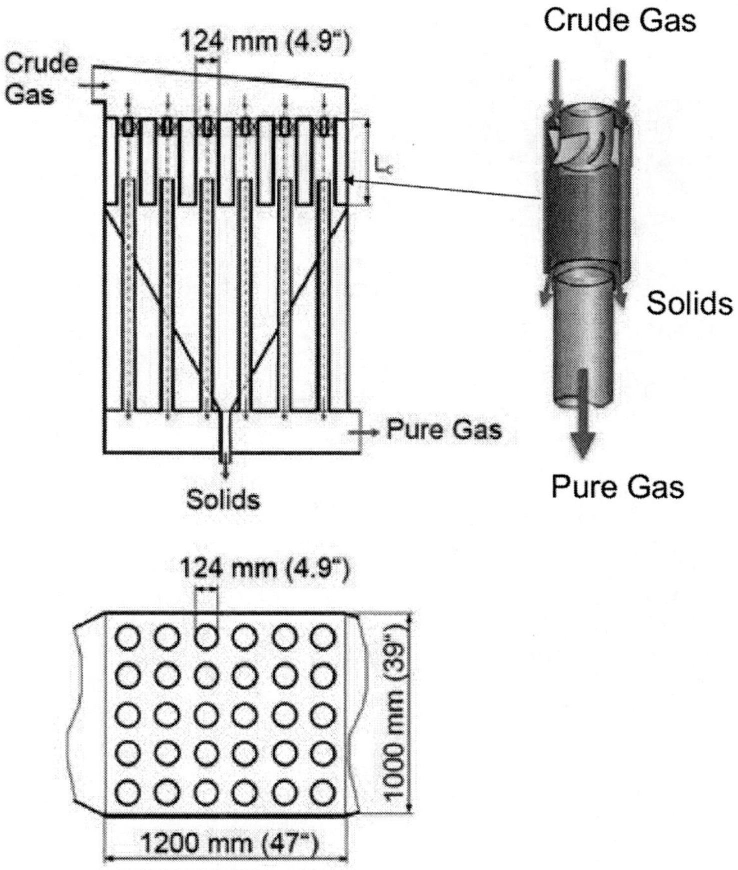


Fig. 8. Multicyclone-Variants ST30-L (left) and ST30-S (right) with 30 Uniflow Cyclones.

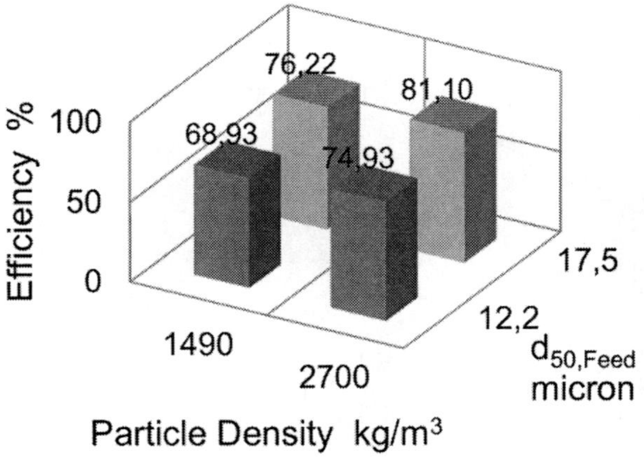


Variant ST30-L:
30 "large" Swirl Tubes (Fig. 7, left)

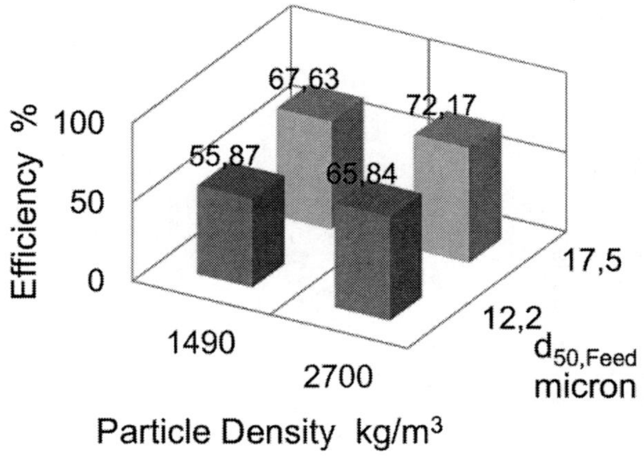


Variant ST30-S:
30 "small" Swirl Tubes (Fig. 7, right)

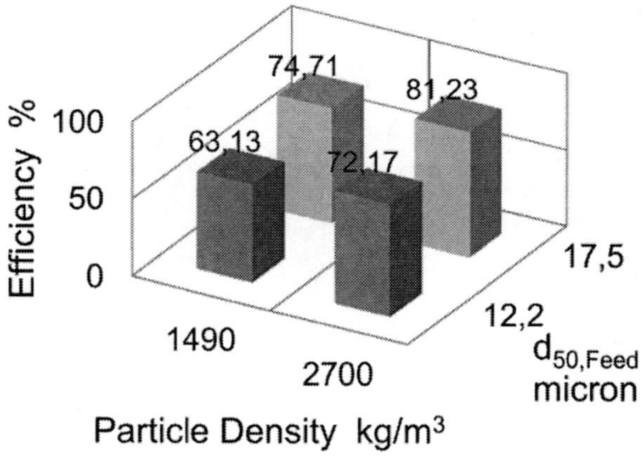


Variant UC30:
30 Uniflow Cyclones (Fig. 8)

Fig. 9. Calculated Separation Efficiencies of Multicyclone-Variants ST30-L (top), ST30-S (center) and UC30 (bottom) with $\Delta p = 1000$ Pa.

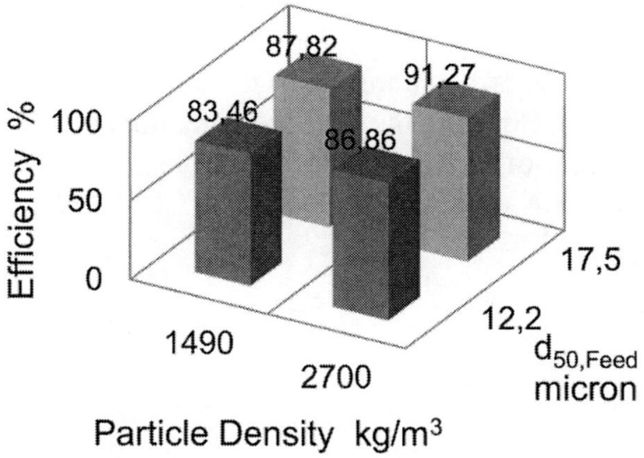


Variant ST30-L:
30 “large” Swirl Tubes (Fig. 7, left)

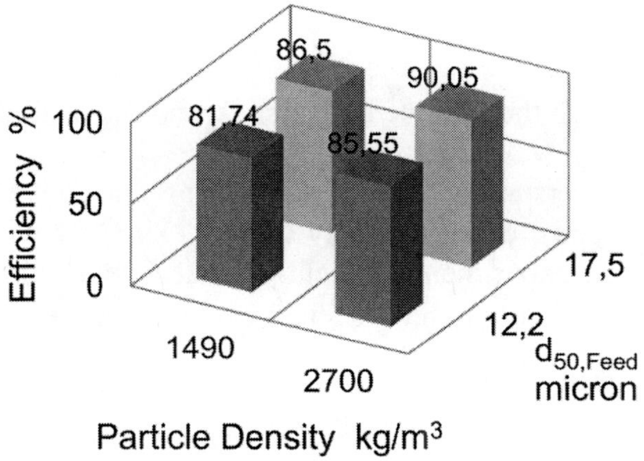


Variant ST30-S:
30 “small” Swirl Tubes (Fig. 7, right)

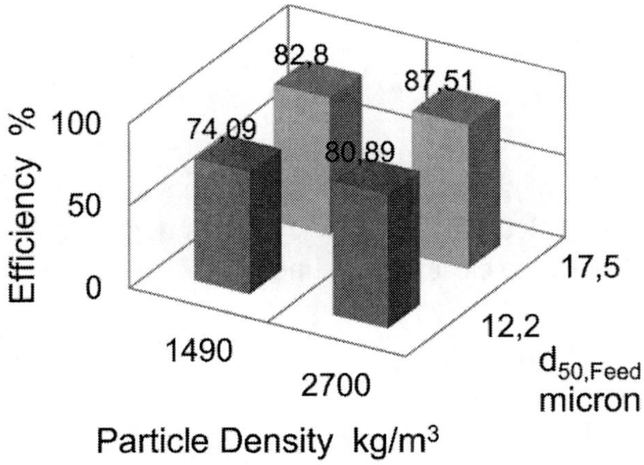


Variant UC30:
30 Uniflow Cyclones (Fig. 8)

Fig. 10. Calculated Separation Efficiencies of Multicyclone-Variants ST30-L (top), ST30-S (center) and UC30 (bottom) with $\Delta p = 3000$ Pa.

## Comparison of the cut-off sizes

To obtain a better understanding of the results described above it is useful to compare both cyclone types based on dimensionless parameters characterizing the cyclone performance. This is done by applying the Stokes number of particles with the cut-off size

$$Stk(d_c) = \frac{\Delta\rho\, d_c^2\, v_z}{18\,\eta_g\, D_c} \tag{2}$$

and the Euler number

$$Eu = \frac{\Delta p}{\frac{1}{2}\rho_g\, v_z^2} \tag{3}$$

following a proposal described in (25). In eqs. (2) and (3)

$$v_z = (4/\pi)\, Q / D_c^{\,2} \tag{4}$$

is the mean axial gas velocity through the cyclone with the cyclone diameter $D_c$ and $\Delta\rho_g$ is the difference between the particle and the gas density.

Figs. 11 and 12 show the Euler number versus the Stokes number of the above described multiclones as well as of uniflow cyclone investigated experimentally in (6,9,11,13,15). Since grade-efficiency curves have been measured only in a few experiments, the cut-off sizes were estimated from the overall separation efficiencies $\eta$, assuming the separator to have a sharp cut at the cut-off size

$$Q_3(d_c) = 1 - \eta \tag{5}$$

In addition the line resulting from the equation

$$Eu\,\sqrt{Stk(d_c)} = \sqrt{12} \tag{6}$$

is plotted in Figs. 11 and 12, which provides a scaling rule for "well designed cyclones" at low solids concentrations according to (26).

As can be seen from Fig. 11 the data points of the calculated swirl tubes are close to the line resulting from eq. (6), whereas the data points of the calculated uniflow cyclones depart from that line at larger Stokes numbers beyond 0.01, in agreement with the experimental points.

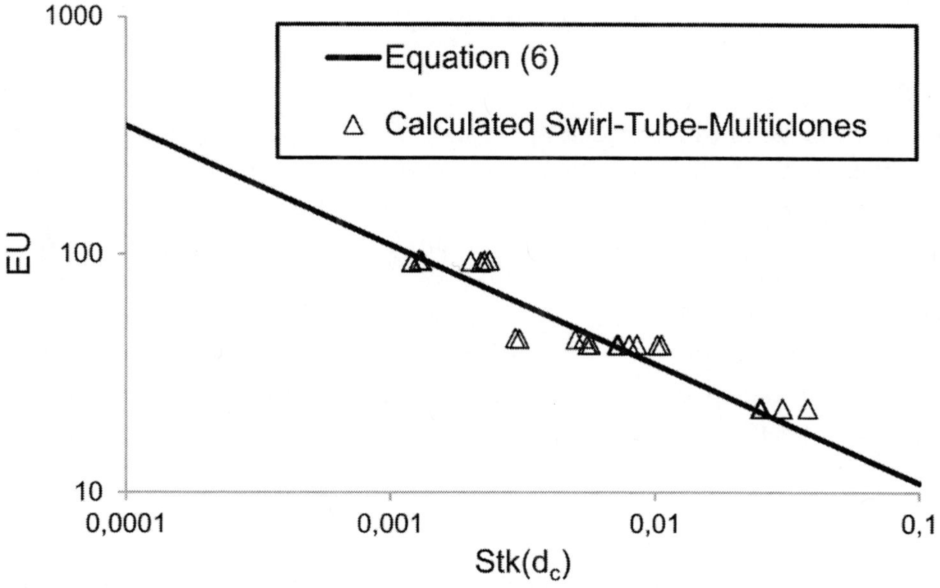


Fig. 11. Euler number *Eu* versus Stokes number *Stk*($d_c$) showing calculated points for the investigated Swirl-Tube-Multiclones and the line for "well designed standard cyclones" according to the equation of Swarovski.

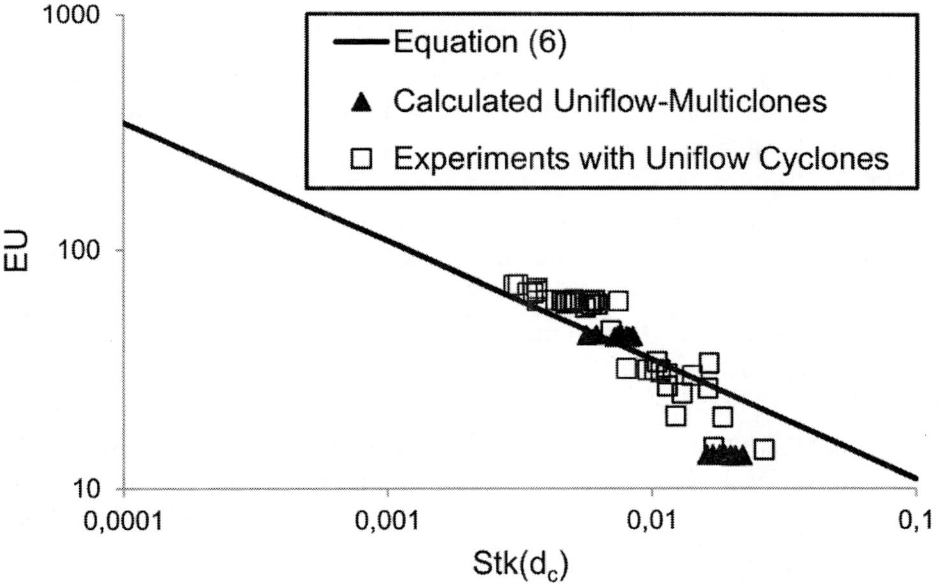


Fig. 12. Euler number *Eu* versus Stokes number *Stk*($d_c$) showing calculated points for the investigated Uniflow-Multiclones and experimental points from uniflow cyclones together with the line for "well designed standard cyclones" according to the equation of Swarovski.

## Conclusion

Recent studies revealed that present state-of-the-art uniflow cyclones are generally slightly less efficient than reverse flow cyclones regarding the separation of a specified gas-solids flow at a specified pressure drop. Nevertheless their more compact construction in comparison to reverse flow cyclones can make them preferable in applications with limited space. The present investigation performed by well proven calculation programs for both cyclone types compares multiclones based on swirl tube cells with those based on uniflow cyclone cells, varying the available space, the number of cyclone cells, the allowed pressure drop, the particle size distribution of the feed and the particle density. The calculations indicate that, under the considered operation conditions, a uniflow multiclone is more efficient than its swirl tube counterpart when the space for the dedusting device is limited onto about the volume needed by the uniflow multiclone for its optimum performance and the available pressure drop is low.

### *References*


1. Hoffmann, A. C., Stein, L. E. (2007), Gas cyclones and swirl tubes. Springer-Verlag, 2nd Edition.
2. Krishnamurthy, S., Koves, W. (2013), Apparatuses and methods for gas-sold separations using cyclones. Uop Llc. Patent US-8419835 B2.
3. Gauthier, T. A., Briens, C. I., Bergougnou, M. A., Galtier, P. (1990),. Uniflow cyclone separation efficiency. Powder Technol. 62, 217-225.
4. Weng, M. (2002), Experimentelle und numerische Untersuchung von Gleichtromzyklonen, Ph.D. Thesis, RWTH Aachen University, Germany.
5. Muschelknautz, U (2002), Separation of coarse particles: New separator with low pressure drop. Proceedings of 7th International conference on fluidized bed technology, Niagara Falls, Canada.
6. Würtl, R. (2007), Experimental investigation of uniflow cyclone performance. Diploma Thesis, MCI Management Center Innsbruck, Austria.
7. Foidl, F. (2008), Entwicklung von Leitapparaten zur Druckverlustminimierung in Gleich-stromzyklonen. Diploma Thesis, MCI Management Center Innsbruck, Austria.
8. Leitner, A. (2009), Einfluß der Staubart auf den Trenngrad eines Gleichstromzyklons. Diploma Thesis, MCI Management Center Innsbruck, Austria.
9. Muschelknautz, U., Pattis, P. Reinalter, M., Kraxner, M. (2011), Design criteria of uniflow cyclones for the separation of solid particles from gases. Proceedings of 10th International conference on fluidized bed technology, Sunriver / Oregon, USA.
10. Kraxner, M., Muschelknautz, U., Karri, S.B.R., Cocco, R., Knowlton, T.M., 2011. Applicability of a Uniflow Cyclone as a Third Stage Separator in the FCC-Process. AIChE – American Institute of Chemical Engineers – Annual Meeting, Minneapolis / Minnesota, USA.
11. Muschelknautz, U., Pattis, P., Reinalter, M., Kraxner, M., Kofler, T., Aigner, M., Eichwalder, G. (2012), An experimental study of uniflow cyclone design parameters. AIChE-Annual Meeting, Pittburgh, USA.
12. Kraxner, M., Muschelknautz, U., Wechner, S., Ackermann, S., Greif, V., Bolda, J., Sommer, K. (2012), Influence of the inlet vane geomatrey on the uniflow cyclone performance. AIChE-Annual Meeting, Piitsburg, USA.
13. Muschelknautz, U., Kofler, T., Kraxner, M., Schretter, P., Pattis, P. (2012), An experimental investigation of uniflow cyclones with a spiral inlet. 62nd Canadian Chemical Engineering Conference, Vancouver, Canada.
14. Kraxner, M., Portenkirchner, M., Brugger, T., Muschelknautz, U. (2012) Effective flow rates for

uniflow cyclone performance. 62$^{nd}$ Canadian Chemical Engineering Conference, Vancouver, Canada.

15. Kofler, T., Kraxner, M., Pillei, M., Lanthaler, R., Ilchmann, C., Goller, R., Muschelknautz, U. (2013), Effect of solids loading on the uniflow cyclone performance at low solids concentrations. AIChE – Annual Meeting, San Francisco, USA.
16. Kraxner, M. (2013), Empirical evaluation of design criteria for uniflow cyclones in multicyclone boxes, PhD Thesis, TU Munich, Germany.
17. Muschelknautz, U. (2012), Analytical approach for calculating the separation efficiency of uniflow cyclones, Chem. Eng. Technol. 35(12), pp. 2099-2106.
18. Muschelknautz, U. (2015), Analytical approach for calculating the performance data of uniflow cycones. 22nd International Conference on Fluidized Bed Conversion, Turku, Finland.
19. VDI-Wärmeatlas Atlas (2018), Zyklone zum Abscheiden fester Partikel aus Gasen. 12th German Edition.
20. Muschelknautz, E., Greif, V. 1997. Cyclones and other gas-solids separators, in Circulating Fluidized Beds (Eds: J. R. Grace, A. A. Avidan, T. M. Knowlton), Blackie Acad. & Profess., London, 181-213.
21. Muschelknautz, U. (2016), Uniflow Cyclones compared with Standard Cyclones. Proceedings of FILTECH 2016.
22. Muschelknautz, U. (2017), Performance comparison of uniflow cyclones and standard cyclones. Proceedings of 12$^{th}$ International conference on fluidized bed technology, Krakow, pp. 421-428.
23. Greif, V. (1996), Reduzierung des Druckverlustes von Zyklonabscheidern durch Rückgewinnung der Drallenergie und Erweiterung der Grenzbeladungstheorie auf kleine und kleinste Staubbeladungen. VDI-Fortschr.-Ber., VDI-Verlag, Düsseldorf, Reihe 3, Nr. 470.
24. Roschek, D., Dereschewitz, M., Reckweg, F., Muschelknautz, U. (2015),. Design of Cyclone Improvements in two CFB Boilers and Operation Experience. 22nd International Conference on Fluidized Bed Conversion, Turku, Finland.
25. Peng, Hoffmann, Dries (2004), Separation Characteristics of Swirl-Tube Dust Separators, AIChE J. , 50, pp. 87-96.
26. Swarovski, L. (1986), “Solid-Gas-Separation” in *Gas Fluidization Technology*, D. Geldart, ed. New York, Wiley.

# 508021 Hydroclyclone Experiments and Design Equations for Produced Water Desanding Operation

**Abstract:**

Hydrocyclones are commonly used for liquid solid separations [1] involving high feed flow rates. The advantages of the use of hydrocyclones for the water treatment in oil extraction operations [2] has been proven, such as for example the attainment of produced water for water injection operation [3]. Nevertheless, little information are available in the literature for the design of hydrocyclone desanders related to water-solid system relevant to water treatment in produced water operations.

In this work, we present the experimental results on a series of tests carried out on a purposely-made pilot unit hosting a single hydro cyclone and able to work at water flow rates up to 2.5 L $s^{-1}$. The setup is able to set independently the feed flow rate and the underflow flow rate. Mixtures of water and Tri-hydrate Alumina particles (1 to 70 Î¼m size) up to 2000 ppm v/v of solid concentrations have been tested. Solid particle density, particle size distribution and shape makes the tested system very similar to real water sand mixtures in produced water desanding operation. These results have been used to find the proper hydrocyclone design equation such as those proposed by Coelho and Medronho [4]. In particular, we have found the proper correlations for the cyclone pressure drops, and the cyclone cut size as a function of the liquid flow rate and the water flow ratio between the underflow and the feed stream.

REFERENCES

[1] L. Svarovsky, Introduction to solid-liquid separation, in: Solid-Liquid Sep., Elsevier, 2001: pp. 1â€“29. doi:10.1016/B978-075064568-3/50025-0.

[2] J.C. Ditria, M.E. Hoyack, The Separation of Solids and Liquids With Hydrocyclone-Based Technology for Water Treatment and Crude Processing, in: SPE Asia Pacific Oil Gas Conf., Society of Petroleum Engineers, 1994. doi:10.2118/28815-MS.

[3] M.S. Choi, Hydrocyclone Produced Water Treatment for Offshore Developments, in: SPE Annu. Tech. Conf. Exhib., Society of Petroleum Engineers, 1990. doi:10.2118/20662-MS.

[4] M.A.. Coelho, R.. Medronho, A model for performance prediction of hydrocyclones, Chem. Eng. J. 84 (2001) 7â€“14. doi:10.1016/S1385-8947(00)00265-5.

# 508322 Validation of CFD-DEM Simulation of a Continuous Tablet Coating Process

**Abstract:**

Tablet coating is a widely used unit operation in the pharmaceutical industry. Traditionally, large-scale drum coater systems are loaded with hundreds of kilos of tablets and rotate at relatively low rotation rates (Fr<<1). This leads to several hours of batch processing in order to obtain sufficient coating uniformity. The ConsiGma™ coater from GEA is a tablet coater which can be integrated into the GEA ConsiGma™ continuous process line. The ConsiGma™ Coater operates with low drum loads (3-7 kg) but very high rotation rates (Fr~1) leading to much shorter cycle times in the range of a few minutes. To achieve better coating results the tablets are jetted away from the rotating drum wall by so-called air knives in order to form a stable, gravity-free cascade. This air flow needs to be considered in a simulation model to accurately predict the coating process in the ConsiGma™ coater. Depending on the desired throughput rates there are three interchangeable sizes of the GEA ConsiGma™ Coaters.Interchangable means that the drum wheel of the next smaller drum will also fit and operate in the next larger housing (80/160mm or 160/320mm).

In this work we present a model to simulate the coating process in the ConsiGma™ 160 mm and 320 mm coater using CFD-DEM. The eXtended Particle System (XPS) is used for the DEM simulation and is coupled to AVL Fire® for the CFD[1]. The validity of the simulation results, as well as our method for modeling the non-sphericity of the tablets, are shown. The simulations are compared to experimental results, including the quantitative comparisons of the tablet flow and bed shape, as well as coating-related parameters, such as average coating mass on the tablets, standard deviation and coefficient of variation.

*This work was funded through the Austrian COMET Program by the Austrian Federal Ministry of Transport, Innovation and Technology (BMVIT), the Austrian Federal Ministry of Economy, Family and Youth (BMWFJ) and by the State of Styria (Styrian Funding Agency SFG).*

**References**

1. Jajcevic D, Siegmann E, Radeke C, Khinast JG. Large-scale CFDâ€"DEM simulations of fluidized granular systems. *Chem Eng Sci*. 2013;98:298-310. doi:10.1016/j.ces.2013.05.014.

# 507281 Modelling of High Shear Wet Granulation Using a Population Balance Framework

**Abstract:**

Wet granulation processes are difficult to reproduce and thus are often operated inefficiently which leads to high recycle rates in a continuous processes and high rejection rates in batch processes. A modelling approach can help to improve the efficiency of wet granulation processes significantly by understanding the effects of process parameters and material properties better.

In this study, a compartment-based model is used to model the wet granulation process, employing a one-dimensional population balance framework. For this modelling approach, the mixer is decomposed in three regions spray region, impeller region, and chopper region. Laboratory scale granulation experiments are conducted to estimate unknown parameters for empirical rate expressions. Finally, the model developed is validated using experimental data at different process scales. In order to assess the performance of the granulator model, the granulation end-point is predicted and compared to additional experimental results.

# 507217 Prediction of Pressure Filtration Performance in Systems with Pharmaceutical High Aspect Ratio Crystals

**Abstract:**

Filtration is a key product recovery process in the pharmaceutical industry as it facilitates the separation of suspended solid particles, such as Active Pharmaceutical Ingredients (APIs) from a liquid. Filtration is an intermediate processing step between crystallisation and drying, and it typically involves several washing cycles aiding the removal of impurities and the manipulation of the mother liquor solubility [1]. Pressure filtration is the most common separation technique and its performance is strongly influenced by the cake structure and formation, which in turn are highly dependent on particle properties such as shape and size distribution. Cake filtration in systems with wide particle size distributions (PSDs) and asymmetrical high aspect ratio particles is frequently quite time consuming at scale. Rising the system's pressure drop can accelerate the process, but it may also be counterbalanced by an increase in cake compressibility and/or particle breakage, both leading to denser packings [2]. The aim of this work is to study the cake filterability considering systems with high aspect ratio particles and various PSDs.

Lab scale filtrations have been conducted using a 200 ml, 35 mm diameter pocket filter using nitrogen pressure. Experiments have been conducted using dicalcium phosphate and mannitol. These materials were chosen because as pharmaceutical excipient the materials are commercially available with grades of different particle distributions. Filtration rates were compared for beds with different PSDs, at different applied pressures and for a settled bed or from a slurry. This approach was repeated using an API with high aspect ratio crystals that was wet milled to create three distinct PSDs and the filtration behaviour characterised in the same manner.

Modelling the filter cake structure is of crucial importance, since it allows for the accurate determination of cake properties. DigiDEM, a commercial Discrete Element Method (DEM) software, specially designed to deal with non-spherical shapes, with Lattice Boltzmann (LBM) built-in for drag force calculations [3], is used in this study to model the cake structure and consequently, evaluate the cake voidage as function of PSD and particle shape. DigiDEM is also used for the estimation of the particles' terminal velocities due to gravitational sedimentation. With validated cake voidage and terminal velocities predictions at hand, the improved Ruth equation (an expansion of Darcy's law) [4] is used in conjunction with lab-scale filtration data for the estimation of cake properties such as medium resistance, specific cake resistance and compressibility index. The estimated cake properties are then employed within the filtration model to predict the system response at higher pressures. Such a predictive model can be ultimately used for the successful system scale-up and control, and bring us a step closer to industrial adoption of a new Digital Design paradigm for pharmaceutical product and process design.

The authors gratefully acknowledge the support of the Advanced Manufacturing Supply Chain Initiative through the funding of the 'Advanced Digital Design of Pharmaceutical Therapeutics' project (Grant No. 14060).

## References

1. Rushton, A., A.S. Ward, and R.G. Holdich, ***Solid-liquid filtration and separation technology.*** 2nd ed. 2000: WILEY-VCH Verlag GmbH.
2. Bourcier, D., J.P. Féraud, D. Colson, K. Mandrick, D. Ode, E. Brackx, and F. Puel, ***Influence of particle size and shape properties on cake resistance and compressibility during pressure filtration.*** Chemical Engineering Science, 2016. **144**: 176-187.
3. Structure Vision Ltd 2015, viewed 11 July 2017, http://www.structurevision.com/.
4. Ruth, B.F., G.H. Montillon, and R.E. Montonna, ***Studies in filtration – I. Critical analysis of filtration theory***. Industrial & Engineering Chemistry, 1933. **25**: 76-82.

# 508074 Development of Continuous Manufacturing of Pharmaceuticals and Its Monitoring and Control


**Abstract:**

A novel continuous manufacturing system of pharmaceuticals with its monitoring and control systems was developed. The system consists of several unit operations such as mixing, twin screw granulation and spiral drying system. Performance of each unit operations as a function of each operating parameters was investigated experimentally. Also, a DEM (discrete element method) computer simulation was used to analyze the mechanism of twin screw granulation and a CFD (computational fluid dynamics) was used to analyze the mechanism of spiral drying system. Moisture content of granules in the twin screw granulation and the spiral drying was continuously measured and uniformity of API (Active Pharmaceutical Ingredient) content was also continuously monitored with a developed NIR (Near Infrared) sensor. Performance of the automatic manufacturing with feedback control systems of each unit operations was also investigated.

## 8th World Congress on Particle Technology

# Algorithm to Enhance Performance of Dilute Phase Pneumatic Conveying Systems

**Jonathan O. Thorn M.S.**

## Background

The nature of a dilute (lean) phase pneumatic conveying system suggests that it will be designed with an excess of air volume in order to insure that it operates sufficiently when put into service. A minimum air velocity in the pipeline is required initially to achieve entrainment of the particles. Therefore, that corresponding airflow (or higher) will cause the system operate sufficiently. In the design stage the minimum entrainment velocity will most likely be identified but then several layers of safety factor will be implemented that result in elevated airflows. In Figure 1 we see a simple pressure dilute phase system configured with one feed source and one destination. More complex systems are common with varying numbers of sources and destinations. The excess of design or operation is the following:

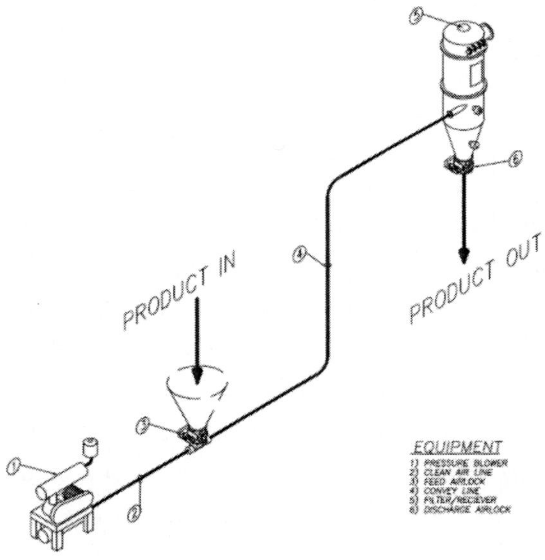


Figure 1. Typical pressure dilute phase convey system

First, the common practice of operating a dilute phase blower at a constant speed necessitates that some level of conservatism be considered. A robust system design will accommodate temperature changes (daily, seasonal, etc) and potential upset conditions without interruption to the process. A system designed with excess airflow will be less susceptible to changes in the process and deemed more reliable operationally. This excess air is then designed into the air movement device. Positive displacement blowers are the most common dilute phase pumps and these are usually packaged with the motor to operate at certain speed using belts and pulleys. Therefore, with only one opportunity to design the drive package a conservative blower speed selection is preferred by users. Once the system is implemented and working satisfactorily, it is very unlikely for the blower speed to be changed.

Secondly, the flexible nature of the dilute phase conveying system means a single system can be used for several different duties. Service variations may include but are not limited to:

**Conveying Rate** – The design rate of the system may not be the rate actually transported through the pipe. It may be that the process was designed for excess capacity that is not yet feasible. It may be that feed issues or other process restrictions are preventing the full flow rate of material from reaching the convey line. Whether intentional or not, operating at any convey rate below the maximum requires less airflow from a design and operational perspective.

**Distance** – The use of diverters allows the piping to reroute to shorter and longer distances as needed to service multiple destinations. In these cases, the longest or more difficult pipe run will be used as the design basis. Even with single destination systems, the distance the system was designed for may not reflect the actual system implemented. In any case, conveying a distance shorter than the maximum design length of the system requires less airflow from a design and operational perspective.

**Material** – A system may convey multiple materials (with varying properties) or the material properties may differ from the original design. Assuming the material properties influence the dilute phase conveying requirements, the system will be organized to handle the material with the most difficult properties.

These operational conditions combine in multiple scenarios to create a pattern of inefficient operation. In the best case, the typical safety factors cause the blower to operate moderately faster than is required to physically transport the material. In the worst case, several factors combine to cause the blower design to greatly exceed the required operating speed of the blower causing dramatic inefficiency and excessive velocity in the convey line. The former is not desired but is usually undetected because of the lack of comparative analysis (each situation is unique). The latter can cause unwanted velocity related impacts such as excessive abrasion or material damage. Regardless of whether the condition negatively impacts the process or is even detected, it can be said that most every dilute phase conveying system operates the blower faster than is required to do the specific duty for which it is engaged.

Figure 2. Contributions of design and operation to excessive blower speed.

Inefficient operation of the dilute phase conveying system can be tied directly to operating speed of the blower as material is being transported. The compression work that occurs in the blower is a function of the air volume being compressed (Q) vs the differential pressure (P2/P1) between the inlet and outlet of the air movement device. The work to adiabatically compress the gas can be given by equation 1 where k=1.4 for air.

$$\dot{W} = \frac{QP_1k}{k-1}\left[\left(\frac{P_2}{P_1}\right)^{\frac{k-1}{k}} - 1\right] \tag{1}$$

where the power to operate the motor is given by the overall efficiency ($\eta$) of the blower package.

$$P = \eta \cdot \dot{W} \tag{2}$$

The overall efficiency of the blower package includes the efficiency of positive displacement compression ($\eta_{PD}$), the efficiency of the drive package transmission ($\eta_D$) and the electrical efficiency of the motor ($\eta_m$). For the purpose of these experiments we will be concerned with overall efficiency.

$$\eta = \eta_{PD} \cdot \eta_D \cdot \eta_m \tag{3}$$

While there is a relationship between blower speed and overall efficiency, the largest contributor to increasing the power consumption is from air volume (Q) and the resulting pressure drop. In Figure 3 we see a typical Zenz diagram plotting air velocity vs pressure drop per length for a certain material. Starting from the pressure minimum, as the velocity (and air volume) increase in the convey line, the resulting pressure drop to perform a convey duty (rate and distance) increases. This indicates less efficient conveying as more energy is being consumed through the combination of increased airflow volume and increased pressured drop. In this way a 10% increase in conveying velocity the system can experience a 12-15% increase in power consumption for the blower. As the velocity is decreased from the pressure minimum, we see the pressure drop start to rise. However, a power reduction is still realized as the air volume is decreasing faster than the influences of pressure rise. Therefore operating the system at a velocity below the point of pressure minimization is the most efficient performance of the system from the perspective of blower power consumption.

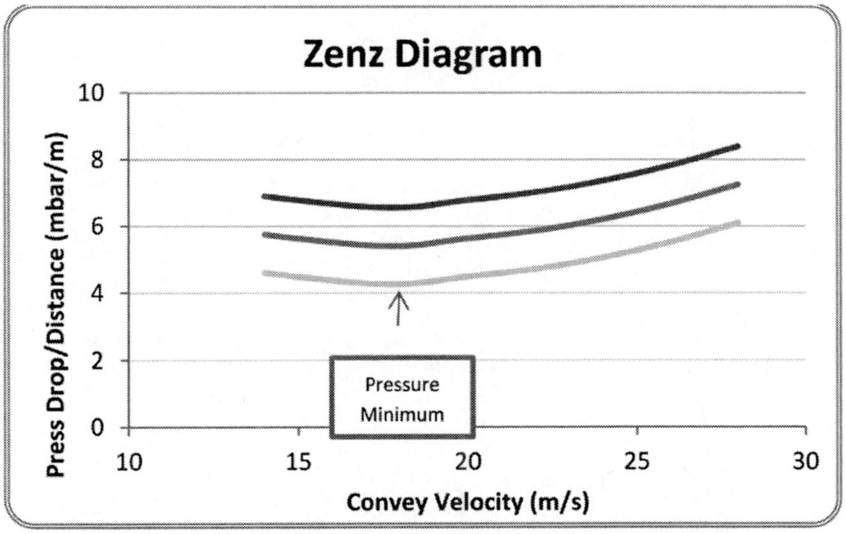


Figure 3. Zenz Plot for fiberglass filled polyethylene pellets

The velocity at which the pressure minimization occurs is widely considered the minimum entrainment velocity of the system. At lower velocities, particles begin to drop out of suspension (saltation) and the pressure rises due to the increased friction of the dragging material with the pipe wall. In a typical system, the increased pressure drop would induce additional saltation through airlock leakage and lower pick-up velocity thereby marching the system toward a plugged condition. In order to support conveying after the initial pressure rise, an algorithm would be required to sense the increase in pressure and make the necessary correction to the blower speed in order to replace the lost air. This would be an enhancement to the typical operation of a dilute phase system and will be referred to Enhanced Dilute Phase for the coming discussion.

At a particular conveying distance, the system operation will be considered in (3) manners:

*Standard Dilute* – Single speed operation with conservative design inputs and safety factors

*Optimized Dilute* – Single blower speed operation with blower speed optimized (VFD)

*Enhanced Dilute* – Blower speed controlled by an algorithm reacting to field conditions

Theoretically, some operational efficiency is gained with each mode as airflows and pressures are reduced. If we introduce a second distance (destination) into the discussion, the blower can now be operated in (5) different ways which include permutations on keeping the same blower speed (from the longer distance) or "resizing" the blower requirements for the shorter distance and operating the system at the revised blower speed (See Figure 4). Introducing a third, even shorter, distance magnifies this point between the longest and shortest runs and significant power reductions can be realized between the two extremes. Primarily the shorter distance requires less pressure drop for the blower to overcome in order to operate the system. For Enhanced Dilute Phase, pressure could be used as an input to indicate to what degree the length has been reduced as opposed to length itself.

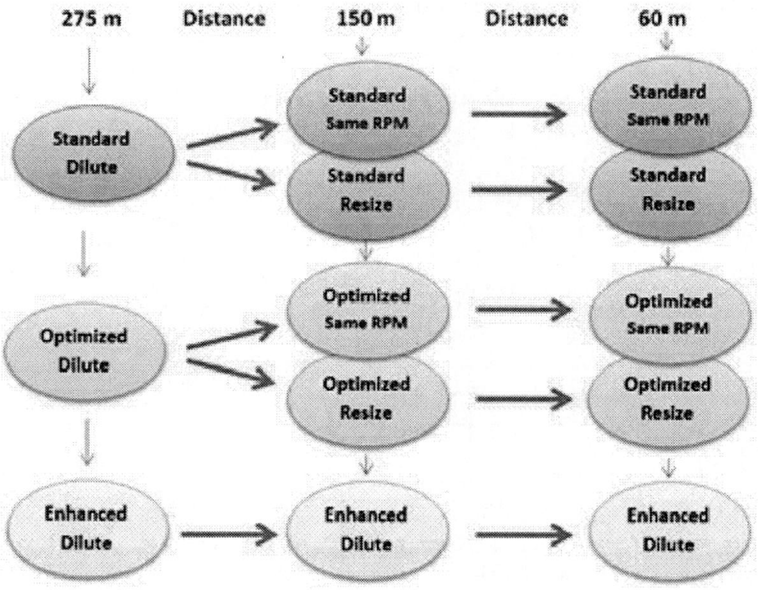


Figure 4. Design and operational variability as convey distance is reduced

Within each of the operational conditions shown in Figure 4, there is an additional variable of convey rate to be considered. It is assumed in most cases that the system will be normally operating at maximum design rate. However, as previously noted there are several common situations that cause the actual convey rate to be less than the design rate. There is also the case where the convey rate goes to zero: the blower is operating but the flow of material is halted in order to satisfy a process requirement. In any constant blower speed system, the lower the actual convey rate the less efficient the system will operate with the least efficient being zero rate.

Conversely, an algorithm that would sense the reduced convey rate passing through the convey line would be able to adjust the blower speed to match the new conditions. While using mass flow rate of material as an input could be an effective approach, the measurement is not normally available in a typical system. However, the fluctuation in mass rate is captured in the pressure drop of the system and can be compensated indirectly in the system pressure.

## Enhanced Algorithm

It has been established that an algorithm to control the speed of the blower would be able to make reductions in power consumption associated with distance, feed rate and excess velocity. Pressure is the common variable that ties the factors together in a tangible form. Pressure enters into the overall system design by strongly influencing the base convey air volume, rotary airlock leakage and blower performance (See Figure 5). The pressure in the convey line specifically compresses the convey air volume and sets the pickup velocity of the system. A longer system or higher convey rate will create more system pressure drop which in turn requires more uncompressed volume to maintain the same pick-up velocity. Higher pressure will also create more leakage across the rotary valve which must be generated by the blower in order for the system to operate. Finally, the differential pressure the blower must overcome to pump the required (total) air volume directly influences the blower speed. The enhanced dilute phase algorithm can then dramatically reduce the blower speed while allowing the system to operate reliably. The reduction in blower speed for the convey duty then creates efficiency.

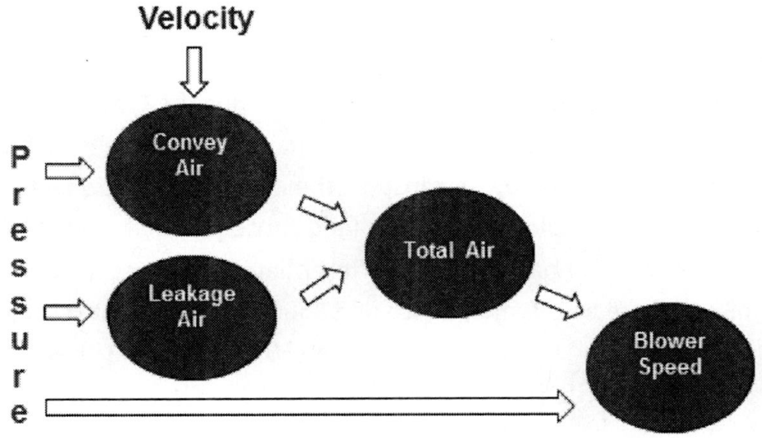


Figure 5. Influences of pressure on an enhanced dilute phase algorithm

## Experiment

An experiment was designed to quantify the benefits of using an algorithm to control the speed of the blower versus operating at a constant speed. Because there is a gradient from minimum blower speed to operate the system to maximum safe operating speed on the blower, (2) reference points for constant speed are used. *Standard* blower design is a blower speed that incorporates the longest distance, highest rate and appropriate safety design factors. *Optimized* blower design incorporates a resize of the blower speed for the shorter distance with much of the safety factor removed. All tests were conducted using pressure dilute phase flow.

A convey loop consisting of 100 mm stainless steel pipe was used that can be configured for multiple distances. Considering horizontal, vertical and elbow fittings, pipe runs with equivalent lengths of 60 m, 150 m and 275 m were used. A 45 KW positive blower capable of 1200 $M^3$/hr provided the conveying air volume and was controlled by a variable speed drive (VFD).

Tests were conducted on fiberglass filled polyethylene pellets with a bulk density of 600 kg/$m^3$. A gravimetric weigh belt feeder was used to set the convey rate of 8 TPH maximum rate and 5 TPH reduced rate. A rotary airlock valve of 250 mm diameter flanges and 17 L/rev internal volume was used deposit the material into the pipeline.

The experiment begins by operating the system at the longest distance (275 m) and the highest rate (8 TPH) and setting the blower at a standard design speed. Pressure, airflow and blower VFD power are recorded at steady state and averaged across the time frame. The rate is then reduced (5 TPH) and the operating data recorded at steady-state. Finally the material feed was turned off (0 TPH) and the blower was allowed to run with clean air only while the data was again recorded. This data set (operational data at 8, 5 and 0 TPH) was then repeated for multiple scenarios. First the data set was repeated at the same distance using an Optimized blower speed followed the Enhanced control algorithm. Next the distance was reduced to the shorter lengths of 150 m and 60 m using different permutations of Standard and Optimized blower speeds (see Figure 4) to get a data array for this material.

## Results

The results for the above experiment are listed in Table 1 showing the blower speed, convey pressure and blower power for each scenario. It should be noted that convey pressure differs from blower pressure in that it only includes the losses to transport the material. Additional pressure loss in the clean air line exists and impacts the blower pressure (and subsequently the blower power). The values for blower power are colored by tonnage in order differentiate the power usage in each data set. Values for standard dilute phase at 150 m distance and 3500 RPM were not recorded in the original testing but deemed useful in the analysis. Therefore the 150 m values are interpolated from the 275 m and 60 m values which were recorded.

| 275 m | 8 TPH | 5 TPH | 0 TPH | |
|---|---|---|---|---|
| Std Dilute | 3500 | | | RPM |
| | 538 | 385 | 200 | mbar |
| | 34.1 | 28.2 | 21.3 | KW |
| Optimized | 3300 | | | RPM |
| | 525 | 373 | 180 | mbar |
| | 30.9 | 25.0 | 18.4 | KW |
| Enhanced | 3150 | 2770 | 1950 | RPM |
| | 525 | 365 | 83 | mbar |
| | 29.0 | 19.1 | 5.1 | KW |

| 150 m | 8 TPH | 5 TPH | 0 TPH | |
|---|---|---|---|---|
| Std Dilute Same Speed | 3500 | | | RPM |
| | 365* | 370* | 140* | mbar |
| | 27.6* | 23.9* | 19.4* | KW |
| Std Dilute Resize | 3000 | | | RPM |
| | 345 | 255 | 110 | mbar |
| | 20.9 | 17.9 | 13.3 | KW |
| Optimized Same Speed | 3300 | | | RPM |
| | 350 | 260 | 130 | mbar |
| | 24.2 | 21.1 | 16.4 | KW |
| Optimized Resize | 2800 | | | RPM |
| | 385 | 250 | 95 | mbar |
| | 20.4 | 15.8 | 11.3 | KW |
| Enhanced | 2790 | 2490 | 1790 | RPM |
| | 375 | 240 | 48 | mbar |
| | 19.3 | 12.0 | 4.1 | KW |

* values interpolated to complete analysis

| 60 m | 8 TPH | 5 TPH | 0 TPH | |
|---|---|---|---|---|
| Std Dilute Same Speed | 3500 | | | RPM |
| | 240 | 185 | 95 | mbar |
| | 23.0 | 21.1 | 17.9 | KW |
| Std Dilute Resize | 2600 | | | RPM |
| | 220 | 138 | 55 | mbar |
| | 13.2 | 10.8 | 8.6 | KW |
| Optimized Same Speed | 3300 | | | RPM |
| | 225 | 170 | 83 | mbar |
| | 19.8 | 17.9 | 15 | KW |
| Optimized Resize | 2400 | | | RPM |
| | 235 | 150 | 55 | mbar |
| | 12.0 | 9.7 | 7.1 | KW |
| Enhanced | 2350 | 2110 | 1650 | RPM |
| | 250 | 160 | 28 | mbar |
| | 11.0 | 7.8 | 3.3 | KW |

Table 1. Blower speed, convey pressure and blower power for various test scenarios

As we take the power data from the various scenarios, a large division grows between the typical system operation (standard dilute phase) and other ways to operate the system in question. In Table 2 we see % reduction in power as compared to operating at the original blower conditions (3500 RPM).

| | 275 m | | | 150 m | | | 60 m | | |
|---|---|---|---|---|---|---|---|---|---|
| | 8 TPH | 5 TPH | 0 TPH | 8 TPH | 5 TPH | 0 TPH | 8 TPH | 5 TPH | 0 TPH |
| Std Dilute (Same Speed) | 0% | 0% | 0% | 0% | 0% | 0% | 0% | 0% | 0% |
| Std Dilute (Resize) | * | * | * | 24% | 25% | 32% | 43% | 49% | 52% |
| Optimized (Same Speed) | 9% | 11% | 14% | 12% | 12% | 16% | 14% | 15% | 16% |
| Optimized (Same Speed) | * | * | * | 26% | 34% | 42% | 48% | 54% | 60% |
| Enhanced | 15% | 32% | 76% | 30% | 46% | 79% | 52% | 63% | 82% |

Table 2. Power reduction percentage compared to Standard Dilute Phase operation

Plotting the power consumption vs convey rate for each distance we see the power differential.

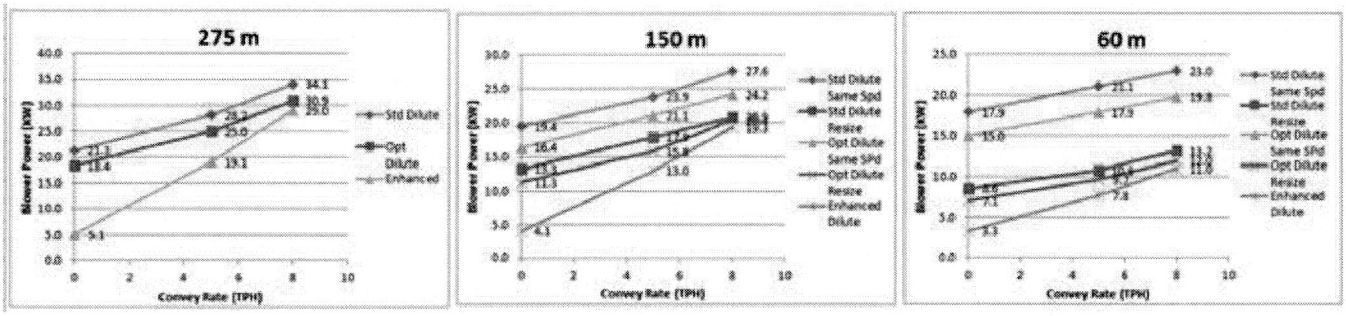


Figure 6a, 6b, 6c. Plots of the power consumption for each distance.

## Analysis

It can be seen from the experiment that blower speed plays a large role in the efficiency of the operation of a (pressure) dilute phase pneumatic convey system. It can also be seen that the blower speed can vary widely depending on how the system was designed and whether it is expected to perform other duties. The influences of distance, rate and design conservatism affect the volume of air needed and the resulting blower discharge pressure to operate the system.

For the same distance and rate, the enhanced dilute phase algorithm reduced the power consumption by 9-17% for the systems operating at full rate. Therefore this would be the minimum expected savings by controlling the blower speed compared to standard design methods.

Within each data set, the rate of material feed was reduced from 8 TPH to 5 TPH. A standard operation would not make any corrections for these typical occurrences. However the enhanced control algorithm will reduce the blower speed in order to account for the reduced material feed rate. The experiment saw a reduction in the comparative power consumption by 32-63% on the basis of rate alone.

As the conveying distance was reduced, the needed blower speed to operate the system could theoretically be reduced. As the distance was shortened the algorithm instructed a 20% and 33% reduction in speed which resulted in 30% and 52% reduction in power compared to standard methodologies.

The greatest efficiency gains could be seen when the material feed was ceased and the blower was pumping air only through the convey line. The expectation is that the power consumption of the blower will drop off considerably when the material feed is halted. However, due to the excessive airflow and resulting pressure drop the power only falls 22-38% from its peak value. The enhanced control algorithm reduces the air only power consumption 70-83% compared to its peak conveying value. The energy reduction for the comparative full speed operation is then 76-82%.

After process efficiency, air velocity can be considered. In the case of excessive blower speed, the system will experience excessive air velocity in the convey line. In the uncontrolled examples the pickup velocity of the system could rise from 10% - 80% higher than the required minimum in a controlled state. Control over the velocity in the convey line is a less tangible benefit but important to the aspects of material degradation and system abrasion.

## Summary

An algorithm that will control the speed of the air source (positive displacement blower) in order to adapt to changing conditions in the convey line can dramatically improve the operating efficiency with respect to power consumption. By measuring pressure in the system, an algorithm can make corrections for material flow rate, convey distance and design safety factors in order to reliably operate a dilute phase conveying system with 20-50% less power on average. Almost as important is the impact on air velocity reduction which can positively affect material degradation or abrasion.

# 509580 Investigations on Heat Transfer in Dilute Two-Phase Flow during Pneumatic Conveying


**Abstract:**

Multi-phase flows are present in a large amount of technical unit operations, especially in chemical, food and power industries. For many of these processing steps using hydrodynamic calculations exist good models to describe multi-phase-flow properties in e.g. heat exchangers, reactors or dryers. Especially for fluidized bed reactors many publications show the effort to describe heat-transfer phenomena for the disperse phase system gas-solid where essential progress took place within the last decades. Several mechanisms for heat-transfer differentiated in particle-particle, particle-gas and particle-wall contact have been well evaluated.

Less attention was spent on heat transfer phenomena within dilute pneumatic conveying systems, although this operation step is essential for material handling. Approaches from related disciplines like fluidized bed reactors are not directly transferable because of obvious increased air velocities up to the point of turbulence. Drying a product, followed by pneumatic conveying into a hopper or filling station, it is necessary to know the exact product temperature as possible to prevent caking or degradation of the bulk solids in which the particle temperature is not measurable during conveying.

Within the scope of this present work, experiments and approaches were carried out to determine the heat-transfer of a two-phase flow during pneumatic conveying with focus on heat convection and heat conduction. The experiments were carried out for different types of material and particle sizes respectively mass loadings varied over a wide range of scale. Furthermore, air volume flow rate was constant for several conveying gas velocities, with Reynolds numbers from non- to turbulent regime. A novel measurement device was developed capable to detect the heat transfer and successfully implemented into the testing setup. In addition, during the measurements it was possible to detected changes in flow profile as well as sedimentation effects. Finally based on data analysis, an empirical correlation is determined to predict the heat conduction of two-phase flow for the design of pneumatic conveying systems with attention to heat transfer.

# 509619 Effect of Flow Aids on Calcium Carbonate Flow Rate in the Dilute Phase Pneumatic Conveying System

**Abstract:**

Calcium Carbonate (CaCO3) is one of the difficult materials to convey in bulk solids handling system. Based on the angle of internal friction, angle wall friction, and flow function, CaCO3 needs steeper hopper angle and wider outlet dimension for mass flow, which may not be possible in many situation due to headroom requirement. In that situation, it is common to use flow aids such as fluidizer, impactors, vibrators and moving bottom. In this study, effectiveness of these flow aids fluidizer and impactors were analyzed. The vacuum dilute phase pneumatic conveying experimental set-up for the study was selected with 125 feet horizontal and 25 feet vertical line in a 4â€ sch 10 carbon steel pipe. Three target flow rate of 3000, 6000 and 9000 lb/h and three target pick-up velocities of 5250, 5750 and 6250 FPM were selected. The following flow aids fluidizer, and impactors were selected. The pressure drop of negative (vacuum) pressure of 5-7 inches of Hg were observed. Different combination of impactors (pulses/min) with and without fluidizer were studied and analyzed.

# 507118 The Effect of Fluid on Particles Conveying in Pipes


**Abstract:**

Pneumatic conveying (PC) and hydraulic conveying (HC) have significant applications in various industries. These two ways of conveying differ only by the type of the carrier fluid, one is compressible and the other is not. Even so, these two ways of conveying considered very different attracting different research groups rarely having a dialog. Therefore, one of the main objectives of our ongoing research is to compare both systems and bridge the gap to found common analysis for both PC and HC.

Pneumatic conveying is basically analyzed by showing a measured phase diagram or state diagram (Zenz diagram). The diagram presents the steady-state pressure drop per length of pipe versus the superficial gas velocity. Each line presents a constant particle mass flow rate. Hydraulic conveying of larger particles is called heterogeneous flow or settling flow where the flow is treated as two-phase flow as for PC. In such a case, it is common to analyze the flow by a similar phase diagram. However, there are a number of differences:

1. Instead of pressure drop, head loss is presented.
2. Instead of superficial fluid velocity, mix velocity is presented.
3. Instead of lines of constant solid flow rate, lines of constant volumetric concentration are presented.

In order to bridge the gap between PC and HC we designed a new feeder for HC based on PC technology, thus, a rotary valve feeder. This enabled to control the particle mass flow rate and to measure the liquid superficial velocity. This presentation will present the new design and compare phase diagrams for the same particulate materials conveyed in various fluids (air, water and brine).

# 508860 Particles in Contact: The Key Challenge in Solids Processing

**Abstract:**

One of the grand challenges in Particle Technology is how to characterize the contact mechanisms and contact forces between particles. Mechanical particle properties are the key parameters in discrete element models (DEM) which have been developed during the last decades and gradually reach "predictive power". Here and in many other aspects of Particle Technology, it is helpful to distinguish between process and materials functions. The process functions describe the effects of external parameters such as energy input, rotor revolution, machine geometry on the motion and stresses acting on particles in the process. As a result, the particles in the process or in the machine (such as mills, mixers or granulators) are stressed by these external forces and react accordingly depending on their material properties. In order to make DEM truly predictive, intrinsic material properties such as Young's modulus, hardness, plastic deformation or fracture toughness of the particles must be known quantitatively. For instance, we have shown how to model packing densities of glass beads as a function of particle size in dependency of intrinsic material properties. However, measured particle properties (and their distributions) are only known in a few rare cases.

The (nano-)mechanics of particles across all length scales down to the nanoscale are highly relevant for many applications including adhesion, friction, powder flow, comminution and tribology. Particular interesting new fields are additive manufacturing and 3D printing which "promise" a revolution in industrial manufacturing based on powder technologies. Nanomechanical elastic and inelastic particle properties have been measured systematically in a novel nanoindentation device installed within a scanning electron microscope to obtain simultaneously force-displacement information and optical images of deformed and broken particles. Quantitative data of intrinsic particle properties (and their respective distributions) are derived with high statistical relevance from force-displacement curves of single oxide, metal and polymer particles. The force-displacement of well-defined model particles can be used to obtain so far unknown distributions of stress energy and stress number distributions in mills and other solids processing devices. Finally, we show how novel 2D materials such as graphene, $MoS_2$ or BN can be produced by scalable mechanical delamination. Here, the challenge is also to characterize the delaminated nanosheets which is accomplished by advanced ultracentrifugation techniques. All these results are discussed in the context of unifying principles of product design and solids processing.

# 508411 Impact Behavior of Microparticles on Component Surfaces Considering the Micro-Topography: Experiment and DEM Simulation

**Abstract:**

The Discrete Element Method (DEM) is widely used to simulate particulate processes. The walls of apparatuses and tolls in these simulations are often calculated as smooth surfaces, and the micro-topography, for example the surface roughness, is not considered. However, with decreasing particle size the influence of the surface asperities is becoming more and more important, because the deformation of single asperities and adhesion forces in the contact area can become dominant compared with the entire deformation and adhesion of the particle. On the micro scale the contacts with asperities take place often oblique whereas the impact on the macro scale is in normal direction to the wall surface. Furthermore there is no continuous contact area and the contact force during impact is distributed over a many single asperities. This leads to an additional energy dissipation and influences the rebound angle and rotation of the particle. Therefore it is necessary to describe the influence of the surface micro-topography on the particle impact behavior depending on the particle size.

In this work the impact behavior of microparticles with component surfaces was investigated by the simulation using DEM and experiments. The main focus was on the influence of the surface micro-topography on the rebound angle and the coefficient of restitution, which depends on the ratio between kinetic energies of the translational movement after and before particle impact.

The impact tests of single microparticles with different surfaces were performed. For these experiments a novel setup was developed. The collision of spherical particles with a diameter varied in the range of 100-700 μm was captured three-dimensionally with three high-speed cameras coupled with the microscope. The particle was dropped vertically to the surface to obtain a normal impact. The trajectory, angles and velocities of the particle were measured directly during the impact and rebound. The tests were performed with a polished and an unpolished titanium surface and smooth polystyrene microparticles. The experiments were repeated several times for each particle size and surface to measure the distribution of rebound angle and coefficient of restitution depending on the micro-topography.

To study the micromechanisms of the collision the DEM simulation was used. The titanium samples investigated by experiments were measured by Scanning Probe Microscopy (SPM) imaging with a nanoindenter (TI Premier, Hysitron) to obtain the micro-topography of surfaces. The SPM imaging was performed with a diamond Berkovich tip (with a tip radius of 100 nm) at a constant normal force of 2 μN. A total area of 2 mm x 2 mm was scanned by 40 x 40 consecutive areas with a size of 50 μm x 50 μm. The total area was reconstructed by merging the measured areas of the SPM imaging. A CAD model of this area with the resolved micro-topography was obtained by reverse engineering for the polished and unpolished surfaces. The point cloud of the surface obtained from the SPM imaging was meshed in an unstructured triangulated surface and implemented directly in the DEM simulation environment. The mesh resolution of the surfaces was varied to investigate the influence of the triangle size on the simulation results. With coarsening and refinement of the triangulated grid a mesh convergence could be obtained. Additionally an ideal smooth surface was simulated. The roughness of the polystyrene particles was neglected in the simulations, because it is by a factor of three to five smaller than the roughness of the investigated titanium surfaces. The required parameters for the contact model, like the coefficient of restitution and elastic modulus, were obtained by the impact tests and nanoindentation of the samples and particles. The impact point of the particles on the microstructured surface was randomly varied in the simulation to obtain the distribution of the parameters which describes the influence of the surface roughness. The simulation results for the different surfaces (unpolished, polished and ideal smooth) were compared with the experiments using the mean values and distributions of the coefficient of restitution and the rebound angle.

# 512547 Keynote Presentation: The Advancement of Fluid/Particle Separation for Environmental Protection


**Abstract:**

Environmental and health protection efforts continue to be key challenges and opportunities for the scientific community. Whether the challenge is air pollution caused by the particulate matter (e.g. PM2.5), fine particle removal for wastewater discharge/recycling, or sufficient dewatering of a solid for disposal, fluid/particle separation technologies have been playing key roles and evolving over the years. Recent technology developments and the future outlook are reviewed in this presentation. For clean air, filter media development is the major focus. Advanced filtration media allow for less fuel consumption in energy generation and water purification process and, as a result, reduces emissions.

For effluent water, improvements to traditional solid/liquid separation technologies, such as clarifier design and the use of membrane filters, have allowed industries to meet more stringent regulatory requirements. In the area of solid waste reduction, innovative technologies are also being developed. A new chemical amendment for difficult to dewater tailings is highlighted as an example.

# 506343 Atomic Scale Engineering of Polymer Fibers for Enhanced Filtration Performance

**Abstract:**

Manipulation of atomic scale structures of molecules in fibers provides a method to control and modify material properties, including short range forces near fiber surfaces, for practical applications. In this work mats of small PVDF electrospun fibers are polarized. The fiber structure is observed on a stabilized high resolution SEM with indications of the atomic structure of the fluorine atoms as bright spots on the SEM image. The polarized mats are evaluated for surface charge and tested for performance in the capture of nanoscale salt particles from flowing air stream.

# 507262 Wet Classification of Fine Particles Using Crossflow Filtration

**Abstract:**

1.Introduction

Due to the increasing use of particulate products with a fine and narrow particle size distribution, processes are needed to generate such particle systems. Particles below 10 µm are used for pigments, polishing compounds or pharmaceutics. Especially for these products, it is necessary that not a single coarse particle occurs. Conventional production processes like precipitation, crystallization or comminution of coarse particles cannot guarantee that all particles are smaller than a few µm or have a narrow particle size distribution at a small size. Additionally, conventional wet classification processes like hydro-sieves and centrifuges cannot separate fine particles in highly concentrated suspensions with a high selectivity. Disadvantages of these processes include the suspension concentration limit and high energy consumption.

In this work, the crossflow classification is investigated as a method to classify particles in a particle size range down to 1 µm. Crossflow classification utilizes the principles of the crossflow filtration. A suspension flows tangentially across the surface of a membrane. The pore size of the membrane is smaller than the particle size. Due to a higher pressure on the concentrate side, pure liquid permeates through the membrane (permeate flux). Meier et al. [1] suggest a new method of wet classification with the crossflow filtration. They recognized a selective deposition of fine particles. Altmann and Ripperger [2] developed a model for the particle deposition, which is based on a hydrodynamic model. The permeate flux creates a drag force on the suspended particles which transports them towards the membrane surface. The velocity gradient in the boundary layer of the crossflow creates a lift force on the particles, which transports them back into the bulk flow. In the present work, the model of the lift force by Rubin [3] is used. Both the drag and lift forces depend on the particle size. By regulating the crossflow rate and permeate flux, the separation size can be adjusted. The particles smaller than the separation size are deposited on the membrane, while the larger particles remain in the bulk flow. After rinsing of the system with a pure liquid, the particles deposited on the membrane can be resuspended by backwashing and discharged from the system. To find an optimal operating point, the influences of velocity and concentration were investigated.

2. Experimental setup

For the initial experiments, a lab-scale system. containing a pipe module with tubular membranes, which have a good resistance against backflushing, was constructed. The polypropylene (PP) membranes used have a nominal pore size of 0.2 µm. The effective membrane area was adjusted with different modules between 0.008 and 0.086 $m^2$. An automatic regulated pump produced a defined constant overflow velocity. By modulating the transmembrane pressure, the flux through the membrane was adjusted. To prevent decreasing flux caused by a particle layer on the surface, the transmembrane pressure was increased.

3. Experiments

A study of crossflow mean velocity on the separation size and selectivity was performed. The overflow velocity was varied between 0.1 and 2.5 m/s. The models for the lift force were adjusted for non-spherical particles and the influence of particle concentration and validated by experiments with different particle systems at multiple concentrations. The resuspension and discharge of the deposited fine particles was optimized. Factors affecting the stability of the resuspended particles after the classification such as particle charge were investigated.

4. Results

The experiments demonstrate successful classification of particles smaller than 5 µm. The influence of the pH on the stability of the suspension is shown. By adjusting the pH, resuspension of agglomerates into primary particles can be achieved.

[1] J. Meier, G.-M. Klein, Volker Kottke, Crossflow filtration as a new method of wet classification of ultrafine particles, Separation and Purification Technology 26 (2002) pp.43-50

[2] J. Altmann, S. Ripperger, Particle deposition and layer formation at the crossflow microfiltration, Journal of Membrane Sciense 124 (1997) 119-128

[3] G. Rubin, Widerstands- und Auftriebsbeiwerte von ruhenden, kugelförmigen Partikeln in stationären, wandnahen laminaren Grenzschichten, University of Karlsruhe, 1977

# 514884 Study of Dry Beneficiation of Coal Particles in Air Dense Medium Fluidized Bed using CFD-DEM Method

**Abstract:**

Dry beneficiation of coal particles is an effective method of removing the ash, sulfide minerals and other harmful substances from raw coal. In this study, CFD-DEM method was adopted to study the dry beneficiation of coal particles in dense bubbling fluidized bed, where coarse-grained model is employed to simulate the motion of large amount of Geldart B magnetite powders and DEM is employed to model the motion of coal particles. In the CFD-DEM model, the weight function method and the center-count method are used to calculate the solid fraction and velocity of coal particles and coarse-grained magnetite powders, respectively. It was shown that (1) the simulated separation process of coal particles is in a good agreement with the experimental data under both of the superficial gas velocities of $1.5U_{mf}$ and $1.75U_{mf}$; (2) The light particles with densities of 1400kg/m$^3$ and 1700kg/m$^3$ are easily floated to the bed surface while the dense particles with densities of 2300kg/m$^3$ and 2500kg/m$^3$ settle down slowly to the bed bottom at the superficial gas velocity of $1.5U_{mf}$. But for the higher velocity of $1.75U_{mf}$, the particles with middle densities of 1700kg/m$^3$, 2000kg/m$^3$ and 2300kg/m$^3$ cannot be separated well due to the entrainment of large bubbles; (3) The force analysis indicates that the dense particles of 2300kg/m$^3$ and 2500kg/m$^3$ have a greater resistance force from dense medium particles in separation process. In summary, the CFD-DEM method is an effective method for studying dry beneficiation of coal particles.

# 507737 A Compartmental Population Balance Modelling of Continuous Twin Screw Wet Granulation Using Mechanistic Process Kernels

**Abstract:**

Wet granulation is a critical process to create larger stable granules from fine powders in pharmaceutical manufacturing. Twin screw granulation (TSG) becomes increasingly popular as a method of wet granulation in the pharmaceutical industry as part of continuous manufacturing flowsheet due to its customized configuration, short residence time, relative insensitivity to changes in formulation properties and ability to produce porous granules suitable for downstream compression. TSG is very different from traditional batch high shear wet granulation (HSWG) and there is evidence that the balance of rate processes within is quite different. For example, breakage of large wet granules largely controlled by screw element geometry, and subsequent layering of unwet fines is a very important process in the TSG. Thus, rate expressions used in population balance models (PBM) for HSWG, or even for ball mills, may not be applicable to the TSG. This paper presents a compartmental population balance modelling of twin screw wet granulation with a mechanistic understanding of the rate processing kernels. To develop and validate the rate expressions, we use experiments explicitly designed to study a single rate process in a single type of screw element. As granule breakage is a key mechanism in controlling the granule size, breakage-isolated experiments are carried out in the conveying and distributive mixing elements. The dominant breakage mechanism in the conveying and distributive mixing elements is chipping and fragmentation respectively. Based on the breakage results in different elements, the breakage kernels including the selection function and the breakage function are mathematically formed. Other rate process kernels were also clearly specified in each compartment via mechanistic approach. Model results from the compartment based PBM are compared with experimental results for TSG of a model formulation over a range of liquid to solid ratios. Avenues to improve the predictive capacity by coupling PBM with DEM (Discrete element method) are also discussed.

# 508166 Understanding Powder Transfer in Force Feeders of Tableting Machines: A Comparison


**Abstract:**

Solid dosage forms, in particular tablets constitute the most frequent application form of active pharmaceutical ingredients (API) to patients. During tablet production, the three different processing stages, namely die filling, powder compaction, and tablet ejection, influence the tablet quality attributes to a different extent. The first stage, the die filling, is mainly defining the dose and its uniformity as it comprises the powder transfer from the blending hopper through the feeding system of the tableting machine into the dies of the rotating turret. For decades tablet press manufacturers have established a unique experience in pharmaceutical materials processing and tablet machine design however their "gut feeling" lacks a scientific basis. Although each of them have their own features and design components, only two main mechanisms for powder transfer from the hopper into the dies exist: gravity and force feeding. The latter mechanism is more popular and consists of one or more rotating paddle wheels in the feeder, which force the powder into the dies.

Supported by the increase of computational power, numerical methods that simulate the behavior of granular matter have gained increased attention also in the pharmaceutical industry within the last ten years. The *discrete element method* (DEM) is a numerical model that tracks the positions and velocities of individual particles through Newton's second law of motion thus providing a high resolution of particle based systems. Although some groups investigated the powder flow in force feeders through the DEM and improved process understanding, some limitations still exist. Those include the consideration of large particle sizes and material properties that are not resembling pharmaceutical powders and the neglect of the poly-disperse particle size nature of tableting blends. In this study, light is shed on the powder flow in force feeders of different tableting machines by means of the DEM. An iterative calibration procedure allowed the computation of a model direct compression formulation. First, the powder flow for two different sets of paddle wheel shapes in one tableting machine (FETTE) is studied. Afterwards the performance of this machine is compared with another one (KIKUSUI) to identify the setup that provides best product quality in terms of API content and tablet mass.

The computations were carried out using the open source DEM code known as LIGGGHTS. The material properties mimic a low dose strength (5% (w/w) API content) direct compression formulation including one API and six excipient particle sizes. The micro-mechanical particle properties were calibrated through an iterative process whereby the powder flowability of various commonly used direct compression diluents was compared with the numerical blend. Two different lab-scale tableting machines ("FETTE": FETTE 1200i, Fette compacting GmbH, Schwarzenbek, Germany) and ("KIKUSUI": KIKUSUI VIRGO, Kikusui Seisakusho Ltd. Kyoto, Japan) were compared. In the FETTE, the hopper is located in the middle of the feeder on top of the first paddle wheel, known as dosing wheel. The dosing wheel is transferring the powder to two paddle wheels below, namely the filling wheel which is the first filling station and which is rotating in opposite direction of the turret, and the reverse dosing wheel. The latter constitutes the second die filling station which is rotating in same direction as the turret with 24 punch stations. The reverse dosing wheel has angular shaped blades whereas for the dosing and filling wheel two different configurations are available. The first set includes rod like shaped blades that are alternating arranged and penetrating half of the powder bed height. The second set differs in two ways. First the blade shape is angular and second the blade is of the same height as the geometry demarcation. The KIKUSUI with 18 punch stations contains only two opposed rotating paddle wheels that are, in contrast to the

FETTE paddle wheels, overlapping. The feeding hopper is located above the left paddle wheel and the feeding plate contains an additional volume above the paddle wheels.

The numerical setup was adapted according to experiments which included an iterative filling of the system. Afterwards the actual high-speed tableting started by increasing the turret speed to obtain a throughput of 57,600 tablet per hour (40 rpm for FETTE and 53.3 rpm for KIKUSUI) at a paddle wheel speed of 30 rpm. The computations were carried out until in total 240 dies have been filled.

First of all, the powder flow in the force feeder of FETTE will be compared for the two different sets of paddle wheel shapes followed by a tablet quality comparison of the tableting machines FETTE and KIKUSUI.

The results show that the filling of the FETTE system itself caused particle size segregation for both paddle wheel sets that is then also found during steady state tableting. During steady state the following descending order of API content was found in different locations: for the rod like shaped paddle wheel combination: reverse dosing wheel (11.3%) > hopper (4.8%) dosing wheel (4.2%) > filling wheel (3.4%), for the angular shaped paddle wheel combination: reverse dosing wheel (14.2%) > dosing wheel (4.9%) > filling wheel (4.2%) > hopper (3.4%). These API content differences were also transferred into the die as follows.

In case of the rod like shaped paddle wheels, 85.3 mg of API mass entered the die underneath the filling followed by an additional 125.6 mg during reverse dosing wheel zone. Afterwards 5.3 mg API were scraped during dosing cam in which the lower punch was moved upwards by 3 mm to obtain a more confined powder bed that would subsequently be subjected to the compaction process. In contrast, for the angular shaped dosing and filling wheel combination more than double the API amount (175.6 mg) entered the die from filling and 120.8 mg from reverse dosing wheel zone. In other words, in the reverse dosing wheel, which was the same for both paddle wheel shape sets, a similar amount of API was transferred into the die. After scrapping of 16.8 mg API mass during dosing cam, 36.1% more API mass (279.6 mg vs 205.5 mg) compared to the rod like shaped paddle wheel set were in the filled dies. These differences between API masses can be explained by two phenomena. First a different API content in the paddle wheel zones (confer above) was found and the paddle wheel shape influenced the extent of the force feeding effect. As mentioned in the shape description, the angular shaped filling wheel was transporting the complete height of the powder bed during rotation in contrast to the rod like shaped wheel in which the blades only penetrated half of the powder bed height. Thereby more mass (angular shape) could be transported from the force feeder into the dies. As a consequence the blend mass flow rates were 32.8 g/s and 25.7 g/s for the angular and rod like shaped filling wheels, respectively. Although the final tablet masses were only of minor difference (4.3%), the API contents difference of 21.7% exceeded this value unproportionally, indicating that incorrect API dosing might constitute a quality risk. However further analysis is necessary to verify this observation in long term production (> 240 dies).

In case of the KIKUSUI tableting machine, particle size segregation during steady state tableting revealed significant different API contents throughout the feeder with 5.0%, 1.2%, and 4.3% in left, and right paddle wheel zones, and hopper, respectively. The first particles entered the die underneath the right paddle wheel zone resulting in 133.3 mg API mass. As a significantly higher API content was found in left paddle wheel zone, a 1.6-fold more API amount (216.2 mg) compared to right paddle wheel zone was filled in the second die filling half. The API particles entered the die later thereby the majority of them were located in the top portion of the die thus they were scraped to a considerably high extent during dosing cam (67.4 mg or 19.3%). At the end, a final API mass of 282.1 mg was obtained which was higher compared to both paddle wheel sets in FETTE.

The comparison of all tableting machines gave the following order in terms of tablet mass: KIKUSUI 5.0 +/- 0.1 g, FETTE angular 4.9 +/- 0.1 g, and rod like shaped paddle wheel set 4.6 +/- 0.2 g. With respect to API content 5.5 +/- 1.0 %, 5.6 +/- 0.8%, and 4.6 +/- 1.1% were obtained for KIKUSUI, FETTE angular, and rod like shaped paddle wheels, respectively. Hence most significant differences between the tableting machines could be revealed for API dosing.

These distinctive differences in tablet mass and API content showed that the powder transfer from the hopper into the dies was significantly influenced by the force feeder design, in terms of number of paddle wheels and their shape in addition to the housing geometry. This study shows that the tablet quality is not only affected by the formulation and process conditions (e.g. operating speeds) but the tableting machine exerts a major impact. In next studies the powder flow mechanism will be analyzed more thoroughly by particles' velocity and coloring, residence time, and size segregation comparison to provide a more detailed understanding.

# 508215 A Development of Combined PBM-RTD Technique to Simulate Continuous Wet Granulation Systems for Pharmaceutical Manufacturing Processes

**Abstract:**

The healthcare requirements of the developing countries in the world are expected to grow in keeping pace with their burgeoning populations. In efforts to meet the current and future projected supply-demand figures, there is a paradigm shift in pharmaceutical manufacturing towards producing the drug products (most commonly tablets) in a continuous operation as opposed to traditional batch methods of manufacturing in the industry. Granulation is a key unit operation that is responsible for the reduction in the segregation of the components, increase in the blend composition and, is responsible for producing particles with homogeneous particle size distributions. In an effort to predict the outlet compositions of the particles of a continuous granulator, the study of residence time distributions (RTDs) is crucial in order to predict the effects of variabilities in the inlet parameters such as feeder fluctuations and blend inhomogeneity on the outlet granule particle sizes and compositions. Population Balance Models (PBMs) are developed and extensively employed in modelling Wet Granulation systems in order to track the particle size distributions (PSDs) and porosities of the particles. Therefore, this work has initially studied two (2) different RTD models, compared them, and has applied them to a continuous granulation system. The RTD models are being combined with a PBM model in order to develop a model that quantifies the processes and attempts to describe the phenomena in as mechanistically as possible.

RTD models in granulation literature have been modelled in similar vein as the models for continuous liquid reactor systems [1]. The models account for plug flow volume fractions (conveying of material), number of continuous stirred tanks (back-mixing within mixing and kneading sections), and dead volume fraction (stagnant material within the unit operation). RTD models for hot-melt extrusion process (HMEs), which occur in similar equipment as granulation in polymeric media, have been given similar treatment in literature. The HME models too are described as reactor trains, and have even been modelled as cascading series of stirred tanks with recycle flows added from the subsequent reactor to the preceding tanks [4].

The present study has initially done a case study of two kinds of cascading reactor models for granulation operation. The first one, that has been published and studies for HME systems assumes that the ratio between the amount of material recycle to the previous tank to the amount of fresh feed coming into the current tank is constant for all the tanks in the cascade. This model holds true when the flow of material in either direction is uniform in any particular local mixing section, but the degree of mixing decreases gradually as the material is conveyed forward. The other approach that was developed in this work models the system such that the amount recycle to a previous stage is a fraction of the total inlet coming into the section. This accounts for the case when the degree of mixing is uniform in all the sections of the operation but the amount of material exchanged is different.

The two different models are being developed and studied in detail. They are applied to an experimental case of continuous granulation whereby, the model predictions and their closeness to experimental results are being compared.

An initial experimental study on a twin screw granulator has been performed using a blend of caffeine, microcrystalline cellulose (MCC), lactose and polyvinylpyrrolidone (PVP) across a DOE consisting of variation in powder feed rate, liquid content and screw rotation speed. The RTD experiments were carried out on the DOE using the pulse method of tracer study using caffeine (one of the components of the formulation). The exit concentration was measured in real time using an inline near infrared probe (NIR). The NIR was calibrated with blends containing the aforementioned components in different combinations of proportions in order to measure real time changes in caffeine concentration due to addition of pulse tracer. The initial test results showed that the mean residence time (MRT) decreased with an increase of either the screw rotation speed (RPM) or increase in the feeder powder flow rate.

Further refined studies were carried out using the same components and experimental methods on the full DOE and the data is being fitted to the various RTD models. It is hoped that the RTD model would rely less on the similarity and approximation to the various TIS models developed for traditional liquid reactor systems and would be more inherently mechanistic and derived from basic principles applicable to granulation and pharmaceutical material systems.

In order to achieve the above mentioned goals, it is envisioned that the traditionally developed PBM equations [3] would be employed for estimating the particle size growth and porosity change as the primary powder particles are transformed into granules. The PBM equations take into account the binary aggregation of the smaller particles, breakage of larger particles and granules into smaller ones due to shear from the screws, liquid addition due to binder spraying, liquid evaporation due to hot air drying, and gas consolidation due to compression of the particles. The aggregation rate is generally considered to be proportional to the number of particles and the fractional area of the wetted region in the colliding particles. The breakage rate is primarily proportional to the screw speed and a power law factor to the particle size. These four elementary particulate rates are responsible for the changes in particle sizes and porosity.

Combining PBM with RTD models hasn't been reported yet in pharmaceutical granulation modelling literature. However, a combined approach has been employed in modelling the rate processes of colloidal aggregation of beta-lactose globulin in calcium chloride solutions [2]. Therefore, it is envisioned that a similar approach would lead to a combined PBM/ RTD model for simulating continuous pharmaceutical granulation. It is hoped that the model developed would be as mechanistic as possible that would lead to reduction in dependence on experimental data and parameter fitting/ assumptions often involved in the current semi- empirical models.

Such models would then be employed in the simulation of a continuous pharmaceutical process line with integrated process control models. The integrated models would then be useful for designing quality manufacturing setups which would provide the advantages of lower equipment footprint, reduced manufacturing cost and reduced human intervention in the entire pharmaceutical manufacturing process thereby, leading to high- quality pharmaceutical products at affordable prices to the uninsured and under-insured consumers in the developed and developing countries.

References:

[1] A. Kumar *et al.*, 'Conceptual framework for model-based analysis of residence time distribution in twin-screw granulation', *European Journal of Pharmaceutical Sciences*, vol. 71, pp. 25–34, Apr. 2015.

[2] N. Erabit, F. T. Ndoye, G. Alvarez, and D. Flick, 'Coupling population balance model and residence time distribution for pilot-scale modelling of β-lactoglobulin aggregation process', *Journal of Food Engineering*, vol. 177, pp. 31–41, May 2016.

[3] A. Chaudhury, 'Mechanistic modeling, simulation and optimization of wet granulation processes', Rutgers University - Graduate School - New Brunswick, 2015.

[4] J. Puaux, G. Bozga, and A. Ainser, 'Residence time distribution in a corotating twin-screw extruder', *Chemical Engineering Science*, vol. 55, no. 9, pp. 1641–1651, May 2000.

# 508511 Development and Application of Coupled CFD-DEM Models for Complex FLUID Particle Systems (FIBROUS MATERIAL, SLURRIES ETC...) in Industrial Scale Processes

**Abstract:**

Many processes in the chemical industry involve the both fluids and particle flow.. Slurry mills, cyclones, desanders, pneumatic conveyors, grains drying and sorting, fluidized beds and catalytic reactors are just a few examples of granular-fluid systems. In order to predict the behavior of such systems accurately using a first principle modeling tool, it is important to model the effect of fluid flow on the particles. Design and scale-up, as well as optimization of such processes, require a deep understanding of the thermo-hydrodynamics, which is determined by the particle-level interactions between the fluids, particles, and boundaries.
The complexity of the fluid-solid flow present in these systems makes modeling them a challenging task. The primary source of difficulty is the wide range of length scales existing in these system, as the largest structures are a few meters whereas the momentum, heat and mass transfers occur at the micrometer scale of the interactions between particles and fluid and particles. Although, there are some inherent challenges in developing CFD-DEM models, such as large computational run times, the advent of fast GPU hardware and advanced algorithms are mitigating these limitations.

In this work, details of the DEM-CFD integration between two commercial softwares, Rocky® (DEM) and ANSYS Fluent® (CFD) will be addressed. In particular, the coupling formulation for non-spherical particles shapes and a wide variety of application that includes flexible fibrous material, slurry flows etc... will be evaluated. In addition, a few application examples, including both one-way and two-way coupling cases will also be presented. These models capture the discrete nature of the particle phase while maintaining the computational tractability. This is accomplished by not solving the detailed fluid field at the particle level, which enlarges the range of equipment and processes that can be studied with numerical simulations.

# 507311 Investigation the Complex Gas-Solids Flow Characteristics in a Fluidized Bed with Wurster Tube By Process Tomography and CFD Simulation

**Abstract:**

Wurster type fluidized beds are widely used in pharmaceutical industry for pellets coating. They are considered to be the best choice for pellets coating as they can produce a superior film compared with other coating strategies. For a Wurster type fluidized bed, there exists a concentric cylindrical tube which is called Wurster tube inside a conical chamber, and the flow region is mainly divided into two parts, i.e. the coating zone inside the Wurster tube and annulus zone outside the Wurster tube. A dual-fluid nozzle is placed in the bottom of the Wurster tube for spraying the coating solutions. In this research, process tomography sensors including electrical capacitance tomography (ECT) and microwave tomography (MWT) are designed and used to investigate the complex gas-solids flow characteristics in a Wurster type fluidized bed for coating and drying. One set of 12-4-8 combined electrical capacitance tomography (ECT) sensor was designed and used to monitor the solid concentration inside the chamber. The ECT measurement was conducted synchronously both inside and outside of the Wurster tube, i.e. coating zone and drying zone, to achieve a complete monitor of fluidised bed process with Wurster tube. The MWT sensor has 16 antennas and the measurement frequency is in the range of 1.0 GHz and 2.5 GHz. A series of tests were carried out by varying the operational parameters including the Wurster tube gap, air flow velocity and particle mass load. Different flow patterns as well as flow stability were evaluated based on the tomography measurement results. To evaluate the flow characteristics inside the bed, fast flourier transform (FFT) was applied to investigate the fluctuation in the coating zone and drying zone. The hydrodynamics of gas-solids flow with different moisture content were revealed by the tomography images and spectrum structure based on FFT analysis.

To verify and optimize the measurement result, computational fluid dynamic (CFD) based on the kinetic theory of granular flow (KTGF) is used to investigate the flow characteristics in the same bed. Different drag models, i.e. the energy-minimization multi-scale (EMMS) model, Wen-Yu and Gidaspow model, are used to investigate the gas-particle interface effect on the flow characteristics. Key process parameters, including the solids concentration, solids velocity and pressure drop, are analyzed and verified by the measured results. Both the CFD simulation and tomography measurement results provide valuable information to reveal the hydrodynamic characteristics in the Wurster type fluidized bed. With the lab-scale fluidized bed, the results of Wen-Yu and Gidaspow drag models have a good agreement with the ECT measurement. The results also indicate that the design of the Wurster tube has significant effects on the flow. Based on the CFD simulation and ECT measurement results, optimized parameters are chosen to improve the solids circulation efficiency and coating quality.

# 507075 Tracking Single Particle Temperature during Hot Melt Coating in the Free Stream


**Abstract:**

Coating plays an important role in the pharmaceutical industry to protect agents or release them in specific parts of the stomach or bowel, but as well in the food industry to protect flavors. The coating is usually applied in a fluidized bed, mainly in a Wurster coater. In this process the particles are transported and dispersed by an air flow in the center of the device while being sprayed on with the coating liquid. After coating in the center region of the Wurster coater, particles are collected again in a part with lower air velocity. The coating process itself, including spraying, hitting of the droplets on the particle, spreading on the surface and solidification or drying influences the final coating quality.

A process of particular interest is coating with hot melt. Natural substances like beeswax, carnauba wax and palm fat are interesting in the food industry, as they do not need to be declared as food additives. Hot melt coating has several advantages including a faster process, as significantly less energy needs to be transferred. Additionally, substances like fats and waxes create a more efficient barrier against vapor compared to water-soluble coatings. The coating process is more difficult to control because of higher viscosity, which leads to larger droplets and lower spreading of the material, and more complex crystallization behavior. The process parameters and the suitability of the coating depend on the crystallization temperature range, which needs to be passed to go from the melt to the solidified coating. An additional challenge is the tendency of the wax coatings to agglomerate.

It is difficult to determine temperatures, drying characteristics and kinetics during the process in the Wurster coater itself. Instead, the coating process was reproduced by fluidizing a single particle in the free stream and spraying on it similar to the work of Link [Link and Schlünder 1997]. The advantage is that the complete information of the spraying process is hence accessible, including the local air velocity.

A test rig was set up to fluidize a single particle in the free stream above a nozzle. Additionally, a second nozzle was installed above the fluidized particle to spray coating liquid. The coating liquid was applied to investigate the influence of isolated parameters on the coating quality. Spraying was done by an ultrasonic nozzle to uncouple droplet size from the nozzle's air volume flow. The coating can hence be applied in a controlled way. In the free stream the air velocity depends on the particle size and density. The particle fluidization was calculated analytically by balancing drag force and gravity force. The energy transport from the air to the sphere was calculated based on temperature and differential velocity.

The temperature of the particle was adjusted by setting the temperature in the fluidization nozzle. Spraying coating dissolved in water at room temperature changes the temperature on the surface of the sphere. The temperature evolution on the surface of the sphere was monitored by an infrared camera (see Figure 1 left and middle). The evaporation cools the particle until the liquid is completely evaporated, allowing to monitor the coating process. It was hence possible to determine the surface temperature and to monitor the process of drying of the coating on the particle. This allowed to visualize zones with applied coating during the process and to determine the duration of drying of the coating. It showed to be in the range of up to 20 s depending on the sprayed coating volume, temperature and on particle size and air flow. The coating yield in the free stream was low due to the large spraying cone compared to the particle surface size, as well as the fluidization air redirecting most of the liquid from the particle. Still this reproduces the Wurster coater process well. It was possible to create defined coating layers in the levitator. The single particle coating with hot melts, which show a complex drying or crystallization behavior, is currently in work.

The hot melt process was reproduced in a Wurster coater. Taking the crystallization temperature range into account, the process conditions can be realized similarly to solvent-based coatings. While coatings like beeswax tend to agglomerate, different melts like stearic acid could be processed well. The produced coating was investigated in micro-computed tomography to evaluate for a closed coating layer (see Figure 1 right). Hot melt coating showed to be less homogeneous compared to solvent-based coating, with delamination effects on the particle despite the low surface tension of waxes.

Figure 1: Thermography image of fluidized particle (left); Temperature evolution during spraying different volumes of solvent-based coating (middle); μCT image of a melt-coated particle (right)

**References**

Link, K. C. and E.-U. Schlünder (1997). "Fluidized bed spray granulation." Chemical Engineering and Processing: Process Intensification **36**(6): 443-457.

# 507556 The Design and Operation of Multi-Point Dust Collection Systems

**Abstract:**

It is very common in particulate-producing plants to have a vacuum dust collection system to collect fugitive dust from all manner of unit operations such as container filling, container dumping, size reduction, transfer points, etc. The design methods have been well established in the *Industrial Ventilation* manual published by the American Conference of Governmental Industrial Hygienists [1]. Nevertheless, in practice these systems are not always properly designed or maintained, thereby not providing the dust collection and subsequent worker protection for which they are intended. Following the guidance of the ACGIH manual, we will first outline the philosophy and methodologies to design a multi-point vacuum pick up system. At first glance, this would appear to be an easy task, however when one considers the multiple requirements (saltation velocity, minimum velocity to meet NFPA guidelines, required volumetric flow rate of air for dust collection at each pickup point, etc.), the problem is considerably more difficult. In practice, many systems will either not operate as expected or performance will degrade over time. We will highlight some of the pitfalls in the design and operation of these systems, such as intermittent operation of branches of a system, unintended extension of the system by adding longer lengths of flexible tubing, etc. This highlights the need for the design engineer and the system operator to work together to conceptualize and design a system which provides the level of dust collection needed and can be easily maintained for the life of the process plant.

[1] Industrial Ventilation: A Manual of Recommended Practice, 29th Ed, American Conference of Governmental Industrial Hygienists (ACGIH), 2016.

# 508790 Rejection Mechanisms of ZnS Quantum Dots and Au Nanoparticles and Selection of Membrane Filters for Ultrafiltration and Nanofiltration

**Abstract:**

Dramatic increase in demands of engineered nanomaterials can increase their release [S1] into water environments. Efforts on improving water treatment techniques to prevent the release of hazards to the environments should be made. Membrane processes are considered as a very effective and promising method for water treatments. However, particle removal[S2]mechanisms have not been fully elucidated due to complex interactions between colloids and membranes, especially for very small colloidal particles. In this study, a series of systematic filtration tests using eight different types of membrane filters against 1.7 nm ZnS quantum dots (QDs) and 5, 10 ad 20 nm Au nanoparticles (NPs) was performed to understand the rejection mechanisms. By calculating retention, recovery and adsorption efficiencies, we found that adsorption, electrostatic repulsion and reentrainment of NPs attached on membrane surfaces are the three dominant rejection mechanisms for NPs smaller than pore sizes. Results showed that each test filter had different rejection mechanisms for removing small NPs. Nylon membrane filters had a greater adsorption affinity with Au NPs compared to other membranes with even smaller nominal pore sizes. Moreover, we found electrostatic repulsion played a significant role in removing NPs, but the efficiency was affected by flux conditions, or filtration flow rate, due to hydrodynamic drags. NP and membrane properties such as zeta potential and Hamaker constant did not correlate well with membrane rejection abilities. Different membrane filters with the same nominal pore size of 0.1 μm showed different NP removal efficiencies. Our study provides the detailed insights of rejection mechanisms of various membranes and suggests a way of selecting membrane filters in different filtration purposes.

---

[S1]Your original sentence: "Dramatic increase in demands of engineered nanomaterials can result in their release [S1]into water environments." May not be true, it is not 'must or have to happen event'! This is the common problem of your English writing.

[S2]Removal 'behavior' is not commonly seen

# 511639 Discrete and Continuum Modeling of Granular Flow in Silo Discharge

**Abstract:**

The discharging of granular materials from a flat-bottomed silo is simulated by using continuum modeling and 3D discrete element method (DEM). The predictive abilities of three commonly used frictional viscosity models (Schaeffer, S-S and Î¼(I)) are evaluated by comparing with DEM data. Both the funnel flow pattern (type C) and the semi-mass flow pattern (type B) predicted by DEM simulations can be successfully represented when Schaeffer or Î¼(I) model is adopted, whereas S-S model gives a consistent type B flow pattern. Moreover, all the three models over-estimate the discharge rate when compared to DEM result. By checking the profiles of solids volume fraction and vertical velocity right above the outlet, it is found that the larger discharge rates given by Schaeffer and Î¼(I) model is due to the over-estimation of volume fraction, while the deviation appeared in S-S model stems from the failure in predicting both solid vertical velocity and volume fraction

# 507036 Improving Circulating Dehydrogenation Technology through Optimization of Fluidization

**Abstract:**

The dehydrogenation of light alkanes to corresponding light olefins is industrially an important process. One of the commercial alkane dehydrogenation circulating fluidized bed reactors was underperforming for quite some time. In addition, the catalyst losses were significantly higher than the design value. Several aspects related to fluidization, catalysis and process engineering were investigated to understand and resolve these issues in the Plant. The co-authors of this study focused on the fluidization aspects of the process, and benchmarked equipment design as well as operating conditions against industrial best practices. The team extensively studied the design of feed distributors, cyclones, internals, strippers, etc. Furthermore, attrition properties were also assessed for different types of catalyst used in the unit. Several recommendations and improvement modifications were proposed by the team in order to increase olefin production and to reduce catalyst losses significantly. This study provides insights into the improvement of fluidization behavior and systematic approach followed by the team to determine underlying root causes in the circulating fluidized bed unit.

# 508509 Hydrodynamic Characteristics of High Pressure High Temperature Counter-Current Moving Bed Chemical Looping System Circulated with Geldart Group D Oxygen Carrier Particles

**Abstract:**

The chemical looping combustion processes developed at the Ohio State University (OSU), which uses Fe-based oxygen carrier and a counter-current moving bed reducer design, have the capability of achieving nearly 100% carbon capture and the flexibility of co-generating high purity hydrogen and electricity. To accommodate the moving bed operation, the oxygen carriers are classified as Geldart Group D particles with an average particle diameter of 1.5 mm and a density of 2500 $kg/m^3$. The system consists of a reducer operated as counter-current moving bed reactor, and a combustor operated as a fluidized bed reactor, to perform reductionâ€“oxidation reaction cycles for converting fuels carbon dioxide ($CO_2$). An oxidizer can also be integrated into the system, between the reducer and combustor reactors, to produce a nearly pure stream of hydrogen ($H_2$). The oxidizer is operated in the counter-current moving bed mode similar to the reducer reactor. An L-valve is used in the chemical looping systems to transport the solids between the moving bed reactors and the fluidized bed combustor where the aeration gas flow rate is directly correlated to the global solid circulation rate. An iso-kinetic device is installed on the standpipe of the moving bed side to measure the solids circulation rate of the system. Thousands of hours of operations with different solids circulation rates and gas processing capacities have been performed on sub-pilot and pilot scales of the systems under high operating pressures (>10 atm) and temperatures (>1,000 °C). The hydrodynamic characteristics of oxygen carrier particles operated in the systems, such as pressure drops and fluidization regimes, under different operational conditions are obtained. The operational ranges of gas flow rates for the reducer and the combustor under different pressures and temperatures are obtained. The characteristic relationship between solids circulation rate and aeration gas of L-valve under different temperatures is investigated. The gas demands for fluidizing the combustor and transporting solids through the riser under varing operating temperatures and pressures are also investigated. Fluidization regimes of the combustors are identified. The slugging behavior of the combustor with high height-to-diameter ratio under elevated temperature conditions is also investigated with the aid of ECVT measurement.

# 509052 Â Defluidization Behaviour of Industrial Reactive Powders at High Temperature

**Abstract:**

Fluidized bed reactors are used in a wide range of industrial chemical processes, and they are particularly useful in high temperature systems thanks to their ability to provide high heat transfer rates and rapid solids mixing which lead to isothermal and more controllable conditions. Their applications spread the environmental, chemical, energy and process industries. For example, they are a key technology in the petroleum industry, catalytic chemicals synthesis processes, combustion and gasification of solid fuels (coal, wastes and biomass), metals productions and many more.

Operation below incipient fluidising conditions in combustion â€“ type operations can lead to particle agglomeration and sintering which change the characteristics of the bed, leading to possible bed defluidisation unscheduled process downtime and additional costs. However, the details of the physical processes under such conditions are still not well understood. Despite the countless researches that have been done so far, such processes still disclose unrevealed features and challenge the researchers worldwide.

The objective of this work is to assess the effect of high temperature on the flow properties and fluidization behaviour of such systems using carbon coke and two different titanium ores. To this end, a simple approach for the early detection of agglomeration and defluidisation in the regime below incipient fluidization - when the carbon combustion initiates the defluidization- will be followed. Such an approach is based on the simultaneous measurements of local temperatures and the pressure drop across the particle bed.

# 508544 Fluidization of Cohesive Materials Using Modified Solids Mechanics Models

**Abstract:**

Fluidization of cohesive materials is difficult to achieve. These materials tend to form channels during fluidization. They also form structures within the fluid bed that lift surround bulk material and can develop pressures and pressure gradients that are larger than the hydrostatic gradient normally experienced in fluid beds. We want to approach this phenomenon from a solid mechanics point of view. One can model the channel formed in a fluid bed as a pipe or a rathole in material that has strength and treat it as a slope stability problem â€“ except that the bulk material has pressure gradients that modify the stability of the channel surface. The basis of these equations is part of a previous work. The size of stable ratholes with gas pressure gradients can be computed using this past work. However, the motion of material in the channel during fluidization can induce a shear that is not compatible with the traditional assumptions for calculation of rathole stability from a solids mechanics point of view. Traditional theory assumes an unconfined state at the surface of the rathole and implies unique conditions on principle stress at this location. This assumption is not valid when a shear stress exists on this surface. We have reformulated the rathole stability equations with the assumption of a local shear stress at this boundary and also included gas pressure gradients. The resulting equation was designed to predict the stability of a channel formed in a bulk solid subject to movement in the channel and local gas pressure gradients. This is closer to the conditions that one would expect in the fluidization of a cohesive material.

We constructed a fluid bed and measured the standard fluidization curves as well as the size of channels formed in the bed. We also measured the unconfined yield strength of the material at low solid contact pressures and then compared this data to the values predicted from the modified rathole theory, including movement of material. An FEM analysis of gas pressures and solids stresses surrounding the channel was coupled with the results of the rathole equations to predict the channel size. We also measured the settlement time of cohesive materials as a function of the unconfined yield strength, and then created a settlement model that included gas loss through channels that formed. Finally, we correlated this behavior with the expected aerated rathole stability predictions. The results are presented in this paper.

# 507721 Dry Fine Grinding: Aspects of Particle Stabilization in Mill Classifier Circuit

**Abstract:**

In many industrial applications the demand on minerals with increasing product fineness rises. The production of these materials is often performed in closed mill-classifier circuits which enable the production of materials with high product fineness and constant product quality. Thereby, the use of dry operated stirred media mills is a promising and emerging option for the energy efficient production of fine particles due to their high stress intensity as well as high stress frequency compared to conventional ball mills. Generally, in dry grinding processes fine powders are not easy to produce and to handle since attractive particle-particle forces become more decisive with decreasing particle size. These forces are leading to a variety of challenges, in particular caused by a high state of agglomeration, a decrease of the powder flowability as well as material adherences on the machine and plant equipment. Obviously, these aspects may complicate the comminution within a mill-classifier circuit on various levels, simultaneously: As the attractive particle-particle forces determine the particle and bulk properties, they have a strong impact on the stress mechanism inside the mill, the particle classification and final product filtration as well as on the material transport and retention time inside the single devices. In order to control these forces, so-called liquid grinding aids which reduce the particle-particle forces by adsorbing on the surface of the product particles are often used as stabilizing additives. As the name suggests, these additives are especially known for improving the grinding performance inside the mill. However, there is only very little information available how far liquid grinding aids affect the process in mill-classifier circuits. Therefore, this study examines the application of grinding aids in mill-classifier-circuits with stirred media mills on different levels:

In a first step, the stressing behavior inside a dry operated stirred media mill was investigated in dependence of the grinding aid as well as the most important process parameters like stirred tip speed, bead size and bead density. Therefore, limestone was ground in a batch operated 1.6 liter laboratory stirred media mill. It became clear, that those mill parameters which determine the stress energy of the grinding beads have a strong impact on the grinding performance inside the mill. Beyond that, an enhancement of the grinding process as well as its control is effectively achieved by an appropriate application of grinding aids. Thereby, a crucial role of the powder flowability was identified: As the grinding aid determines the flow behavior of the product powder and thus the bulk flow as well as particle capturing between the grinding beads, it influences different important values like the number of stressed particles at each grinding bead collision as well as the resulting stress intensity. Hence, the grinding result strongly depends on the applied grinding aid. Additives that show different impacts on the bulk properties require different optimal process settings. Hence, those parameters need to be adjusted to the applied grinding aid very carefully in order to ensure effective fine grinding. As a rule of thumb, lower stress energies and higher collision frequencies are more favorable when adding powder flow assisting additives.

In the second step, similar experiments were carried out in a continuously operated 6.6 liter horizontal stirred media mill. Thereby it was examined, how far findings from the batch grinding experiments may be transferred to continuous grinding. It is shown that the batch results may give good indications for optimizing continuous stirred media milling. However, an overlapping impact caused by the different powder flowabilities was identified: Since the flow behavior of the product powder determines the axial material transport through the mill, grinding aids may imply a strong impact on the material hold-up as well as retention time within the mill. Thereby, a high powder flowability is not necessarily leading to a lower mill retention time. This circumstance is mainly caused by a deflector wheel in the end of the mill which does not only hold back the grinding beads but also big particles that have not been comminuted satisfactory. Thus, the particle stabilization achieved by grinding aids is affecting the grinding result on different levels simultaneously. All of these issues need to be taken into account when designing dry fine grinding processes in stirred media mills.

In addition to the grinding experiments, the impact of different grinding aids on the classifying performance in a dynamic air classifier was investigated independent of the mill-classifier circuit. Within a systematic investigation, the process parameters of the air classifier were examined in dependence of the grinding aid species, grinding aid concentration as well as fineness of the limestone fraction. As the grinding aids strongly affect the attractive particle-particle forces and thus the agglomeration behavior of the product powder, they also influence the classification performance in dynamic air classifier. Consequently, grinding aids also work as dispersion additives or classifying aids. It is shown how several classification characteristics like cut size, separation sharpness and incorrect discharge depend on the grinding aid application. Thereby, an ideal operating range with regard to the particle stabilization was identified.

Finally, grinding results from a mill-classifier circuit consisting of a dry operated horizontal stirred media mill and a deflector wheel air classifier are presented. It is shown how far the grinding aids affect the product quality, product throughput and specific energy consumption. Thereby, the results are correlated with the findings from the investigation of the single units shown above. For an even more comprehensive understanding of the pilot plant, a material circulation factor is considered which describes the mean number of material circulations inside the plant. It enables a more realistic description of the residence of the particles within the pilot plant. Again, effects of the different grinding aids are identified and are correlated with the grinding results.

# 508125 Influence of Filtration on Particle Size and Shape of Crystalline Material

**Abstract:**

The production process of pharmaceuticals and fine chemicals often includes a biotechnological synthesis. Subsequent, the target molecule has to be separated from the surrounding liquor selectively and with high purity. Therefore the crystallization is a common step in downstream processing. After the formation of the solid crystalline form, the material has to be separated from the broth using a solid-liquid-separation-step. During the separation the product is exposed to mechanical stress, which can lead to a modification of the outer appearance of the crystals.

In literature the impact of crystallization conditions, for example agitator speed or additional compounds, on the shape and agglomeration of crystalline particle systems is described [1,2]. Besides the formation of the solid phase, also the subsequent processing is influencing the particle form. Events which influence the shape of the particles can be either breakage of crystals or the formation of solid bridges due to solved compounds in the remaining liquid after mechanical dewatering. These changes in particle structure can have a high impact on solubility, bioavailability and further formulation properties of the final product [3,4]. By choosing the right separation device as well as the appropriate machine parameters, unfavorable changes in product shape can be decimated or in best case avoided. Therefore, the intensity of the following formulation steps can be reduced by adjusting the required product characteristics during the separation step.

In the following work, the influence of the separation on the crystal appearance is investigated. As criterion of the product quality the grade of agglomeration is evaluated. As model system for pharmaceutical and fine chemicals amino acids are used. For the quantification of the agglomeration grade of the crystalline systems an automated image analysis and classification was used based on literature [5,6]. The different classification methods of agglomerated and single crystals is applied to different system of crystalline amino acids.

Since this work focuses on the filtration as separation method, different types of filtration, for example pressure filtration and vacuum filtration are compared regarding the impact on the agglomeration degree as well as the particle size distribution. Automated image analysis was used to classify the crystals regarding their appearance. Therewith the grade of agglomeration before and after filtration was analysed and set into relation.

With this information the choice of the appropriate separation device can be complemented by the knowledge of the influence of the separation on the particle shape.

References

[1] D. Wieckhusen, *CHIMIA* 2006 *60 (9)*, 598. DOI: 10.2533/chimia.2006.598.

[2] E.M. Ålander, M.S. Uusi-Penttilä, Å.C. Rasmuson, *Ind. Eng. Chem. Res.* 2004 *43 (2)*, 629. DOI: 10.1021/ie0301280.

[3] S. Schmidt, D. Havekost, K. Kaiser, J. Kauling, H.-J. Henzler, *Eng. Life Sci.* 2005 *5 (3)*, 273. DOI: 10.1002/elsc.200500116.

[4] N. Blagden, M. de Matas, P.T. Gavan, P. York, *Advanced drug delivery reviews* 2007 *59 (7)*, 617. DOI: 10.1016/j.addr.2007.05.011.

[5] L.-M. Terdenge, S. Heisel, G. Schembecker, K. Wohlgemuth, *Chemical Engineering Science* 2015 *133*, 157. DOI: 10.1016/j.ces.2015.03.002.

[6] D.R. Ochsenbein, T. Vetter, S. Schorsch, M. Morari, M. Mazzotti, *Crystal Growth & Design* 2015 *15 (4)*, 1923. DOI: 10.1021/acs.cgd.5b00094.

# 506374 Prediction of Granule Critical Quality Attributes Using Twin Screw Granulation

**Abstract:**

Wet Granulation is a commonly used unit operation for particle size enlargement of fine powders to mitigate issues such as blend segregation, dust formation, and poor flowability. Twin Screw Granulation is a robust and continuous approach towards performing wet granulation at various scales of operation. A predictive understanding of the granule properties in Twin Screw Granulation is becoming increasingly important in order to ensure a Quality-by-Design approach in the wet granulation and downstream processes.

Granule breakage is found to be the dominating rate process in Twin Screw Granulation. We have performed breakage isolating experiments to understand the influence of formulation properties, process parameters, and screw design on granule breakage characteristics, specifically, granule breakage probability and daughter size distributions. A model has been proposed for the breakage rate process in Twin Screw Granulation, based on the results from the breakage isolating experiments. We have used screw pullout experiments to understand the development of the critical quality attributes of the granules as a function of the axial distance from the end of the granulator. These experiments provide further insight into the granulation mechanisms in different screw element designs. The results from the screw pullout experiments are consistent with the granule breakage model deduced from the breakage isolating experiments.

# PARALLEL GPU-BASED MONTE CARLO TECHNIQUES FOR THE FLOW-SHEET SIMULATION OF SOLID PROCESSES

*Gregor Kotalczyk, Krishan Lambach and Frank Einar Kruis, Faculty of Engineering, University Duisburg-Essen, Duisburg, Germany*

## Abstract

We describe two parallel algorithms (the deterministic 'merging' algorithm and the stochastic 'random removal' method) for the simulation of particulate flow between single units in the context of flow-sheet simulations. These algorithms are tested on an exemplary flow-sheet system modelling typical particle production processes. The mixing of several particulate instreams into single units as well as the separation of particles in screen units according to their size are included into this test system. The proposed methods are validated by comparison with numerical solutions of the corresponding sets of differential equations. Although the 'random removal' technique can be used in the context of coagulation or nucleation, we find that it leads to very high levels of statistical noise in the here presented transport implementation. These high noise levels make the method nearly inapplicable for problems of the same complexity as the here presented one. The presented introduced parallel merging algorithm shows less statistical noise and a very good reproduction of the benchmark results at the cost of greater computational complexity and thus computing times.

## Introduction

The description of particle production processes by means of a flow-sheet simulation allows a simple framework for performance analysis and optimization. The modularization of a whole process into single unit operations allows the investigation of specific settings of single units and consequences of these settings for the overall process. The solution of the corresponding equations, however, poses a subject of great difficulty, hence mostly several particle properties have to be taken into account for the complete characterization of the particulate material. The corresponding particle dynamics may exhibit strong non-linear dependencies on these multi-variate properties and are mostly described by the population balance equation (PBE) (see e.g. [1,2]). The Monte Carlo (MC) technique poses a solution method for the PBE, which allows to take several particle properties into account. Such multivariate modelling requires normally a large amount of computational resources, if a sectional method is used for the solution of the PBE (see e.g. [3]).

The modelled particle size distributions (PSD) can describe differences in particle concentrations which can span several orders of magnitudes. The computational efficient description of such large differences of particle concentrations becomes possible by the application of weighted MC simulation particles [4,5]. Recent algorithms for the nucleation, coagulation [6], condensational growth/evaporation [7] and breakage [8] of particles by means of weighted MC simulation particles have been formulated specifically for graphic processing units (GPU). It has been shown that the parallel computation of the GPU architecture could be exploited for the accelerated solution of the PBE, making thus the incorporation of this stochastic technique for the simulation of flow-sheet processes suitable.

Production processes on plant scale can be mostly formulated as a sequence of coupled unit operations (with possible tear-streams), which in turn can be described by the PBEs. There exist several approaches, to solve such a system, next to the sequential modular approach, a (simultaneous) equation-oriented approach can be chosen to describe such a coupled system (see e.g. [9,10]). The parallel formulation of the solution of the PBEs by means of the MC method is best suited for the equation-oriented approach. We present in the following a parallel simulation algorithm for particulate transport, which allows the incorporation of these several mechanisms in an operator splitting technique (see e.g. [11]), this in turn allows the analysis and optimization of a complex manufacturing process, like it is described in the next section further below.

## Simulated system

**Table 1.** Properties of the initial particle populations and the particles within the feeds.

| **Location of particle population with unit number () or Feed symbol** | **Initial population in Disperser (1)** | **In Feed 1** $F_{2,\mathrm{in}}^{(\mathrm{ext})}$ | **In Feed 2** $F_{3,\mathrm{in}}^{(\mathrm{ext})}$ |
|---|---|---|---|
| Particle concentration $n_{\mathrm{F},i}$ / kg (carrier gas) | - | $5\cdot10^{15}$ | $2\cdot10^{15}$ |
| Initial number of particles in Disperser / resp. total number of particles inserted by feeds in 28 s | $10^{10}$ | $8\cdot10^{11}$ | $3.2\cdot10^{11}$ |
| Initial particle mass in Disperser / resp. particle mass inserted by feeds in 28 s / ng | 5.98 | 6.28 | 0.07 |
| Mean geometric diameter $d_\mathrm{g}$ /nm | 50 | 10 | 3 |
| Geometric standard deviation | 1.2 | 1.46 | 1.46 |

We describe a transport algorithm for a particle production process as it is shown in Figure 1. The initial particulate material represented by large particles of the size of ca. 50 nm is dispersed into a carrier gas. These initial particles are mixed in two subsequent units with pulsed aerosol feeds carrying particles of different materials, representing size ranges of around 3nm and 10 nm, the exact properties of the simulated particles can be found in Table 1. In a real application, the particles would grow due to agglomeration and the screen would separate large agglomerated product particles (these are collected by the following filter) from smaller particles (these are directed to a mixer unit for further agglomeration). Hence the agglomeration process is not considered in this work, the screen is operating in a different size range (defined further below) than the one presented in Figure 1.

***Transport equations for the carrier gas***

The net amount of carrier mass $M_i^{(\mathrm{C})}$ (units kg) changes over time for each unit $i$ and is modeled by the following set of equations:

$$\frac{\mathrm{d}M_i^{\mathrm{C}}}{\mathrm{d}t} = \sum_{j\neq i}\left(f_{i,\mathrm{in}}^{(j)}\cdot M_j - f_{i,\mathrm{out}}^{(j)}\cdot M_i\right) + F_{i,\mathrm{in}}^{(\mathrm{ext})}, \qquad \text{with: } f_{i,\mathrm{in}}^{(j)} = f_{j,\mathrm{out}}^{(i)}\ \forall i\neq j, \tag{1}$$

where $f_{i,\text{in}}^{(j)}$ denotes the relative inflow rate (units 1/s) into unit $i$ from unit $j$ and $f_{i,\text{in}}^{(j)}$ the outflow from unit $i$ to unit $j$. The exact values are listed in Table 2. The pulsed feeds $F_{2,\text{in}}^{(\text{ext})}$ and $F_{3,\text{in}}^{(\text{ext})}$ are described by the following formula:

$$F_{2,\text{in}}^{(\text{ext})}(t) = F_{3,\text{in}}^{(\text{ext})}(t) = \begin{cases} F_{\text{in}}^{(\max)} = 0.01\text{g/sec} = & ,\text{if } \lfloor t/\tau_{\text{F}} \rfloor = 0,2,4,6,\ldots \\ 0 & ,\text{if } \lfloor t/\tau_{\text{F}} \rfloor = 1,3,5,7,\ldots \end{cases}, \quad \text{with: } \tau_{\text{F}} = 4\,\text{sec}. \tag{2}$$

The floor function $\lfloor x \rfloor$ denotes thereby the largest integer number smaller than the rational number $x$.

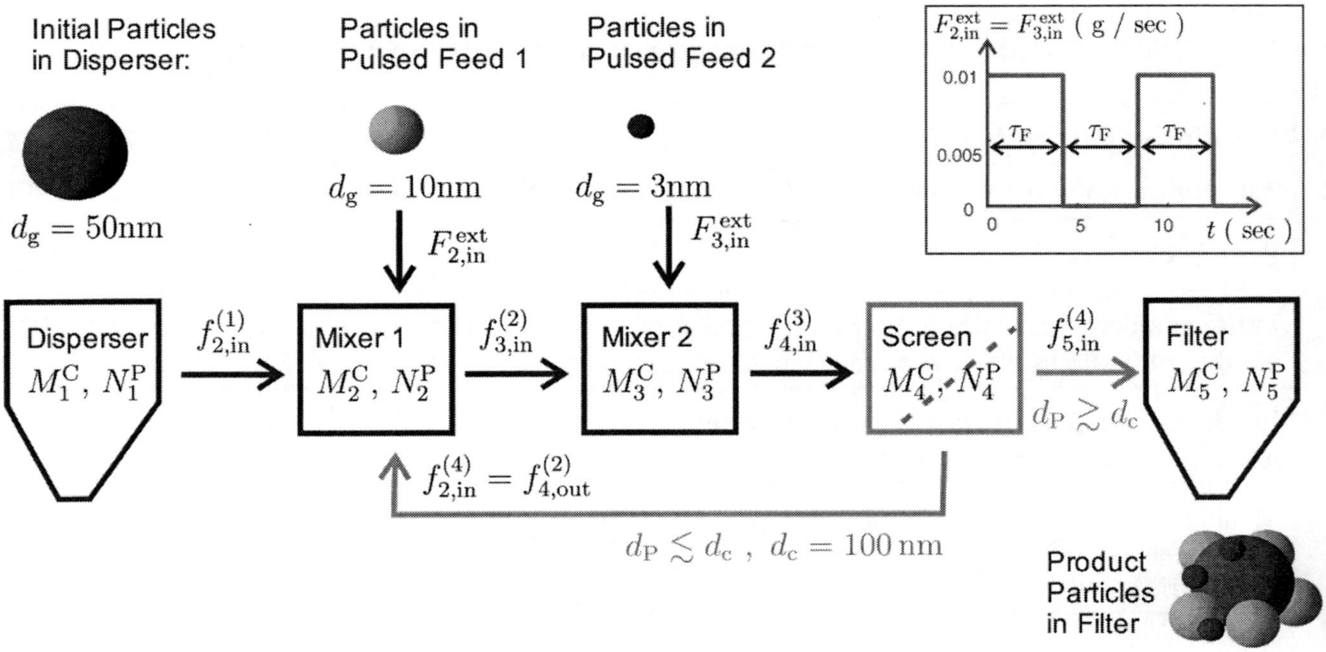


**Figure 1.** Flow-sheet of the simulated system.

**Table 2.** Initial Conditions and flow rates for each unit in Figure 1. The properties of the particle populations are listed in Table 1.

| **Unit**<br>**number in equation (1),(3)** | **Disperser**<br>**(1)** | **Mixer 1**<br>**(2)** | **Mixer 2**<br>**(3)** | **Screen**<br>**(4)** | **Filter**<br>**(5)** |
|---|---|---|---|---|---|
| Initial particle populations | Yes | No | No | No | No |
| Initial carrier gas masses $M_i^{\text{C},(0)}$ /mg | 100 | 0.1 | 0.1 | 0.1 | 0 |
| Outflow rates / $s^{-1}$ | $f_{1,\text{out}}^{(2)} = 0.2$ | $f_{2,\text{out}}^{(3)} = 1$ | $f_{3,\text{out}}^{(4)} = 1$ | $f_{4,\text{out}}^{(2)} = 0.7; f_{4,\text{out}}^{(5)} = 2$ | No |
| External pulsed feeds maximal rates / mg /s | - | $F_{2,\text{in}}^{(\text{ext})} = 10$ | $F_{3,\text{in}}^{(\text{ext})} = 10$ | - | - |

***Transport equations for particles***

We assume that the particles are mixed homogeneously and instantaneously upon arrival in each unit, so that a constant concentration $n_i^{(C)}$ (units / kg) of particles can be found throughout each unit. For the number of particles $N_i^{P}$ (dimensionless number, $N_i^{P} = n_i^{(C)} \cdot M_i^{C}$) within each unit, the following equations result for the change of the number of particles due to the transport:

$$\frac{dN_i}{dt} = \sum_{j \neq i} \left( f_{i,\text{in}}^{(j)} \cdot N_j - f_{i,\text{out}}^{(j)} \cdot N_i \right) + F_{i,\text{in}}^{(\text{ext})} \cdot n_{F,i} \text{ for } j \neq 4, \tag{3}$$

with the relative flows $f_{i,\text{in}}^{(j)}$, $f_{i,\text{in}}^{(j)}$ and absolute flows $F_{i,\text{in}}^{(\text{ext})}$ used in Equations (1) and (2) and tabulated in Table 1. The values for the particle concentrations $n_{F,i}$ within the feeds are listed in Table 1.

For the description of the screen (unit 4, $j = 4$), it is assumed, that the total outflow of the carrier mass $\left( f_{4,\text{out}}^{(2)} + f_{4,\text{out}}^{(5)} \right)$ corresponds to the correct decrease of the particle number, $N_i \cdot \left( f_{4,\text{out}}^{(2)} + f_{4,\text{out}}^{(5)} \right)$, but that the particles are either found in the stream to the second unit or to the fifth unit, depending on their volume $v$, resulting in the following modified particle flow rates:

$$f_{4,\text{out}}^{P(2)}(v) = \left( f_{4,\text{out}}^{(2)} + f_{4,\text{out}}^{(5)} \right) \cdot S(v), \quad \text{and: } f_{4,\text{out}}^{P(5)}(v) = \left( f_{4,\text{out}}^{(2)} + f_{4,\text{out}}^{(5)} \right) \cdot \left(1 - S(v)\right) \tag{4}$$

, where $S(v)$ is a separation function describing the screening process. The resulting equations (4) have to be used instead of $f_{4,\text{out}}^{(2)}(v)$ and $f_{4,\text{out}}^{(5)}(v)$ in Equation (3) for the screen unit. In this work, the following simplified separation function is applied:

$$S(v) = \begin{cases} 1 & , \quad v < v_S = \pi \cdot d_S^3 / 6, \qquad \text{with: } d_S = 40 \text{ nm} \\ (v_E - v)/(v_E - v_S) & , \quad v_S < v < v_E = v_S + 2 \cdot (\pi \cdot d_C^3 / 6 - v_S), \qquad \text{with: } d_C = 50 \text{ nm}. \\ 0 & , \quad v > v_E = \pi \cdot d_E^3 / 6 \qquad \text{with: } d_E \approx 57.8 \text{ nm} \end{cases} \tag{5}$$

## Simulation Methods

***Transport of weighted Monte Carlo (MC) particles***

The application of weighted MC particles allows to render the PSD in a more precise way [4]. Each MC particle $k$ in the unit $i$ is assigned an additional property, the statistical weight $W_k^{(i)}$. This approach allows alternative formulations to the stochastic determination, which particles are selected to leave a compartment (or unit) and to be inserted into the next one (specific transport events are defined by [12]; [13] allows multiple simulation particles to leave within one time step and formulates binomial probabilities for these events).

If weighted MC particles are simulated, exact calculation of the removed weight becomes possible. The following simple Euler approach summarizes this for the MC particle $k$ in the unit $i$:

$$W_k^{(i)}(t + \Delta t) = W_k^{(i)}(t) - W_k^{(i)}(t) \cdot \sum_{j \neq i} f_{i,\text{out}}^{(j)} \cdot \Delta t = W_k^{(i)}(t) \cdot F, \qquad \text{with: } F = 1 - \sum_{j \neq i} f_{i,\text{out}}^{(j)} \cdot \Delta t. \tag{6}$$

For each connected unit $j$ which is connected to $i$, new MC particles are added with the same properties as particle $k$ in the unit $i$ and the statistical weight:

$$W_{\text{new},k}^{(j)}(t + \Delta t) = f_{i,\text{out}}^{(j)} \cdot \Delta t \cdot W_k^{(i)}(t). \tag{7}$$

A scaling factor $F$ can be used to keep track of the depleted statistical weight for all particles, if the flowrates are independent on the particle size, like proposed by [14]. However, in case of a unit operation like the screening process, the flow rates depend on the particle properties (the volume $v$ according to Equation (4)), Equations (6) and (7) have to be evaluated for each individual particle. This might be computationally disadvantageous in the case of a serial implementation, but it is very efficient if a parallel implementation in combination with a graphic processing unit (GPU) is applied. The new particles with the weights $W^{j}_{\text{new},k}$ can be inserted into the simulation by means of parallel merging algorithms or by the random removal of existing particles.

***Merging***

Following the approach presented in [6], two particles $i$ and $j$ with the statistical weights $W_i$, $W_j$ and the volumes $v_i$ and $v_j$ are merged with each other to a new particle with $W_{\text{new}}$ and $v_{\text{new}}$ with the merge error $E_{\text{m}}$ defined via:

$$W_{\text{new}} = W_i + W_j, \qquad v_{\text{new}} = (W_i \cdot v_i + W_j \cdot v_j)/(W_i + W_j), \qquad E_{\text{m}} = \left(\left(v_i - v_j\right)/\min\left(v_i, v_j\right)\right)^2. \qquad (8)$$

While the parallel algorithm in [6] dealt with the problem to include one nucleated particle into an existing particle population by finding merge partners with low statistic weights and the lowest merge errors of a sample, the here presented work merges two entire particle populations with each other. This is done by a large number of parallel comparisons. The algorithm compares for two particles, whether the merging error is smaller than a predefined value $E_{\min}$. The pair is merged if the condition is true. The value $E_{\min}$ is increased in the course of the simulation.[1]

***Random removal?***

The random removal algorithm has been introduced in the framework of a MC simulation of coagulation, in order to ascertain a constant number of simulation particles and avoid the decrease of computational accuracy [15]. We use this algorithm with some modifications for weighted simulation particles (like described in [6]). MC particles of the streams are included into the holdup by removing either the existing particles from the holdup or the particles from the stream and redistributing the removed mass on the rest of the particle population by multiplication with a constant factor. Random numbers are used for the selection of the particles which are to be removed.

***Fixed pivot technique***

For the validation of the transport code, a fixed pivot technique is used – although defined for the simulation of coagulation [16], it can be easily applied to the given problem. A fine one-dimensional grid of 1000 pivot points is set up, spanning particle sizes from 1 nm (first pivot point) to 10 µm (last pivot point), the sizes of two adjacent pivot points increase by a factor of $10^{(1/12)} \approx 1.028$. The pivot points

[1] An increase of $E_{\min}$ by a factor of 4 is done after every 64 parallel comparison steps – if the program is not terminated earlier because of the successful merging of all populations. During each parallel comparison step, 512 MC particles from each stream are compared with 512 MC particles in the hold up. The compared pairs change with each loop iteration, so that after 512 parallel loops (7 increments of $E_{\min}$) each of the 512 MC particles in the holdup unit would have been compared with every particle from every stream, if no particles could have been merged. The data representing the particles is divided in blocks, where each block operates on its 512 particles for each stream and the holdup of the unit (if 10240 MC particles are used for the simulation, then 20 data blocks are used in parallel). We observed, that an initial setting of $E_{\min} = 10^{-4}$ leads to the successful merging of all streams long before the end of the theoretically possible 512 loops – the last loop iteration would be done with the largest value for $E_{\min}$, which is: $E_{\min} = 1.64$.

are initiated with concentrations corresponding to the particle size distributions defined in Table 1, and for each point the transport equation (3) is solved numerically.

## Results

Figure 2 shows exemplary PSDs resulting from the prolonged application of the merging technique as well as the random removal technique. The PSD in the disperser can be approximated by both methods exactly, hence no particles have to be inserted into this unit. It can be seen, that the random removal technique fails to predict the PSDs of the other units correctly. Both simulation techniques preserve the total mass of all particles in all units in the scope of the floating point precision, like it is shown in Figure 3a. The total mass in all units resulting from the pivot method, $M_{piv}$, is compared in the Figure with the total mass $M_{MC}$ resulting from MC simulations. Similar results can be seen for the number of particles ($N_{piv}$ for the pivot method and $N_{MC}$ for MC particles) in the screen unit, which is shown in Figure 3b. It can be seen, that the accuracy of the simulation can be tuned by the choice of the Euler step size and that this accuracy increases for smaller step sizes. Smaller step sizes lead, however to larger computational times, which are summarized in Table 3.

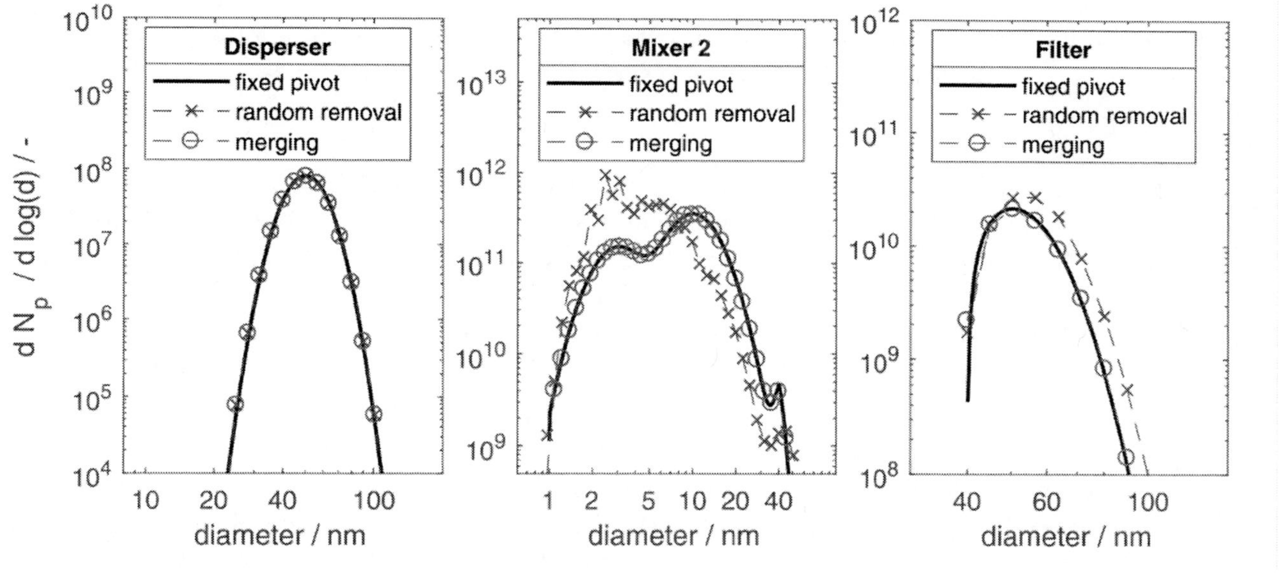


**Figure 2.** PSDs for 3 units in Figure 1 (the screen operates for particle ranges defined by Equation (5)) for a simulation time of 28 seconds, 10 240 MC particles and an Euler time step of $\Delta t = 0.1$ .

**Table 3.** Computational times for 10 240 MC particles in seconds using a NVIDIA GTX 980 Ti GPU.

| **Euler step size** | $\Delta t = 0.1$ | $\Delta t = 0.01$ | $\Delta t = 0.001$ |
|---|---|---|---|
| Merging (10 parallel simulations) | 198.9 | 2 012.7 | 19 928.6 |
| Random Removal (100 parallel simulations) | 49.5 | 128.7 | 1 181.5 |

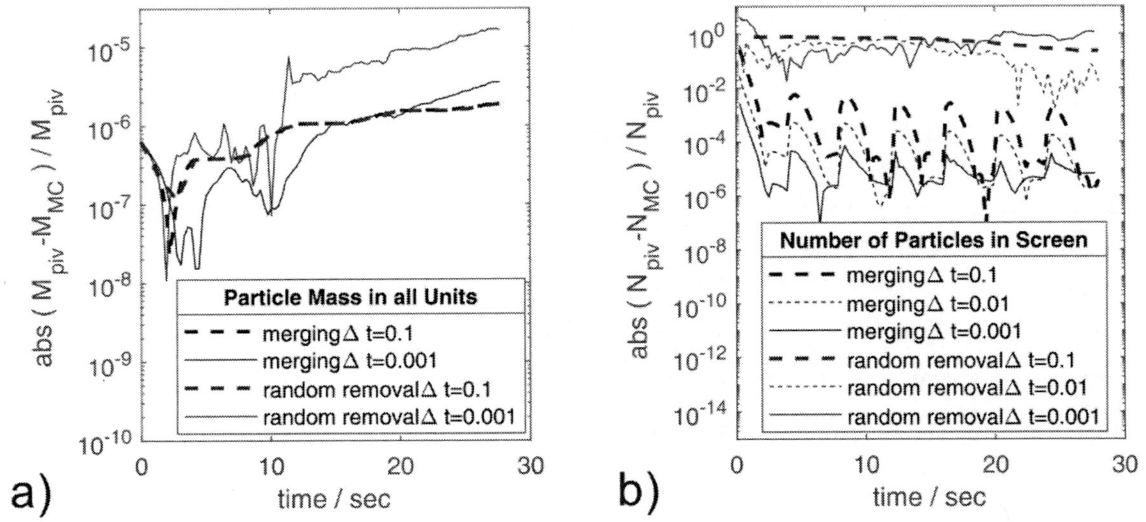


**Figure 3.** The deviations of the MC simulations from the pivot results for different values of the Euler step $\Delta t$. a) the total mass of the system is compared. b) the number concentration in the screen unit is compared. The results are shown for 10240 MC particles.

It should also be mentioned, that the random removal methods exhibit a large amount of statistical noise. The arithmetic standard deviations for the values of the number-concentration of the particles (or the geometric mean diameter or the geometric standard deviations of the PSDs) are higher than 10 % of the arithmetic mean of these values – this very high level of statistical noise disqualifies the here presented random removal technique for many flow-sheet applications.

## Conclusions

We introduce and investigate Monte Carlo (MC) simulation techniques for the particulate transport in a network of units (or compartments). The techniques are benchmarked by comparison to simulations based on the fixed pivot method. The application of parallel algorithms, designed to run on graphic processor units (GPUs) has been tested for the simulation of an exemplary production setup with recycle streams, mixers and a screen unit. The simulation of the transport alone poses a challenging task, and it could be shown, that the simulation of a deterministic outflow of particles from single units coupled to a stochastically simulated inflow (simply termed 'random removal' in this work) is not able to reproduce the benchmark results with a high level of accuracy. A deterministic method, a massive parallel merge algorithm has been introduced, and the Euler step size has been identified as a parameter, which allows to tune the accuracy at the expense of the required computational time. This algorithm proves to reproduce the benchmark results with a high accuracy and is a promising candidate for a computationally efficient simulation of a multivariate flow-sheet problem, which also includes other non-linear particle processes, like coagulation, growth, evaporation, breakage, nucleation and others.

***Acknowledgement***

This work is supported by the DFG (Deutsche Forschungsgemeinschaft) within the priority program SPP 1679 “Dynamic flowsheet simulation of interconnected solids processes”.

***References***

1. Ramkrishna, D. and M. R. Singh (2014), "Population balance modeling: Current status and future prospects", *Annual Review of Chemical and Biomolecular Engineering*, 5, pp. 123–146.
2. Ramkrishna D. (2000) ." Population balances: Theory and applications to particulate systems in engineering", New York,Acad. Press.
3. Xiong, Y. and S. E. Pratsinis (1993), "Formation of agglomerate particles by coagulation and sintering-Part I. A two-dimensional solution of the population balance equation", *Journal of Aerosol Science*, 24(3), pp. 283–300.
4. Zhao, H., Kruis, F. E. and C. Zheng (2009), "Reducing statistical noise and extending the size spectrum by applying weighted simulation particles in Monte Carlo simulation of coagulation", *Aerosol Science and Technology*, 43(8), pp. 781–793.
5. Patterson, R. I., Wagner, W. and M. Kraft (2011), "Stochastic weighted particle methods for population balance equations", *Journal of Computational Physics*, 230(19), pp. 7456–7472.
6. Kotalczyk, G. and F. E. Kruis (2017), "A Monte Carlo method for the simulation of coagulation and nucleation based on weighted particles and the concepts of stochastic resolution and merging", *Journal of Computational Physics*, 340, pp. 276–296.
7. Kotalczyk, G., Skenderovic, I. and Frank Einar Kruis, "A GPU-Based Monte Carlo Technique for the Simulation of Simultaneous Nucleation, Coagulation and Growth Based on Weighted Simulation Particles. In: *Particle Technology Forum 2016 - Core Programming Area at the 2016 AIChE Annual Meeting*. AIChE, 2016, pp. 490–497.
8. Kotalczyk, G., Devi, J. and F. E. Kruis (2017), "A time-driven constant-number Monte Carlo method for the GPU-simulation of particle breakage based on weighted simulation particles", *Powder Technology*, 317, pp. 417–429.
9. Skorych, V., Dosta, M., Hartge, E.-U. and S. Heinrich (2017), "Novel system for dynamic flowsheet simulation of solids processes", *Powder Technology*, 314, pp. 665–679.
10. Hillestad, M. and T. Hertzberg (1986), "Dynamic simulation of chemical engineering systems by the sequential modular approach", *Computers and Chemical Engineering*, 10(4), pp. 377–388.
11. Celnik, M., Patterson, R. I., Kraft, M. and W. Wagner (2007), "Coupling a stochastic soot population balance to gas-phase chemistry using operator splitting", *Combustion and Flame*, 148(3), pp. 158–176.
12. Marshall, C. L., Rajniak, P. and T. Matsoukas (2013), "Multi-component population balance modeling of granulation with continuous addition of binder", *Powder Technology*, 236, pp. 211–220.
13. Irizarry, R. (2012), "Fast compartmental Monte Carlo simulation of population balance models: Application to nanoparticle formation in nonhomogeneous conditions", *Industrial and Engineering Chemistry Research*, 51(47), pp. 15484–15496.
14. Menz, W. J., Akroyd, J. and M. Kraft (2014), "Stochastic solution of population balance equations for reactor networks", *Journal of Computational Physics*, 256, pp. 615–629.
15. Smith, M. and T. Matsoukas (1998), "Constant-number Monte Carlo simulation of population balances", *Chemical Engineering Science*, 53(9), pp. 1777–1786.
16. Kumar, S. and D. Ramkrishna (1996), "On the solution of population balance equations by discretization - I. A fixed pivot technique", *Chemical Engineering Science*, 51(8), pp. 1311–1332.

# 507254 Evaluating the Influence of Dry Granulation Processing Variables on the Rheological Properties of Granules


**Abstract:**

Granulation is often employed in the production of pharmaceutical formulations for oral solid dose delivery. It is most often required when constituent materials in a formulation have poor flow properties, enabling these components to be combined to improve process performance and ensure content uniformity.

Dry granulation, based around the use of a roller compactor and integral mill/screen, can have significant advantages over more common wet granulation techniques. Dry granulation can reduce processing costs and can also be used with moisture/solvent/heat sensitive materials. Unlike a wet granulate, the dry granulate does not require additional drying and milling before encapsulation or tabletting.

Dry granulation is not suitable for all types of material, however, there is little research into which properties make a formulation suitable for this process. In addition, there is little to indicate which process parameters produce optimal granulate quality in order to achieve interruption free processing and high quality final products.

In this study an FT4 Powder Rheometer® (Freeman Technology, Gloucestershire, UK) was used to demonstrate how processing parameters influence the properties of a placebo formulation granulate. The granulate was produced using a roller compactor (MiniPactor, Gerteis Maschinen+Processengineering AG, Switzerland) on which the roll gap and compaction force were varied. The dynamic flow, bulk and shear properties of the resulting granulates were then evaluated using the FT4 Powder Rheometer.

The data clearly demonstrates that several rheological properties, known to influence die filling efficiency and Critical Quality Attributes (CQAs) of the resulting tablets, could be controlled by changes in roller compactor settings.

This study emphasises the importance of assessing powder properties using techniques that closely simulate process conditions and how it is possible to identify Critical Process Parameters to optimise a dry granulation process and ensure high final product quality.

# 507489 Assessment of the Homogeneity of an Alumina Powders Mixture Using Rheological Parameters

**Abstract:**

This study has been carried out in order to improve the comprehension of powder mixing in the nuclear fuel manufacturing process. Thus, a binary mixture of two Alumina ($Al_2O_3$) powders, GE15 and CR6, was used to simulate the oxides powders actually involved in the process. Different mixing technics and parameters were tried and compared in terms of homogeneity of the mixtures produced. The concept of intensity of segregation [1] was used for assessing mixture homogeneity through a sampling protocol and classical statistical analysis tools [2]. One of the difficulties of working with these compounds is that, due to the similarities between CR6 and GE15 powders (same chemical composition, same colour, etc.), the direct measure of the composition of a given sample is not possible. Since homogeneity of a mixture is assessed from the variation of the composition between samples, it is important to find an indirect way to measure the composition of an Alumina powders mixture. Therefore, different correlations between measurable parameters (specific surface area, Sauter diameter, cohesion and flow function coefficient) and the composition of the mixtures were found empirically allowing indirect measurement of the composition of a given sample.

The four studied parameters: specific surface area (, measured by nitrogen adsorption using BET method), Sauter diameter (, measured by laser granulometry, dry transportation with air at 0.1 bar), cohesion and flow function coefficient ( and , measured by FT4 shear test at 9 kPa on a 10 mL cell), were compared in terms of precision, relevance and feasibility.

It appears that the homogeneity assessment methods involving rheological parameters (cohesion and flow function coefficient) show many advantages compared to the other methods. Not only are they the most accurate correlations found but both parameters are measured simultaneously, meaning that it is possible to apply both methods on the same samples while all the tests ($N_2$ adsorption, laser granulometry and FT4) are destructive. Those two methods also allow the estimation of the homogeneity at a more interesting scale (around 7 g of powder), indeed, specific surface area and Sauter diameter methods give a homogeneity at a scale of less than one gram which is not relevant for the end-use properties of the final product.

[1] P. Danckwerts, The definition, measurement and some characteristics of mixtures, Applied scientific research. Section A, 3 (1952) 279-296

[2] S. Massol-Chaudeur, H. Berthiaux, S. Muerza, et J. Dodds, A numerical model to identify the structure of a high-dilution powder mixture, *Powder Technol.*, 128 (2002) 131-138

# 508516 Characterising the Grain Shape: In Search of Size Independent Shape Descriptors

**Abstract:**

Granular materials contain a large number of grains that differ in size and shape. While the characterisation of the size distribution is typically done using sieve analysis, the characterisation of the shape of the grains remains a difficult task and is one of the major bottlenecks in the characterisation of the granular material. In this study, we attempt to characterise the shape of three different types of grains commonly used in iron and steel making industry, namely pellet, sinter and iron ore lump. While the pellet particles can be considered nearly spherical in shape, imposing the spherical shape on the iron ore lumps and highly irregular sinter particles seems a very poor choice. We consider two different size ranges of pellet particles and four different size ranges for sinter and iron ore lumps. Imaging of the grains is done using an iPhone 7 Plus camera and sufficient number of grains of each type in all the size ranges are considered. We perform the image analysis and obtain various boundary parameters (such as projected area, minor and major axis, perimeter, etc.) of the grains using MATLAB image processing toolbox. In addition, we define two additional boundary parameters to characterise the deviation of the projected area from that of a circle and calculate their values for all the different type of grains considered in this study. Using these length scales, we define a total of nine different shape descriptors and obtain the distribution of these shape descriptors for each type of grain for all the size ranges. Interestingly, for a given type of grains, the frequency distribution curves for these nine shape descriptors turn out to be nearly independent of the size range of the grains. This suggests that the chosen shape descriptors indeed represent properties related to the shape of the grains alone, irrespective of the size of the grains considered.

We investigate the correlation among these shape descriptors by evaluating the values of the correlation coefficients between each pair of the shape descriptors using the Pearson method, Kendall method and Spearman method. Our results indicate that out of the nine shape descriptors considered in this study, five shape descriptors are strongly correlated with each other and only the remaining four shape descriptors need to be considered in characterising the shape of the grains. Out of these four uncorrelated shape descriptors, two shape descriptors, namely the aspect ratio of the grains and the root of form factor, have been considered important by researchers in previous studies. The two other shape descriptors proposed in this study, which are used to characterise the deviation of the projected grain shape from that of a circle, are not measured previously to the best of our knowledge. We compare these four shape descriptors for the pellets, sinter and iron ore lumps having the same sieve size range and show that the shape descriptors for these three different types of grains are different.

We next calculate the values of these four shape descriptors for relatively larger coke particles using 2d image analysis. We also perform 3d image analysis of some of these coke particles and calculate the values of the analogous shape descriptors in 3 dimensions. We investigate if the identified 2d shape descriptors (obtained from 2d imaging) can be correlated to the analogous 3d shape descriptors (obtained from 3d imaging) and explore if it is possible to use the 2d imaging technique to estimate the 3d shape descriptors, thus eliminating the need of 3d imaging.

# 508543 Investigating Mechanical Properties of Surfactant Films at the Solid-Liquid Interface Using AFM

**Abstract:**

Corrosion is a well-known problem in the oil and gas sector and also in many other industries. Surfactant based inhibitors have been used for corrosion control in oil pipelines but their inhibition mechanism is still not fully known. Our group is studying surfactant corrosion inhibition mechanism by various characterization techniques (AFM, Electrochemical and Spectroscopic techniques). A better understanding of corrosion inhibition mechanism will lead to the design and development of not only cost-effective but also greener surfactants system for corrosion inhibition purposes.

Atomic force microscopy (AFM) is an effective tool to study the barrier properties of surfactant films at the solid-liquid interface. Determination of mechanical and topographical properties of single and mixed surfactant coatings by atomic force microscopy (AFM) can delineate the role of physical vs. electrochemical barrier by such films in the corrosion inhibition process. . So far, the AFM data, obtained using a closed cell, indicate the effect of surfactant concentration and that of hydrocarbon chains interactions on the mechanical properties of cationic surfactant films. Ultimately, integration of the AFM and electrochemical results will reveal the correlation between the surfactant structures at the solid-liquid interface and their corrosion inhibition efficacy.

# 508545 Deposition Rate Consequences of the Formation of Multi-Spherule Cluster Aggregates in Gases "Role of Momentum Shielding™"

**Abstract:**

For engineering flow systems characterized by particulate suspensions in gases, will the formation of *cluster aggregates* (CAs) in the mainstream accelerate deposition rates on immersed targets or confining walls ---or is aggregate formation more likely to suppress capture rates? Recent research on the drag properties of large cluster aggregates (e.g., stimulated in part by the Sorensen (2011) review) has opened the door to answering these important mechanism- (and pressure-) dependent questions. Sub-100 nm "primary" spherules, which may find themselves in even tenuous 'fractal-like' N-spherule clusters, can take advantage of what we call "momentum shielding"---analogous to the drag advantages experienced by birds who elect to fly in formation. We show here that the extent of this "momentum shielding" can be conveniently quantified (in a dimensionless function we call: $S_{mom}(N;Kn_1;\text{aggregate structure}))$, and used to predict the deposition-rate consequences of aggregation in aerosol flow systems when the cluster deposition mechanism is dominated by either: i) isothermal convective diffusion (C-D), ii) thermo-phoresis (TP-) or iii) inertial impaction (II). As expected, we show that for any type of aggregate structure (including 'fractal-like'), momentum shielding increases with cluster size (e.g. N) and gas pressure (via the reciprocal Knudsen number based on spherule radius $R_1$). We also show how $S_{mom}$ can be calculated for the limiting case of a N>>1 random close-packed-$D_f=3$ cluster ---which is nearly gas impermeable. For fractal-like aggregates, including DLCAs ($D_f=1.8$) and RLCAs ($D_f=2.1$) we have recently shown that the abovementioned function: $S_{mom}(N, Kn_1; D_f)$ is actually sufficient to make useful engineering estimates of CA Brownian diffusion coefficients (Rosner and Tandon (2017b), thermophoretic diffusivities (Rosner and Tandon (2017a) and aggregate "stopping times" (Rosner and Tandon (2017c)----each in terms of their previously known values ($D_1$, $\alpha_T D_1$, and $t_{p,1}$) for isolated individual spherules in the same environment.

This information now enables instructive engineering calculations of the deposition-rate consequences of spherule aggregation to form large cluster aggregates-----i.e. at the same mainstream spherule volume fraction we can now determine how much aggregation will increase or decrease the total spherule acquisition rate on an immersed target. We present here representative results for each of the abovementioned particle transport mechanisms----i.e.: i) isothermal convective-diffusion (C-D), ii) thermo-phoresis (TP-) or iii) inertial impaction (II)----- including a brief account of the consequences of dealing with Kn1-dependent coagulation-'aged' aggregate size distributions (ASDs) in the mainsteam. We summarize our current re-examination of the canonical problem of inertial impaction on a circular cylinder in Re>>1 crossflow (Rosner and Tandon (2017c), dealing now with CA-"projectiles" but in the absence of spherule loss from the much larger target. We conclude by briefly outlining our planned work on the efficient "inversion" of sampled cluster-ASD-data to reconstruct, using the abovementioned $S_{mom}(N,Kn_1;D_f$ information, what ASD must have existed in the mainstream---at least in the absence of appreciable supercritical aggregate break-up.

References:

1. Rosner, D. E. and P. Tandon (2017a). Knudsen Transition Effects on the Thermophoretic Properties of Fractal-like Aggregates; Implications for Thermophoretic Sampling of High-Pressure Flames, Aerosol Science and Technology (in press).

2. Rosner, D. E. and P. Tandon (2017b). Aggregation- and Knudsen Transition Rarefaction- Effects on Particle Mass Deposition Rates by Convective-Diffusion, Thermophoresis or Inertial Impaction; Consequences of Multi-spherule 'Momentum Shielding', Aerosol Science and Technology (in press).

3. Rosner, D. E. and P. Tandon (2017b). Mainstream Aggregation and Polydispersity Effects for Inertially-Dominated Deposition Rates on Cylinders (or Leading 'Edges') in Crossflow"---- Consequences of 'Momentum Shielding' for the Impaction of Large, "Compact" ($2<D_f<3$) Multi-spherule Aggregates, Aerosol Science and Technology (prepared for submission, August 2017).

4. Sorensen, C. M. (2011). The Mobility of Fractal Aggregates: A Review, Aerosol Science and Technology. 45: 765-779.

# 508609 Experimental and Numerical Studies on the Thermal Diffusivity of PackedÂ Powder Beds


**Abstract:**

Several industrial processes, such as calcination and drying, involve thermal treatment of particles. In order to achieve efficient processes and good quality products, the thermal properties of the particles and that of the bulk powder must be well defined. Previously, some models, such as the Krischer and Maxwell-Eucken models, have been proposed to relate particle thermal properties to that of the bulk powder. Though these models have been shown to work well, it is not very clear how to account for changes in the packing density, moisture content, and arrangement of particles. In addition, the particle thermal properties of most materials are not readily available, which limits the application of these models. To get a better understanding of the relationship between the particle thermal properties and that of the bulk powder and also to quantify the effects of packing density and moisture content on the bulk powder thermal properties, we run several experiments and discrete element method (DEM) based simulations. In both the experiments and simulations, a pipe is filled with particles where the temperature of the pipe is kept constant. We measured the temperature change in the powder bed and compute the actual thermal diffusivity of the powder bed. We tested several types of powders and glass beads in the experiments. The packing fraction, the moisture content, and size of particles were varied in the experiments. In the simulations, in addition to the packing fraction and particle size, the thermal properties of the particles were also varied. In this talk, we will present the details of the experimental and simulation results.

# 509556 Determination of Size, Size Distribution and Refractive Index of Artificial and Biological Microparticles

**Abstract:**

For the approval of clinical studies, it is mandatory to validate the production process of artificial oxygen carriers by reliably specifying the properties of the microparticles used like the newly developed blood substitute HbMP-700 (1-3). In particular, besides shape, size and size distribution, the amount of hemoglobin (Hb) and methemoglobin (metHb) content is essential.

We present a method based on the simultaneous measurement of collimated and diffuse transmission spectra in the visible and ultraviolet range and numerical analysis to infer the wavelength-dependent refractive index, size and size distribution of microparticles.

For the validation of the method, we used aqueous suspensions of monodisperse polystyrene microspheres of known concentration, determined by flow cytometric measurements. The related data analysis is carried out for synthetic and experimental data. For homogeneous spheres, Mie theory is applied to compute the solution for the scattering of a plane electromagnetic wave by a particle and to calculate the extinction cross section $C_{ext}$, i.e. the shadow cast by the particle as a function of the vacuum wavelength $\lambda$, the particle size, and the complex refractive index $n(\lambda)$. For dilute suspensions of particles, the measured quantity is an ensemble-averaged extinction cross section $C_{ext}$ . This average is obtained by integrating over the particles' size distribution, which may not be known a priori. From the resulting spectral extinction cross sections, the complex refractive index and the mean and standard deviation of the size distribution can be inferred by means of nonlinear least-squares optimization. In order to reduce the number of free parameters of the problem to a feasible amount, $n(\lambda)$ is represented in a set of harmonic-oscillator-like resonance functions. The refractive index of polystyrene microspheres obtained is in good agreement with literature values for the bulk material, the mean and standard deviation of the distribution of particle *size D* are determined with an uncertainty in the range of a few nanometers. For example, for microspheres of mean diameter mean($D$) = 2μm and distribution width std($D$) = 9.4nm the uncertainty of the mean is $u\{\text{mean}(D)\}$ = 1.2nm and the uncertainty of the distribution width amounts to $u\{\text{std}(D)\}$ = 0.4nm.

In a second step, we applied the method to characterize HbMP-700 microparticles, approximately featuring a prolate spheroidal shape with 700nm length and 400nm width. Besides shape, size and size distribution, the amount of hemoglobin (Hb) and methemoglobin (metHb) content is essential. The relative concentrations of oxyHb, deoxyHb, metHb are accessible from spectral extinction and diffuse transmission measurements, since the different Hb variants exhibit different wavelength-dependent refractive indices. The method to infer size and refractive index described above can be applied to these non-spherical particles, since the deviation of their light scattering properties from those of spheres turns out to be small. The results based on the optical method will be compared to alternative procedures for HbMP particle characterization with respect to their size and size distribution, i.e. electron microscopy, centrifugation and dynamic light scattering. In addition, x-ray spectrometry will be used to detect the ratio between $Fe^{2+}$ (oxy/dexyHb) and $Fe^{3+}$ (metHb) as well as the iron concentration.

**References**

1) Novel hemoglobin particles--promising new-generation hemoglobin-based oxygen carriers. Bäumler H, Xiong Y, Liu ZZ, Patzak A, Georgieva R. Artif Organs. 2014 Aug;38(8):708-14. doi: 10.1111/aor.12331.

2) Nonvasoconstrictive hemoglobin particles as oxygen carriers. Xiong Y, Liu ZZ, Georgieva R, Smuda K, Steffen A, Sendeski M, Voigt A, Patzak A, Bäumler H. ACS Nano. 2013 Sep 24;7(9):7454-61. doi: 10.1021/nn402073n.

3) Hemoglobin-based oxygen carrier microparticles: synthesis, properties, and in vitro and in vivo investigations. Xiong Y, Steffen A, Andreas K, Müller S, Sternberg N, Georgieva R, Bäumler H. Biomacromolecules. 2012 Oct 8;13(10):3292-300. doi: 10.1021/bm301085x.

# 517131 FlowCam NanoÃ,Â® Provides Counts, Sizes and Images of Nano-and Microparticles: Application to a Therapeutic Protein Pumping Study

**Abstract:**

Sub-visible particle characterization is becoming a more important method in assessing drug substance and drug product stability. However, the current USP protocol does not require any characterization of particles below 10 Î¼m. This greatly hampers our ability to properly assess protein stability as a function of particulate formation and obscures most sub-visible particles and various morphological properties.

In this study, we assess the capabilities of the new FlowCam Nano from Fluid Imaging Technologies, in conjunction with a pumping study that focuses on particulate formation generated during fill operations. Initial experiments show that the new FlowCam Nano can generate highly resolved images of nanoparticulates, allowing the simultaneous assessment of particulate morphology for nano- and microparticulates. This is extremely useful, as currently unpublished data (for the pumping study) show that as a function of post-pumping agitation, a significant increase in microparticle concentration coincides with a significant depletion of the nanoparticle concentration. This suggests that agglomeration (and probably not nucleation) is the more likely mechanism for larger microparticulate formation.

The ability to characterize the nano- and microparticulate morphology simultaneously, will allow us to better assess the cause of particulate formation (such as observing the transition of nanoparticles to microparticles and seeing if specific morphological properties of larger nanoparticulates are conserved in microparticulates). In addition, the FlowCam Nano is also sensitive enough to assess differences in the colloidal stability of a drug product in different formulations. During filling operations, and accelerated degradation studies, FlowCam Nano showed the differences in size distributions of IVIG in PBS and 0.25 M glycine buffer. Since IVIG is less colloidally stable in PBS, the size distribution varies significantly more per size bin and the overall mean effective spherical diameter is also ~ 600 percent greater after 24 hours of agitation in comparison to the size distribution data of IVIG in glycine. Thus, Flow Cam Nano has capabilities that are very useful in assessing drugs in the substance (e.g. optimizing drug formulations) and product (e.g. assessing particulates post fill & finish operations) phases.

Numerical and experimental estimation on the normal and tangential capillary bridge force adhered to two spheres

Kazuo Murase*, Keisuke Arai, Takato Ootsuka, Daiki Sakamoto and Futa Egawa,
*Department of Applied Chemistry, Chuo University*
*1-13-27, Kasuga, Bunkyo-ku, Tokyo,1128551, Japan*
**hmurase@kc.chuo-u.ac.jp*

## Introduction

Liquid bridges extensively occur as a binder liquid adhered to solid surfaces in a variety of engineering applications such as crystal growth, ink-jet printing, material processing and powder technology. Especially the strength of adhesion and cohesion of liquid bridges adhered to particles contribute to the consolidation of an agglomerate in wet granular materials. Estimation of the attractive or repulsive force between particles is important in controlling kinetics of agglomeration and handling of granular materials. Characteristics of a liquid bridge e.g. the equilibrium bridge geometry [1], stability [2], static and dynamic binding force [3-6], rupture energy [7,8], and wetting hysteresis [9] etc., have been experimentally and numerically investigated.

The liquid distributions within a particles bed are classified into four types of categories, i.e. pendular, funicular, capillary and slurry states, according to the liquid contents [10]. The static normal capillary force at pendular liquid state have elaborately been investigated as Fisher [11], toroidal model [8] and the exact solution of Yung-Laplace equation [12]. The dynamic normal capillary forces [13] are experimentally estimated with lubrication theory. The tangential force of a liquid bridge [14] also contributes to the primary factor for the granular consolidation with the shear stress of wet granular materials but have not yet substantially been investigated.

This work aimed to estimate experimentally and numerically the normal and tangential capillary force of a static and dynamic liquid bridge adhered to two spheres. The asymmetry static liquid bridge with the shear stress demand for the exact solution of Young-Laplace equation for the equilibrium liquid drop including the effect of gravity.

## 1. Theory

### 1.1 Formulation of a asymmetry static liquid bridge

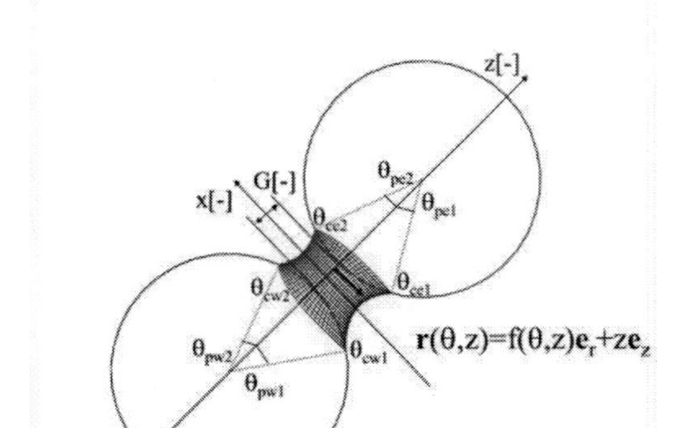


*Figure 1 Geometry of a liquid bridge adhered to two spheres with arbitrary inclination in cylindrical coordinate system.*

The asymmetry static bridge shape demands for the exact solution of the Young-Laplace equation due to the limit of the toroidal approximation. A prevalent geometry of a liquid bridge adhered to two spheres is drawn in Figure 1. The dimensionless position vector of a liquid surface in cylindrical coordinates is described as $\mathbf{r}(\theta,z) = f(\theta,z)\mathbf{e_r} + z\mathbf{e_z}$ divided by spherical radius, R', which $\mathbf{e_r}$, $\mathbf{e_z}$ are unit basis vectors in r-,z-directions. The symbol (') means the dimensional value. Fig.1 indicates the two dimensional front view with four half-filling ($\theta_{Pw1}$, $\theta_{Pw2}$, $\theta_{Pe1}$, $\theta_{Pe2}$) and contact angles ($\theta_{Cw1}$, $\theta_{Cw2}$, $\theta_{Ce1}$, $\theta_{Ce2}$).

The unit outward normal vector of a bridge surface, $\mathbf{n}_L$, is presented as follows [15],

$$\mathbf{n}_L \equiv \frac{r_\theta \times r_z}{|r_\theta \times r_z|} = \frac{f\mathbf{e_r} - f_\theta \mathbf{e}_\theta - ff_z \mathbf{e}_z}{[f^2(f^2+f_z^2)+f_\theta^2]^{1/2}} \quad \ldots(1)$$

,where low suffixes (r,θ,z) describe partial derivatives.

The dimensionless effective potential energy, $E=E'/\sigma R'^2$ with Lagrangian multiplier, $K=K'/\sigma R'^2$, on the constrained condition of a constant volume, $V=3V'/4\pi R'^3$, is described as follows [16],

$$E = \int_{-\pi/2}^{\pi/2}\int_{z_{min}}^{z_{max}} 2\left[f^2\left(f^2+f_z^2\right)+f_\theta^2\right]^{1/2} d\theta dz + \frac{Bo}{2}\int_{-\pi/2}^{\pi/2}\int_{z_{min}}^{z_{max}} (z\sin\beta + x\cos\beta) f^2 d\theta dz$$
$$+ K\left[V - \frac{1}{2}\int_{-\pi/2}^{\pi/2}\int_{z_{cp1}}^{z_{cp4}} f^2 d\theta dz\right] \quad \ldots(2)$$

Bond number (Bo) is a dimensionless number defined by $(\rho_L-\rho_G)gR'^2/\sigma$, where is the ratio of gravitational force to surface tension force equally as Etovoes number (Eoe).

The integral domain of a bridge surface is determined by the trapezoidal section of the upper and lower surface lines, [$z_{w2}<z<z_{e2}$ at $\theta=\pi/2$, $z_{w1}<z<z_{e1}$ at $-\pi/2$] and contact lines, [$z_{w1}<z<z_{w2}$, $z_{e1}<z<z_{e2}$ at $-\pi/2<\theta<\pi/2$] with four contact points settled by four half-filling angles ($\theta_{Pe1},\theta_{Pe2},\theta_{Pw1},\theta_{Pw2}$) in Fig. 1. Two contact lines on the upper and lower spheres were approximated by ellipse lines with parametric aspect ratios, which pass through respectively pairs of two contact points settled with half filling angles, ($\theta_{Pw1},\theta_{Pw2}$) and ($\theta_{Pe1},\theta_{Pe2}$). Filling angles, ($\theta_{PW}(\theta)$, $\theta_{PE}(\theta)$) on the lower and upper spheres have variable properties along the circumferential direction. The formulation of a static bridge shape at equilibrium state results in the extremum problem, which corresponds to the condition that the first order variation equals to zero, $\delta E=0$.

### 1.2 Estimation of the capillary force of a liquid bridge

The dimensionless mean curvature (H) and contact angles ($\theta_c$) of a bridge surface are respectively calculated by eqs. (3),(4) once the outline of a liquid bridge is numerically evaluated.

$$-2H=\nabla_s\cdot\mathbf{n}_L=\left[(\alpha^2+f_\theta^{\ 2})-f(f^2+f_\theta^{\ 2})f_{zz}-f(1+f_z^{\ 2})f_{\theta\theta}+2ff_zf_\theta f_{\theta z}\right]/\alpha^3 \quad \ldots(3)$$

$$,where\ is\ \ \alpha^2=f^2(1+f_z^2)+f_\theta^2$$

$$\theta_c(\theta)=\cos^{-1}(\mathbf{n}_s\cdot\mathbf{n}_L) \quad \ldots(4)$$

,where contact angles also depend on the circumferential angle.

According to the Young-Laplace equation, the dimensionless pressure difference, $\Delta P=\Delta P'R'/\sigma$, across a liquid surface is related with the mean curvature of the surface.

$$\Delta P=(-2H) \quad \ldots(5)$$

The capillary force on a liquid bridge adhered to two spheres must be evaluated only on the contact line as known by the boundary method due to unspecified neck radius of a bridge and asymmetric condition. Dimensionless averaged capillary forces, $F_{Ci}=F'_{Ci}/(2\pi R'\sigma)$ on the upper and lower sphere (i=1,2 respectively correspond to lower, upper sphere) are presented as follows by three terms, which correspond to the surface tension, hydrostatic pressure and buoyancy terms along the longitudinal angle ($\theta$). The sign of the capillary force set to be positive for attractive direction.

$$F_{ci}=\frac{1}{\pi}\int_{-\pi/2}^{\pi/2}\sin(\theta_{Pi})\sin(\theta_{Pi}+\theta_{Ci})d\theta-\frac{1}{2\pi}\int_{-\pi/2}^{\pi/2}\sin^2(\theta_{Pi})\Delta Pd\theta\mp\frac{2}{3}\mathrm{Bo}\cdot V_{bi} \quad \ldots(6)$$

, which $V_b$ is sphere volume submerged in a bridge

Finite Element Method is applied for the solution of the $\delta E=0$ equation. Isoparametric elements of a biquadratic polynomial shape function were selected for the integral domain and unknown radius. The set of non-linear algebraic equations for unknown radius at element nodes and Lagrangian multiplier was resolved by the Newton-Raphson method [9].

## 2. Experiment

### Tensile strength of a static and a dynamic pendular liquid bridge

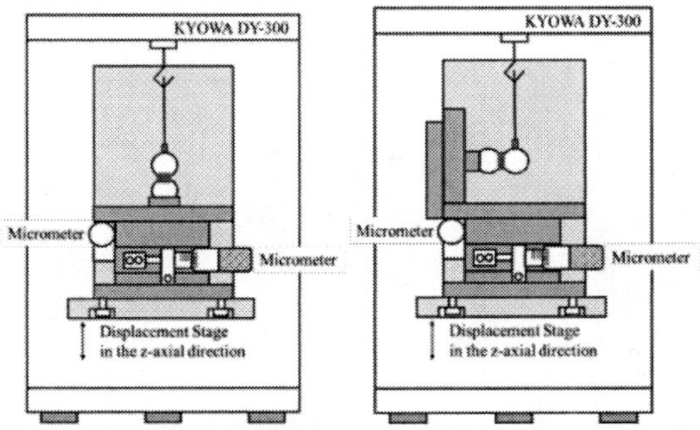


*Figure 2 Experimental apparatus for (a) normal and (b) tangential tensile strength of a liquid bridge adhered to two spheres*

Figure 2 shows the surface tensiometer (Kyowa interface Science Co., Ltd. DY-300) for measuring (a) the normal and (b) the tangential tensile strength of a water liquid bridge adhered to two stainless steel spheres (radius 3.0mm). The rod attached with one sphere is hanged up to the tensiometer and the other sphere fixed on the displacement stage is moved upward and downward by the tensiometer. The discrete and continuous displacement of the stage are performed for the normal and tangential deformation of a liquid bridge for the static and dynamic states. Two micrometers adjust the appropriate position of the lower spheres for the upper sphere. Water droplets are poured into the spheres aperture and substantially wetted by the displacement of the stage. The tensile strength at static and dynamic state and snapshots of liquid droplets by the digital camera and movie are recorded. The outlines of droplets are evaluated by a digitizer and estimate the half filling angles ($\theta_{Pe1},\theta_{Pe2},\theta_{Pw1},\theta_{Pw2}$) and contact angles ($\theta_{Cw1},\theta_{Cw2},\theta_{Ce1},\theta_{Ce2}$) at four contact points. The sphere configuration (radius, gap), the droplet volume, the half filling angles and the contact lines on spheres initiate parameters of numerical simulation.

## 3. Results and discussions

### 3.1 The normal tensile strength of a liquid bridge adhered to two vertical spheres.

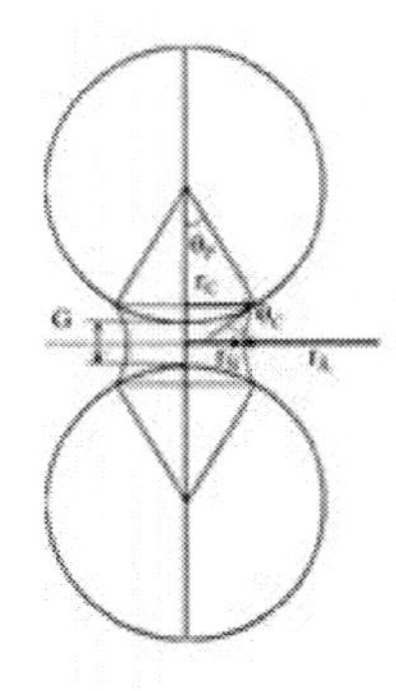


*Figure 3 Geometry of toroidal model*

The symmetric geometry of a liquid bridge adhered to two vertical spheres without gravity has been extensively investigated by toroidal approximation in Figure 3. The boundary method overestimates the capillary for of a liquid bridge rather than the gorge method but is feasibly and distinctly defined for the asymmetric bridge outline. Symmetric condition simplifies eq.(3) for the mean curvature and eq.(6) for the capillary force into eq.(7),(8) [7,16,17].

$$-2H = \nabla_s \cdot \mathbf{n}_L = \frac{1}{\alpha}\left(\frac{1}{f} - \frac{f_{zz}}{\alpha^2}\right), \textit{where} \quad \alpha^2 = 1 + f_z^2 \qquad \ldots(7)$$

$$F_{Boundary} = r_C \sin(\theta_{Pi} + \theta_{Gi}) - \frac{{r_C}^2}{2}(\Delta P) \mp \frac{2Bo}{3} V_{bi}, \quad r_C = \sin\theta_P \qquad \ldots(8a)$$

$$F_{Borge} = r_N - \frac{{r_N}^2}{2}\sin\theta_{cpi}(\Delta P) \mp \frac{2Bo}{3} V_{bi} \qquad \ldots(8b)$$

,where $r_N - r_A = \dfrac{A}{\tan(\pi/2 - \theta_P - \theta_C)} + \sin\theta_P, r_A = \left[A^2 + \{\sin\theta_P - (r_N - r_A)^2\right]^{1/2} A = G/2 + 1 - \cos\theta_P$

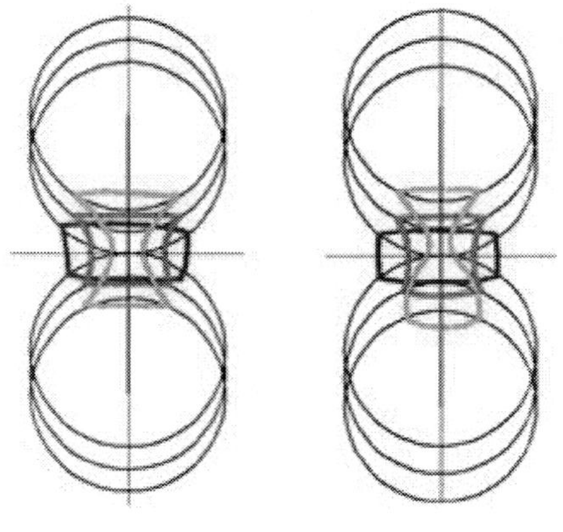

(a) Static state
*G=0, 0.47, 0.91*

(b) Dynamic state
*G=0, 0.59, 1.16*

*Figure 4 Experimental droplet outlines at static and dynamic state, Bo=1.21,V=0.088*

Figures 4 show symmetric outlines of liquid bridges around the vertical axis respectively at the static and dynamic state. Static state results in the shorter rupture gap and the larger dimensionless normal tensile strength from Figures 5a. Sphere gaps at the peak static capillary force depend on liquid volume. The contact angle on the upper sphere corresponds to the advancing contact angle at the higher Bo number due to gravity effect.

Fig. 4b indicates the downward transition of a liquid immediately before the rupture due to the wettability a liquid. The toroidal approximations over-estimate experimental data at the lower sphere gap but the sphere gaps at the peak dynamic capillary force depend on the velocity of stage displacement. The velocity reduces the peak value of dynamic capillary due to the downward transition of a liquid in Fig. 5b. The stepwise profiles of capillary force are due to the movement sensitivity of tensiometer for the stage displacement.

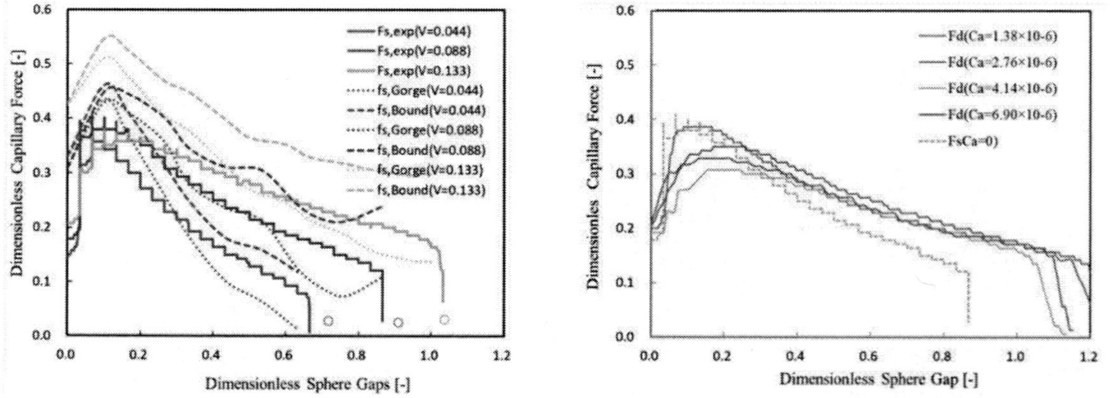


(a) at the static state

(b) at the dynamic state

*Figure 5 Effect of sphere gap on dimensionless capillary force*

### 3.2 Estimation of boundary condition at contact line by LBM

The Young's equation describes the force balance among three interface tensions (solid-liquid, solid-gas and liquid gas) at three phase contact point and also defines the relationship between interface tension and contact angle. The solution of the Young-Laplace equation at the static state demands for one unknown boundary condition of either a contact angle or a contact line. The simulation for the wettability on three dimensional contact line demands for the elaborate process. Our experiment determines the contact lines on spheres under the constant liquid volume and is utilized for the boundary condition. Besides the perfect wettability the smooth transition of three phase contact lines is unexpected due to the advancing and receding contact angle. Lattice Boltzmann Method (LBM) is expected to estimate the boundary condition at static and dynamic state. The geometry of a liquid bridge is numerically estimated by FEM with half-filling angles predicted by 2 dimensional Shan & Chen model of LBM, which is used by D2Q9 model for transition of distribution function. The estimation for the liquid geometry by FEM has been verified. Figures 6 show the predicted liquid outlines by LBM and FEM. at a given liquid volume.

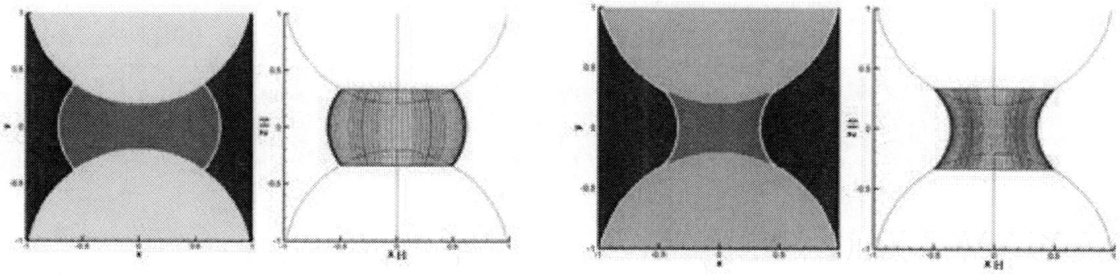

(a) Repulsive interparticle force (b) Attractive interparticle force

*Figure 6 Snapshots of liquid geometries at static state by LBM and FEM*

Figures 7 show snapshots of dynamic liquid outlines at a given liquid volume up to the rapture state by LBM. The transition of boundary conditions, i.e. filling and contact angles, is smoothly described. Gravity and imperfect wettability, i.e. advancing and receding contact angles, are excluded in Figs. 7 but the effect of gravity on the liquid outline is avairable.

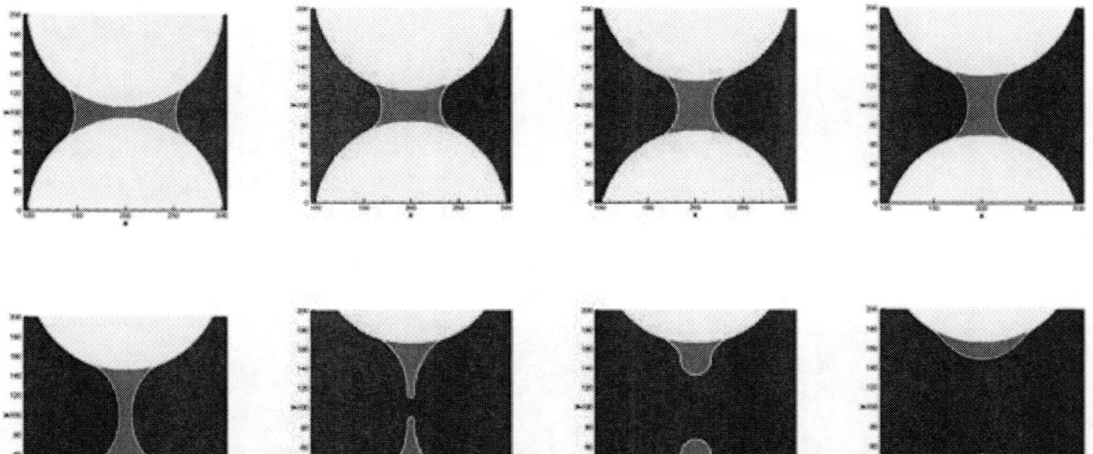

*Figure 7 Snapshots of snapshots of dynamic liquid outlines at a given liquid volume*

### 3.3 The tensile strength of a liquid bridge with shear deformation.

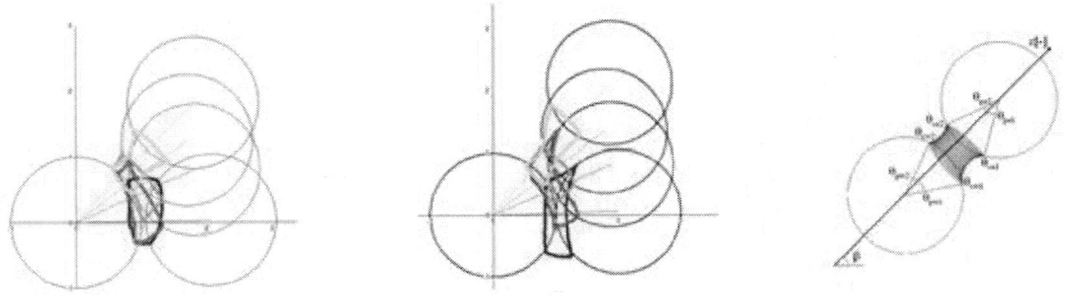

(a) Static state (G=0, 0.59, 0.98) (b) Dynamic state (G=0, 0.59, 0.98) Definitions of half-filling and contact angles.

*Figure 8 Liquid outlines at static and dynamic state ( Bo=1.21, V=0.088)*

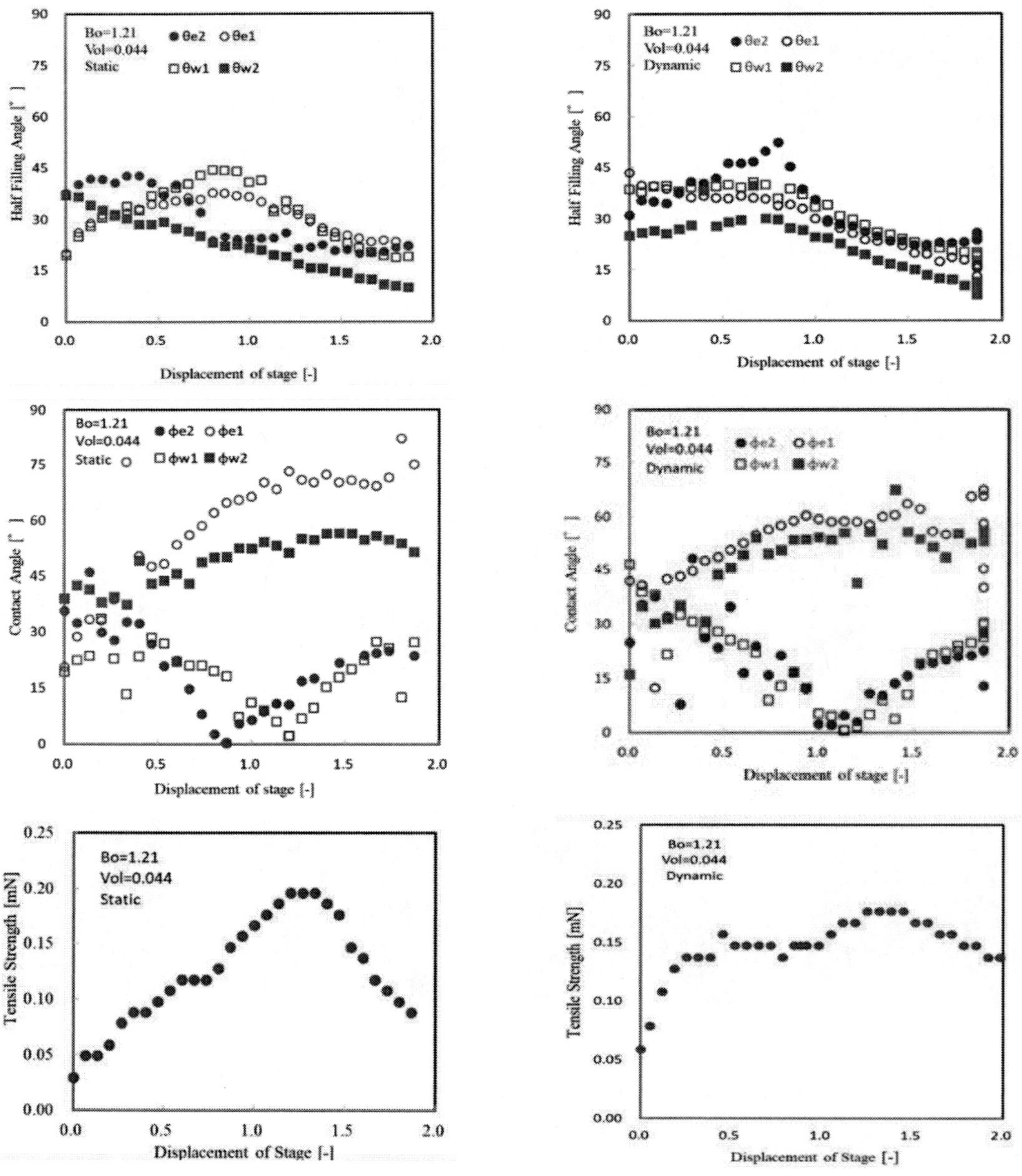


(a) Static state (b) Dynamic state

*Figure 9 Profiles of half-filling angles, contact angles and tensile strength , Bo=1.21, V=0.088*

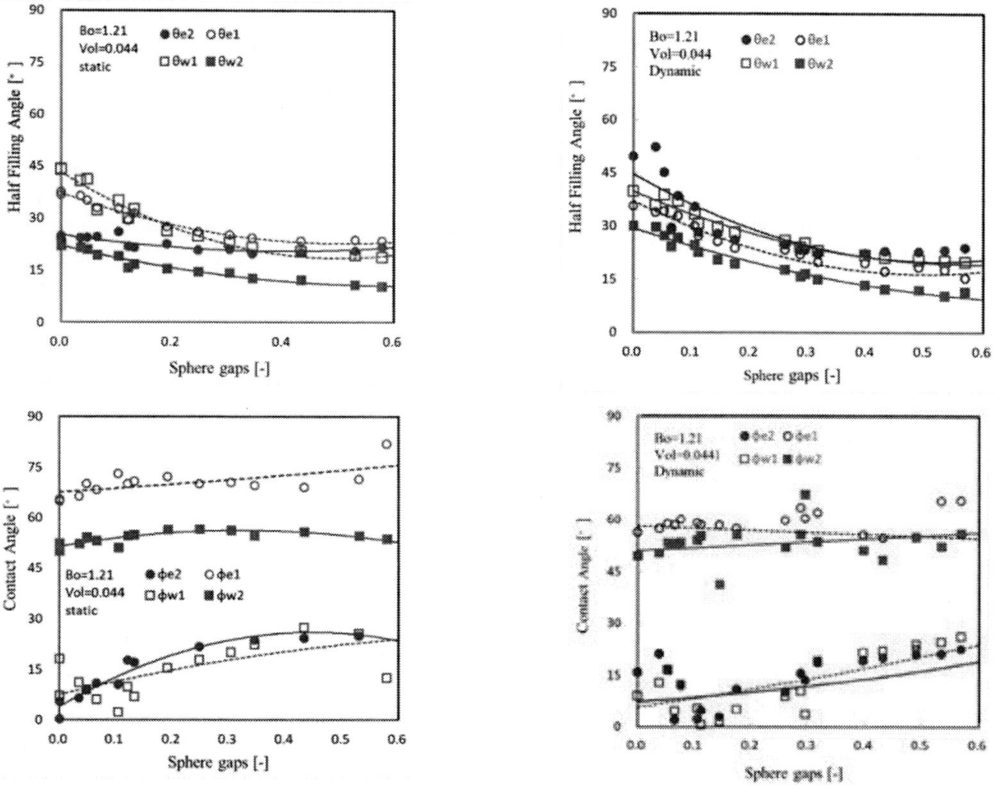

(a) Static state (b) Dynamic state

*Figure 10 Profiles of half-filling angles, contact angles with sphere gaps , Bo=1.21, V=0.088*

Figures 8 show four types of liquid outlines at static and dynamic state. Magenta and green plots are drawn respectively at the extreme values of half filling angles, contact angles and tensile strengths in Figures 9. During the lower stage displacement two spheres keep at the contact state and the half-filling angles reached to the peak values. Half-filling angles sifted with displacements of a stage have the peak values except the static half-filling angle at the top point of the lower sphere ($\theta_{Pw2}$), which monotonously decrease with stage displacement. The increase and decrease of contact angle respectively corresponds to the advancing and receding contact angle. The contact angles at the bottom point of the lower sphere ($\theta_{Cw1}$) and at the top point of the upper sphere ($\theta_{Ce2}$) transit from the receding to the advancing contact angle due to the infinitesimal values. Figures 10 redraw the half-filling and contact angles with respect to sphere gap. Contact angles contribute to the peak tensile strength, which is measured as resultant force ($F_T$) of normal ($F_n$) and tangential ($F_t$) tensile strength in Figure 11. The normal tensile strength conventionally corresponds to the capillary force. The stage displacement at the peak tensile strength minimize the sum of half-filling and contact angle due to the positive hydrostatic pressure across liquid surface. The larger dynamic tangential strength occurs during the sphere contacting due to the asymmetric deformation. The adhesion force of liquid to sphere at the static state contribute to the larger peak tensile strength.

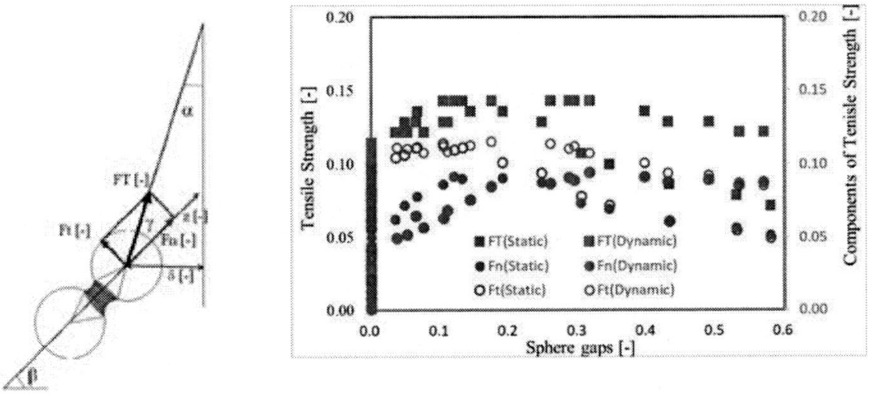

*Fig.11 Normal and tangential components of tensile strength at static state*

### 3.4 Numerical simulation at the static state

The solution of Young-Laplace equation demands for one unknown necessary condition on the contact point, i.e. half-filling or contact angle. Liquid surfaces simulated by FEM are drawn with solid lines in Figures 12. The simulation reveals the asymmetric liquid geometry with half-filling and contact angles along longitudinal angle. Disagreements of liquid interfaces results from the approximation of contact lines on spheres. The tangential strength influences depend on the mean curvature. Especially the second figure at the peak half-filling angle has the negative hydrostaic pressure in the direction of relative sphere displacement and positive pressure in the oppsite direction.

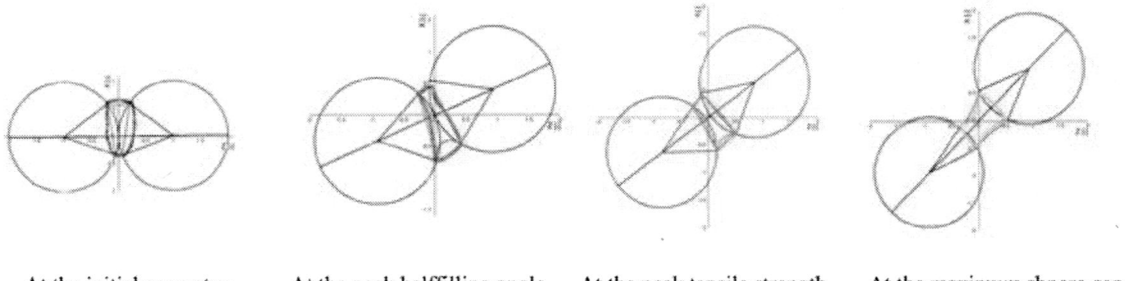

At the initial geometry At the peak halffilling angle At the peak tensile strength At the maximum shpere gap

Figure 12 Numerical estimation at the static state

Figure 13 displays transitions of measured normal tensile strengths ($F_{exp,n}$) and simulated capillary force ($F_{cal}$) to sphere gap. The initial droplet is attached to the above aperture in the direction of sphere movement due to gravity. The stage displacement drifts parts of liquid volume into the lower aperture of spheres. The sphere on displacement stage keep contacting with the other sphere hang up to the tensiometer up to the peak tensile strength. Fcal and FcalB are numerically estimated by outlines of liquid surfaces respectively with Young-Laplace equation and boundary toroidal model. The sign of hydrostatic pressure with deformation of liquid droplets results in the repulsive capillary force. The normal tensile strength corresponds to the capillary force during the range of symmetric liquid bridge. The angle of beta is defined by the inclination of center distance of a sphere pair. The lower normal tensile strength than the capillary force results from the orthogonal component of capillary force at the low range of beta.

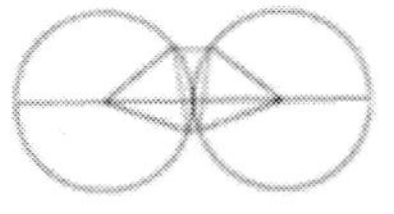

The initial geometry

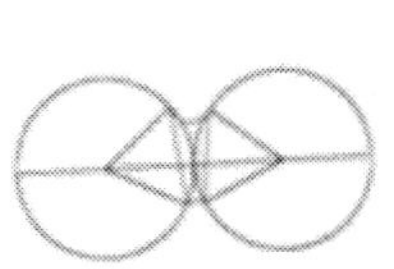

at the peak halffilling angle

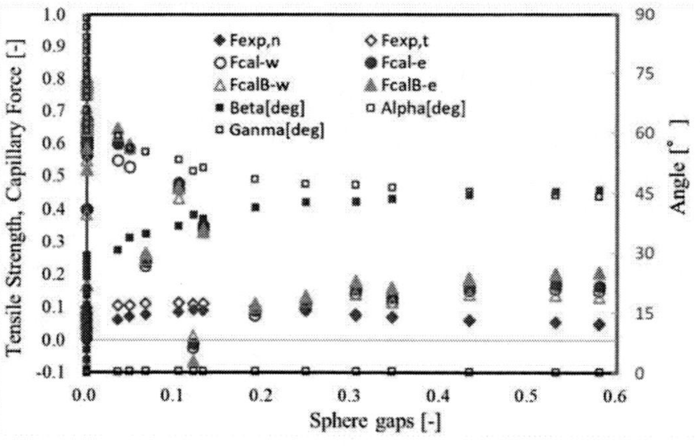


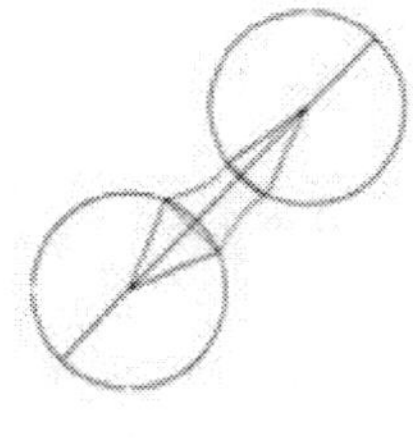

at the maximum shpere gap

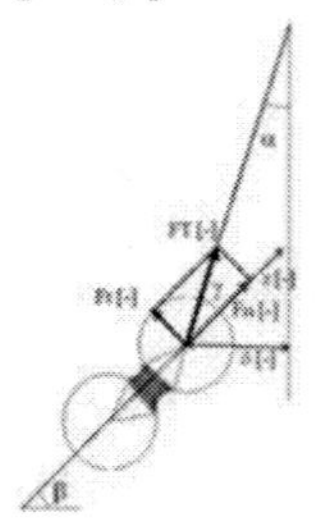

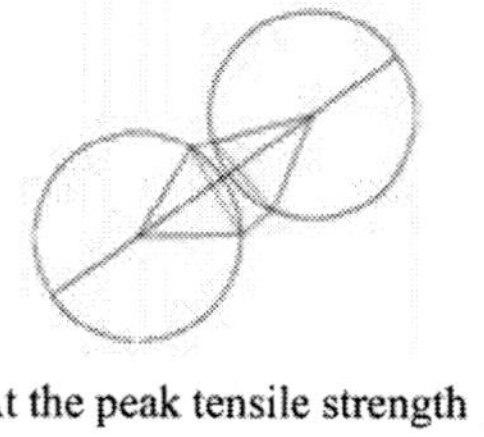

At the peak tensile strength

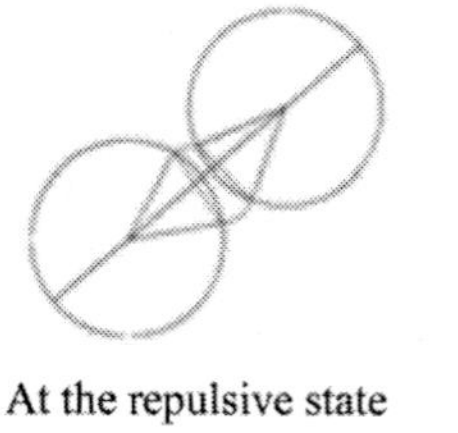

At the repulsive state

Figure 13 Normal tensile strength and numerical capillary force

## Conclusions

Liquid bridges adhered to particles are deformed with the normal and tangential stress in wet granulator. The interparticle force with asymmetric droplet by shear stress must be estimated by the three dimensional solution of Young-Laplace equation. A surface tensiometer and numerical simulation contribute to the normal and tangential tensile strength of liquid bridge adhered to two spheres.

1. The normal strength of symmetric liquid bridge with a surface tensiometer is verified by the conventional toroidal model and the solution of Young-Laplace equation.
2. Sphere gaps at the peak capillary force in the vertical sphere configuration depend on liquid volume and displacement velocity of sphere.
3. The lattice Boltzmann Method is expected to numerically predict the boundary conditions on the contact lines, which are half-filling angle and contact angle at the static and dynamic state.
4. The tensile strength depends on transition of contact angles, which correspond to advancing or receding angles. The larger sum of contact and half-filling angles than the orthogonal angle corresponds to the positive hydrostatic pressure across liquid surface.
5. The normal tensile strength is related with the solution of Young-Laplace equation by FEM. The tangential tensile strength depends on the liquid transition within droplets and the variable mean curvature.

## References


[1] K.Murase, H.Shibata, Y.Yasuda and S.Toyama, *Jpn J. Multiphase Flow,* **5**, 322(1991)
[2] A.E.Saez and R.G.Carbonell, *J. Colloid Interface Sci.,* **140**, 408(1990)
[3] D.N.Mazzone, G.I.Tardos and R.Peffer, *Powder Tech.*, **51**, 71(1987)
[4] C.D.Willet, M.J.Adama, S.A.Johnson and J.P.K.Seville, *Langmuir*, **16**, 9405(2000)
[5] M.J.Adama, S.A.Johnson, J.P.K.Seville and C.D.Willet, *Langmuir*, **18**, 6180(2002)
[6] R.K.Niven, *J Petroleum Sci. Eng.*, **52**, 1(2006)
[7] O.Pitois, P.Moucheront and X.Chateau, *Eur. Phys. J.B*, **23**, 79(2001)
[8] S.J.R.Simons, J.P.K. Seville and M.J. Adams, *Chem. Eng. Sci.*, **49**, 2331(1994)
[9] C.D.Willet, M.J.Adama,S.A.Johnson and J.P.K.Seville, , *Powder Tech.*, **130**, 63(2003)
[10] S.M.Iveson, J.A.Beathe and N.W.Page, *Powder Tech.*, **127,** 149(2002)
[11] R.A.Fisher, *J. Agri. Sci.*, 16,492(1926)
[12] F.M.Orr, L.E.Scriven and A.P.Rivas, *J. Fluid Mech.*, **67**, 723(1975)
[13] O.Pitois, P.Moucheront and X.Chateau, *J. Colloid Interface Sci.,* **231**, 26(2000)
[14] U.El Shamy and T.Groeger, *Int. J. Numer. Anal. Meth. Geomeh.*, **32**, 1763(2008)
[15] F.M.Orr, R.A.Brown and L.E.Scriven, *J. Colloid Interface Sci.,* **60**, 137(1977)
[16] K.Murase, T.Mochida, Y.Sagawa and H.Sugama, *Adv. PowderTech.*, **19**, 349(2008)
[17] C.D.Willet, M.J.Adama,S.A.Johnson and J.P.K.Seville, *Handbook of powder technology*, **11**, chap.28, ed. A.D.Salman, M.J.Hounslow and J.P.K.Seville,Elsevier (2007)

# NUMERICAL SIMULATION OF WIRE-PLATE ELECTROSTATIC PRECIPITATOR - EFFECT OF PARTICLE CONCENTRATION

*Ding Yang*[1], *Baoyu Guo*[2*], *Jun Guo*[1], *Aibing Yu*[2]

[1] *Fujian Longking Co. Ltd, Longyan, Fujian, 364000, China*

[2] *Department of Chemical Engineering, Monash University, Clayton VIC 3800, Australia*

## Introduction

Electrostatic Precipitators (ESP) have been widely used to separate fine particles from large scale industrial exhaust gases, such as coal fired power plants. In-depth knowledge and careful design optimization are critical to meeting the increasingly strict emission standard in terms of $PM_{2.5}$ emission limit, e.g., 5-10 mg/Nm$^3$. Numerical modelling has been employed as a powerful tool in this area. Complicated interactions among electric field, gas flow and particle transport are summarized by Adamiak [1]. To model such a system, various levels of simplifications have been made in published papers.

Among them, Lagrangian particle tracking technique is a widely accepted approach. The mass concentration of ash particles in coal-fire flue gas is typically 10-50 g/Nm$^3$, corresponding to volume fraction in the order of $10^{-5}$. Such a small value is often regarded as sufficiently dilute particle concentration in numerical modeling of multiphase flows. For the case of ESP, complicated interactions among the e-field, gas flow and charged particles are involved, even if the deposited dust layer is ignored. Full coupling between the continuous gas phase and discrete particle phase is computationally expensive to achieve a converged steady solution. To avoid such a high cost, the model for ESP was simplified as so called one-way coupling in many previous works, so that the electric field is solved without the particle phase (Guo et al. [2,3]; Farnoosh et al. [4]). This simplification may be acceptable in the end stage of ESP, but will introduce significant error in the early collection stage. Limited studies reported in the literature have considered the particle charges as a source term in the Poisson's equation of electrostatics. For example, Choi and Fletcher [5] predicted a reduced current for the particle-loaded case ($d_p$ = 1-5μm) compared with the particle-free case, but the corona ion charge density somehow shows an increasing trend along the stream. When fine dust particles are treated, the large specific surface area of the particle phase, combined with reduced collection efficiency, implies high space charge density in the collection region. Its effect was understood as a partial suppression of the corona current [6]. As was experimentally proved by Podlinski et al. [7], space charge of charged particles can contribute to the EHD flow pattern and particle transport. In present work, particle space charge is considered through two-way coupling and interesting results are generated in terms of ESP performance as particle loading changes.

## Integrated ESP Modelling

### Mathematical Equation

Coupled electric field, gas-particle flow and particle charging processes are simulated using

Eulerian-Lagrangian method in Computational Fluid Dynamics (CFD). Major governing equations of vector format are listed below for completeness. For electric field, Poisson equation of potential

$$\nabla^2 V = -(\rho_{ion} + \rho_c)/\varepsilon_0 \quad \mathbf{E} = -\nabla V \tag{1}$$

Current continuity equation

$$\nabla \bullet ((b\mathbf{E} + \mathbf{U}_f)\,\rho_{ion} - D\nabla\rho_{ion}) = S_c \tag{2}$$

where $\rho_{ion}$ is corona charge density, $b$ the ion mobility, $D$ is ion diffusivity, $\rho_c$ is particle space charge density, evaluated by averaging the charges of particles that pass through a control volume (cell),

$$\rho_c = \frac{1}{V_{cell}} \sum_{n=1}^{N} \dot{n}_p q_p \Delta t \tag{3}$$

Where $\dot{n}_p$ is the number flow rate of a parcel of particles tracked, $\Delta t$ is the residence time in the cell. The source term, $S_c$, in Equation (2) is transfer rate of ion charges to particles, which is not considered at the moment, i.e., $S_c = 0$. For the gas flow, the momentum equation is,

$$\begin{aligned}\nabla \bullet (\rho_g \mathbf{U}_g \otimes \mathbf{U}_g) = \nabla(p + \tfrac{2}{3}\rho_g k) + \nabla \bullet [(\mu + \mu_t)(\nabla \mathbf{U}_g + (\nabla \mathbf{U}_g)^{\mathrm{T}})] + \rho_{ion}\mathbf{E} \\ - \frac{3\pi n_p d_p \mu}{\mathrm{C_C}}(1 + \mathrm{Re}_p^{0.687})(\mathbf{U}_g - \mathbf{U}_p)\end{aligned} \tag{4}$$

For the particle flow, the force balance of a single particle is

$$m_p \frac{d\mathbf{U}_\mathrm{P}}{dt} = \frac{3\pi n_p d_p \mu}{\mathrm{C_C}}(1 + \mathrm{Re}_p^{0.687})(\mathbf{U}_g - \mathbf{U}_p) + \tfrac{1}{6}\pi d_p^3 \mathbf{g}(\rho_p - \rho_g) + n_e q_e \mathbf{E} + \mathbf{F}_{dis} \tag{5}$$

k-ω SST turbulence model is chosen for gas flow, and the dimensionless Lawless charging rate equation is used for particle charging [8].

**Model Conditions**

An integrated ESP model developed comprises two parallel modules or submodels, namely, electric field model and coupled gas-particle flow model. One-way coupling between the two modules is used earlier is upgraded to two-way in the present work. The electrostatic field caused by charged particles in suspension is considered in the solution of electric field. The two submodels are coupled through a small under-relaxation factor, in views of strong interaction between the electric field and particles flow, and the stochastic nature of the discrete particle phase. In this way all the volume averaged quantities related to the particle phase (such as the particle space charge density) are smoothed out over time in the simulation.

A single channel, 4-wire and flat plate configuration is considered. The dimensions are 3.5mm in wire diameter, 0.24m for wire-wire spacing, 0.2m for wire-plate spacing. The plate length is 1.2m. The simulation domain is defined as 1.74m long by 0.4m wide box with the 4 holes inside. The grid is smoothly refined at places where higher gradient in electric quantities is expected, such as the proximity of the wires and middle of the adjacent wires. Other base case conditions are as follows.

- Applied voltage on the wire electrode: Va = -60kV. The charge density on the wire surface is determined using an injection law combined with Peek equation (Guo et al.[9]).
- Gas temperature: 120°C. As the inlet boundary condition for the gas phase, the velocity is set as 1 Nm/s; relative turbulence intensity 3.7%, turbulent length scale $0.3D_h = 0.24$m.
- Dust particle size is mono-dispersed at $d_p$=2.5 μm. Particle density 2000 kg/m$^3$; ash dielectric constant of $\kappa_p$=6.6, Particle loading at the inlet, $(\rho_p)_{in}$ , is varied with default value of 10 g/Nm$^3$.

## Typical Results

### Electric Field without Particles

When the inlet/outlet boundaries are moved sufficient long distance away from the electrodes in the simulation domain (referred to as "natural boundary condition"), the electric field around the boundary wires show an asymmetric distribution pattern that differ from that around the inner wires. Electric field-lines/current can go out of the limit of the collection region (Figure 1). Electric field intensity on the collection plate is of interest as it is considered a factor to determine the upper limit of operation before spark-over [10]. Normally the field intensity on the plate is highest at the point that faces exactly the wire center with the least distance. This maximum field intensity on the plate is $3.80\times10^5$ V/m. However, when the natural boundary condition is used, the highest value ($1.06\times10^6$ V/m) predicted in field intensity appears at the endpoint of the plate due to its sharpness. This is a likely point that would trigger electric spark-over prematurely. Practically such a sharp edge should be avoided. Generally, the field strength at the middle wires is smaller than that at a boundary wire, due to the counteraction of adjacent wires. Then the corona charge density at the boundary wires should be greater than that at the inner wires (Figure 2), which is similar to the simulation result of Kasdi [11].

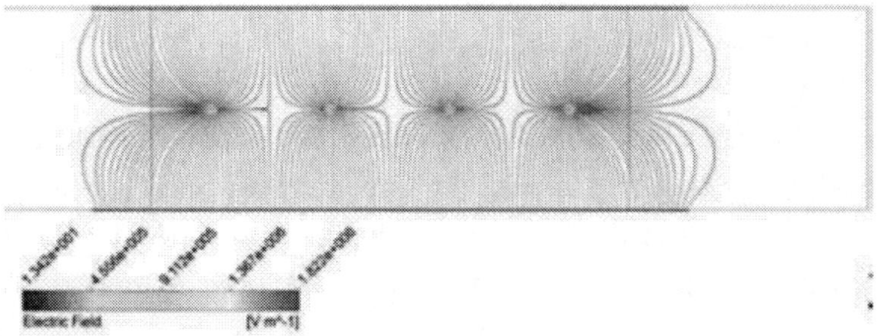

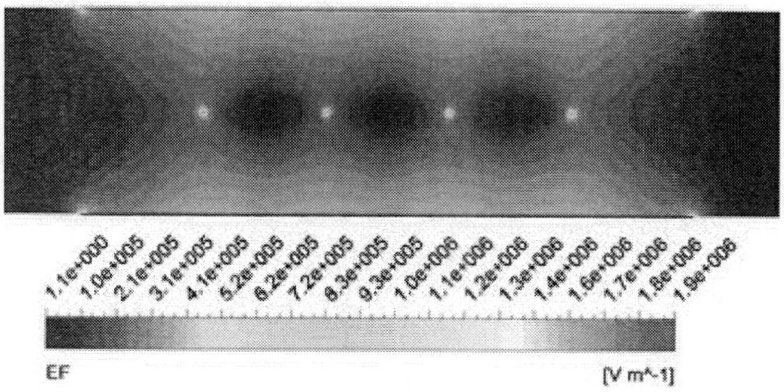


Figure 1. Field streamlines and intensity distribution (no particles).

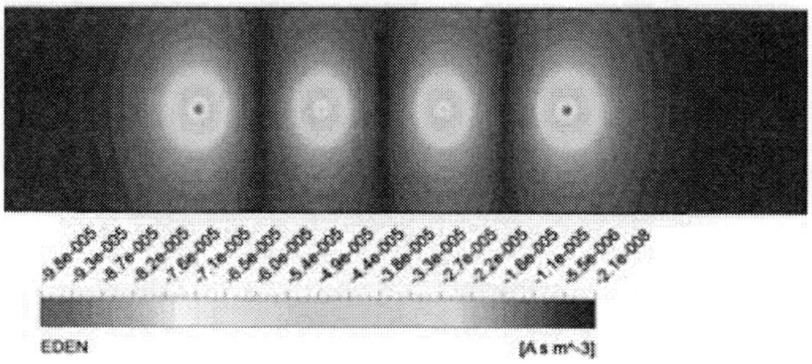


Figure 2. Corona charge density (no particles).

**Effect of Particle Space Charge**

Particles are randomly injected from the inlet plane with uniform probability in locations and no-slip for velocity (same as the gas velocity). In using random number generator, it is ensured that the injection points are not repeated for every coupling cycle. Figure 3 shows sampled particle trajectories with color scaled to particle charge. Figure 4 shows a pattern of particle deposition rate on the plate. It is seen that the end points of the plate, particularly the downstream one, are special spots to accumulate more dust.

Figure 3. Particle trajectories with colour scaled to particle charge.

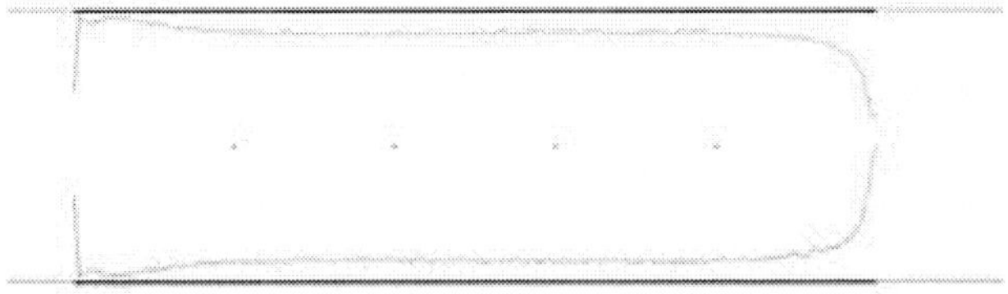

Figure 4. Indicative pattern of particle deposition rate on the plate.

Figure 5 shows the averaged space density of charges carried by particles. Highest space charge density appears near the first wire electrode, where electric current density is highest. It diffuses gradually from the centre line with particles drifting toward the plate while moving forward. Due to the radiating nature of the field-lines around each wire, the lateral component of Coulomb's force in the region close to the centre line may not be as large as that off the centre line. Not collected charged particles of negative polarity are conveyed towards the outlet, lowering the potential downstream outside the collection region (Figure 6), causing distortion of the overall electric field distribution.

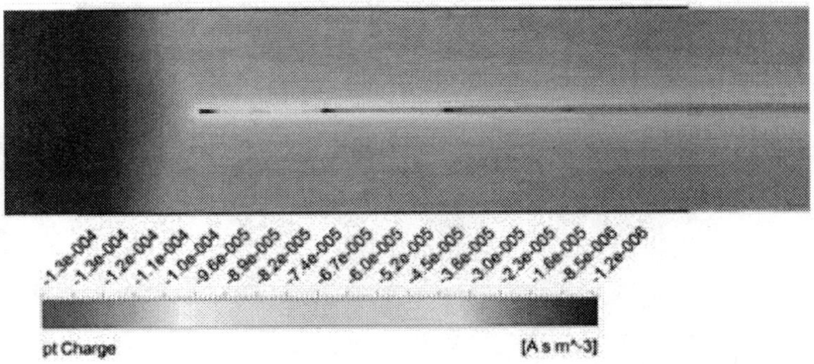


Figure 5. Distribution of space charge density of particles.

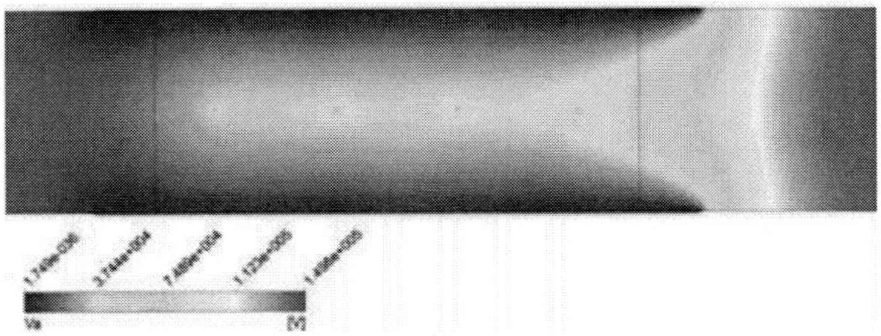

Figure 6. Distribution of electric potential (in absolute value).

## Effect of Particle Loading

### On Electric Current

Figure 7(a) shows the mean corona current, based on unit length of wire. As can be seen, the corona current drops quickly as particle loading increases, although a fixed voltage is applied. This is caused by the secondary electrostatic field created by the space charges of particles that cancels out the primary electric field at the wire surface and subsequently suppressed the corona generation.

### On Particle Collection Efficiency

Predicted particle collection efficiency, as particle loading changes, is given in Figure 7(b). The collection efficiency, increases initially with particle loading (<3 $g/m^3$), peaks at particle loading of 4-5 $g/m^3$ and then decreases as particle loading increases further. When ignoring particle space charge, the electric field would be independent of particle loading. This is the case for very low particle loading, where the fractional collection efficiency is constant, or in other words, the quantity of dust collected as well as that penetrated increases proportionally with particle loading.

The observed enhancement of collection by a slight amount of particles is due to the redistribution of electric field so that the near-plate and volume-averaged field intensity happens to increase, which favor the charging of most particles. The asymptotic lower limit of the curve should correspond to the situation when the corona current is so small that turbulent dispersion dominates over the drift force.

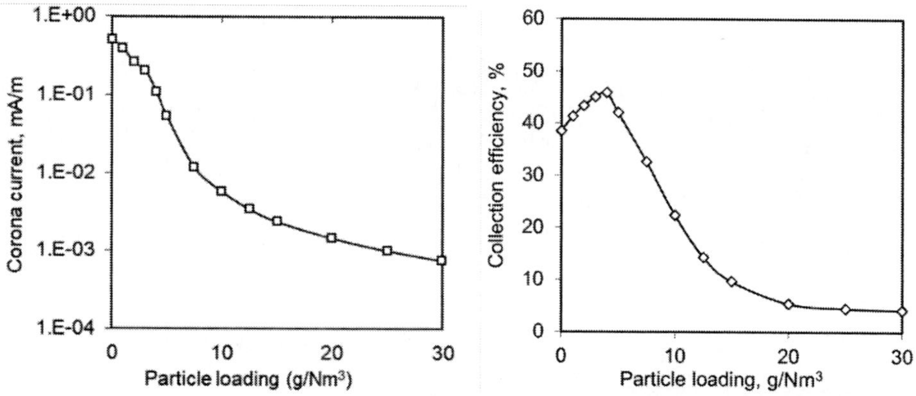


(a) (b)

Figure 7. Mean corona current (a) and collection efficiency (b) as function of particle loading.

## On Electric Field Distribution

Figure 8 visualizes the predicted corona charge density and field-lines, which shows how the electric field distribution evolves with particle loading. As the electric field is distorted starting first from the end side, corona density at 3 of the 4 downstream wires is suppressed faster than the 1$^{st}$ wire. When the particle loading is above 7.5 g/m$^3$, all wires except the first one stop emitting. Meanwhile, reversed field-lines appear around the last wire, and the affected region expands as the loading increases further. Otherwise the field-lines become smoother with increasing loading. Since the field-lines are extended to the outlet, some escaped particles can be pulled back into the collection region, a factor that contributes to significantly greater deposition rate around the end-point of the plate.

Overall, the particle charges affect the particle collection adversely, as they tend to suppress the corona emission or reduce the corona current. Particularly the penetrated particles, which carried significant charges, have prolonged effect as they float along. The corona at the downstream electrodes even extinguish unless the applied voltage is raised individually. When highly charged, more particles precipitate to the plate, reducing the number of particles floating ahead (or less space charge density). The charges carried by these remaining particles exert a negative feedback to the electric field and particle charging, which compromises the collection efficiency.

## Electric Quenching

Corona quenching is a common phenomenon in ESP when fine particles are treated. Riebel et al. [12] experimentally studied the corona discharge characteristics of ESP with glycerol aerosol, which found corona onset voltage increases and the corona current significantly decreases after the glycerol steam was injected into the system. Under fixed voltage in the present simulation, corona currents from all the wire electrodes drop with increasing solid loading. The downstream coronas (from the second one) will eventually extinguish at some point. An interesting phenomenon arises when the particle loading increases above 10 g/Nm$^3$, i.e., the electric streamlines in a small zone around the last wire reverse direction locally. This implies that a charged particle, once falls accidently within the zone, could be trapped and attracted to the same wire, as seen from the particle trajectories in Figure 9. Thus this wire is subject to heavier dust deposition.

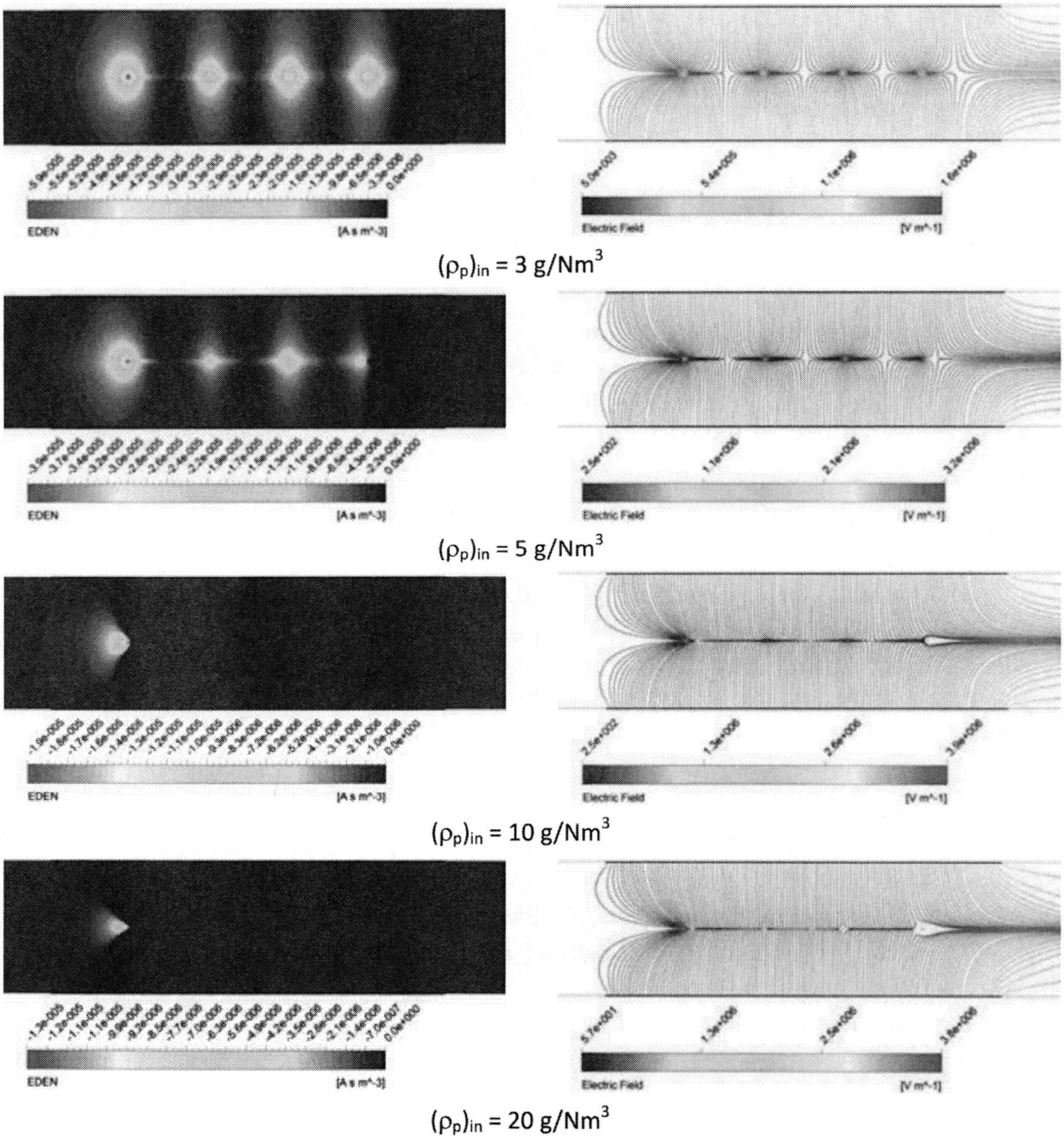


Figure 8. Visualizations of corona charge density and field streamlines for different particle loadings.

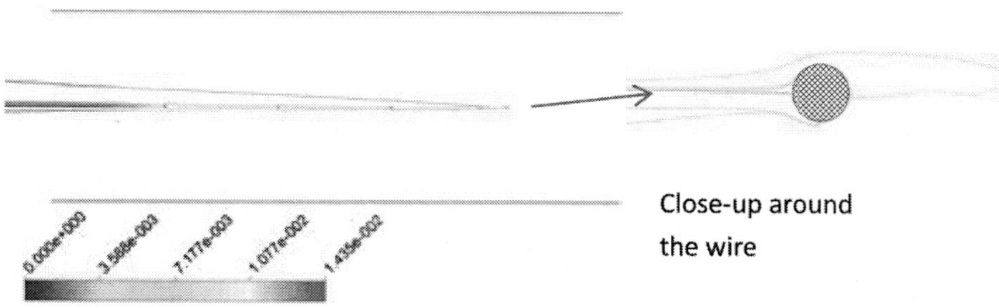


Figure 9. Trajectories of particles falling on the corona wire ($(\rho_p)_{in} = 20$ g/Nm$^3$).

## Concluding Remarks

CFD based ESP model has been advanced, with an enhanced coupling between electric field and particle flow. So the model can be readily applied to the early collection stage of the ESP, where dust particle loading is too high to neglect. Space charge of particles is considered, and its significance is clarified in a broad range of particle mass concentration in the gas. Dust loading significantly affects the dust collection efficiency for fine particles considered. Maximum collection efficiency exists at a critical particle loading that depends on particle size. However, as particle loading increases, collection efficiency decreases and eventually approaches asymptotically to a lower limit. The phenomenon of electrical quenching is observed, where downstream corona is extinguished or one of the emitters is polluted when the particle loading is high.

## References


1. Adamiak K. (2013), "Numerical models in simulating wire-plate electrostatic precipitators: A review", *J. Electrostatics*, 71, pp.673-680.
2. Guo B.Y., Yu A.B. and Guo J. ( 2014), "Numerical modelling of electrostatic precipitation: effect of gas temperature", *J. Aerosol Science*, 77, pp.102-115.
3. Guo B.Y., Yu A.B. and Guo J. (2015), "Numerical modelling of ESP for design optimization", *Procedia Engineering*, 102, pp.1366-72.
4. N. Farnoosh, K. Adamiak, G.S.P. Castle (2010), "3-D numerical analysis of EHD turbulent flow and mono-disperse charged particle transport and collection in a wire-plate ESP", *J. Electrostatics*, 68, pp.513-522.
5. Choi B.S. and Fletcher C.A.J. (1998), "Turbulent particle dispersion in an electrostatic precipitator," *Applied Mathematical Modelling*, 22, pp.1009-1021.
6. Parker K.R., Applied Electrostatic Precipitation, Blackie Academic & Professional, 1997.
7. Podlinski J., Niewulis A., Mizeraczyk J. and Atten P. (2008), "ESP performance for various dust densities", *J. Electrostatics*, 66, pp.246-253.
8. Lawless P.A. (1996), "Particle charging bounds, symmetry relations, and an analytic charging rate model for the continuum regime", *J. Aerosol Sci.*, 27, pp.191-215.
9. Guo B.Y., Guo J. and Yu A.B. (2014), "Simulation of the electric field in wire-plate type electrostatic precipitators", J. Electrostatics, 72, pp.301-310
10. White H.J. (1974), "Resistivity problems in electrostatic precipitation", *J. Air Pollut. Control Assoc.* 24, pp.313–338.
11. Kasdi A. (2016), "Computation and measurement of corona current density and V–I characteristics in wires-to-plates electrostatic precipitator", *J. Electrostatics*, 81, 1-8.
12. Riebel U., Radtke R. and Loos R. (2002), "An experimental investigation on corona quenching", *J. Electrostatics*, 54, pp.159-165.

# Triboelectric Charge of Spherical Glass Particles against Metal Pipeline

*HoSu CHOI [1], Teruo SUZUKI [2] and Kwangseok CHOI [3,*]*
*[1] Christian Academy in Japan, [2] Kasuga Denki Inc.,*
*[3] National Institute of Occupational Safety and Health, Japan*
**choiks@jniosh.go.jp*

## Introduction

Insulating powders are used in many ways such as in the food, pharmacy, and chemical industries. Usually, these powders are highly charged when transported, loaded, dried, and mixed, due to repeated contact and separation between powder particles and the equipment. This results in the powders acquiring electrostatic charges. If not dissipated quickly, electrostatic charge buildup can cause inter-particle cohesion and adhesion to operating equipment, causing a production efficiency decrease. Incendiary electrostatic discharges such as spark discharges, cone discharges, and propagation brush discharges can also take place during industrial operations and may lead to fires and even explosions. Riedewald reported on an industrial fire in a pharmaceutical powder transfer operation [1]. A large number of studies involving electrostatic charge generation on powders, as well as electrostatic discharges and hazards during operation, have been reported. For example, Kaialy reviewed the main factors affecting the electrostatic charging of pharmaceuticals such as particle resistivity, size, shape, and surface roughness [2]. Gibson also reviewed industrial hazards related to electrostatic charges and discharges [3]. Our previous studies [4–6] investigated the electrostatic charging of polymer powder in mixing processes and the ignitability of aluminous, magnesium and polymer powders due to electrostatic discharges involving the effect of the ratio of $N_2$ in the air on the ignitability.

In order to easily measure and confirm the electrostatic charge level of powders, a novel tribocharging system that utilizes the Faraday cup is presented, and the charging behavior of the sample powders is investigated in this paper.

## Experimental Apparatus and method

Figure 1 shows the spiral air tribocharge apparatus presented in this study. The apparatus was divided into four main parts: the air supply part, the powder supply part, the tribocharge part, and powder collection part. The air supply part consisted of a compressor (Hitachi, PO3.7MA) that included an auto dryer (CKD, F3-217199), air controller, electronic valve, and switch. In this study, dried air was used because moisture in the air may have an effect on the electrostatic characteristics of the powder sample. The powder supply and tribocharge parts consisted of a powder port, a spiral air flow device, and a pipeline. The spiral air flow device (photo 1) was attached in front of the sample powder port in order to create spiral air inside the pipeline. The spiral air flow device had 8 holes that were 6 mm in diameter and that opened at a slant (60 degrees), which is very different from conventional models [7, 8, 9]. The straight pipeline was made up of stainless steel metal and was 1300 mm in length and 35 mm in diameter. In order to prevent external noise, the overall powder supply and tribocharge parts were shielded. The powder collection part was made up of a cyclone-type Faraday cup, unlike conventional models [7, 8, 9]. The cyclone-type Faraday cup had a separate metal container (30 mm in diameter) for easy measurement of the amount of collected powders. The other devices included two electrometers (Kethley,

6517A), a data logger (Graphtec, GL900), an electrostatic ionizer (Kasuga Denki Inc., KD-510W), an electrical balance (Shimazu, BL220H), and a timer.

The typical operating procedure for the tribocharge test is summarized as follows:
(1) Adjust the air pressure using the air pressure valve.
(2) Load 1 g of the sample powder into the powder port.
(3) Eliminate the initial charge of the sample powders using an electrostatic ionizer.
(4) Reset the electrometers ($Q$ mode) connected to the pipeline and cyclone.
(5) Switch the air flow on, and run the sample powder through the pipeline where friction occurs.
(6) Measure the indication value when it stabilizes.
(7) Measure the amount of powder collected in the Faraday cup.
(8) Clean out the pipeline and cyclone part using an air spray in preparation for the next test.

The charge to mass ratio, $q$ (nC/g), of the sample powder was obtained by dividing the charge, $Q$ (nC), by mass, $m$ (g), of the sample powder loaded into the Faraday cup:

$$q = \frac{Q}{m} \quad \text{(1)}$$

This experiment is mainly to determine how much the charge to mass ratio, $q$ (nC/g), of the sample powders is affected by the size of the powder and the amount of air pressure. The cyclone-type Faraday cup and the pipeline were suspended electrically with insulators. All experiments were conducted in ambient conditions (air temperature: 22 ℃; relative humidity: 55 %).

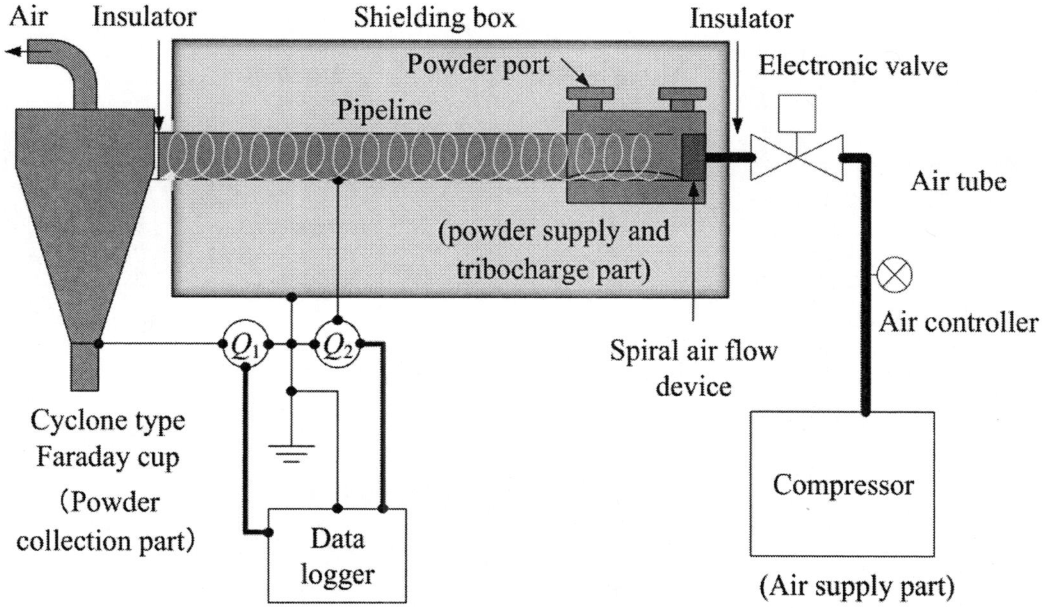


Fig. 1 Schematic of the spiral air tribocharge apparatus used in this study.

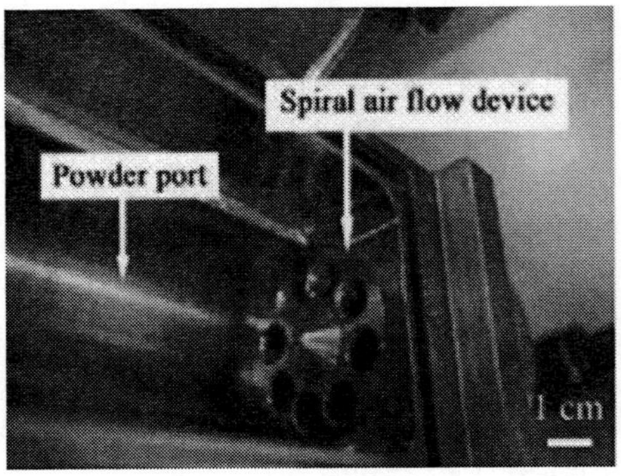


Photo 1 Structure of the powder port and spiral air flow device.

## Sample powders

Glass beads are extremely good light reflectors and thus are widely used as the main component of the coating material for light-reflecting adhesive tape, background paint for traffic control sign boards, and paint for road signs. Four different grades of commercial glass beads were used in this experiment, as shown in Table 1. The particle size was determined using a laser diffraction system (LDA Win 1.21, dry type). Hereafter, they will be called by their abbreviation shown in Table 1. The analysis of the images from microscopy (Neoscope, JCM-500) revealed that the G-4 powder particles were spherical, as shown in photo 2. The other glass bead particles were the same shape as G-4.

**Table 1.** List of powder specimens used in experiments.

| Specimen | Abbreviation | Diameter [μm] | | |
|---|---|---|---|---|
| | | $D_{10}$ | $D_{50}$ | $D_{90}$ |
| Glass beads 1 | G-1 | 22 | 62 | 85 |
| Glass beads 2 | G-2 | 93 | 131 | 175 |
| Glass beads 3 | G-3 | 130 | 180 | 241 |
| Glass beads 4 | G-4 | 162 | 225 | 294 |

$D_X$: X% volume average.

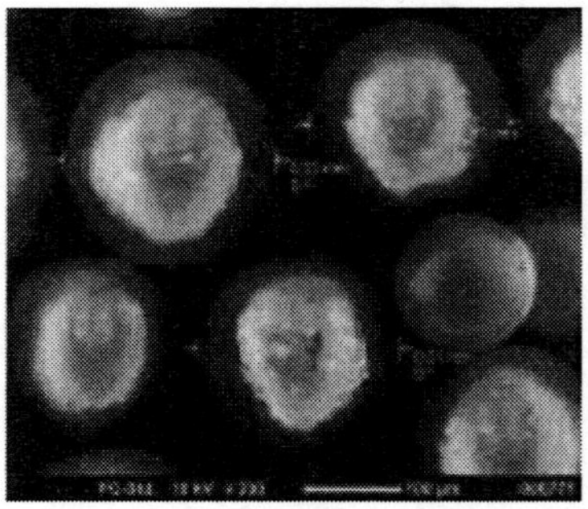

Photo 2 Shape of G-4 powder used in this experiment.

## Experimental results and discussion

Figure 2 shows the typical change of charge detected over time when the G-3 particle was used and the air pressure was 0.15 MPa. It should be noted here that when no particle was used and only air was passed through the pipeline, the charge was detected as almost zero. The charge Q1 of the powder was found through the cyclone-type Faraday cup and then was confirmed through the charge Q2 obtained from the pipeline. As the powder particles passed through the pipeline for about 0.82 seconds, as seen in the Fig. 2, the charge clearly increased. Q1 and Q2 had almost the same trend and absolute values. This result confirmed the reliability of the test and showed that there was no disturbance by external noise.

In this study, all of the glass beads (G-1 to G-4) were negatively charged when stainless steel metal was used as the contact surface. The charge polarity of the materials relates to the work function (or ionization potential) of each material. The contact potential difference between two substances is defined by the difference between their fermi levels, which correlate with the work function. In general, materials with higher work functions tend to appropriate electrons from materials with lower work functions. This point will be carried out in our further study.

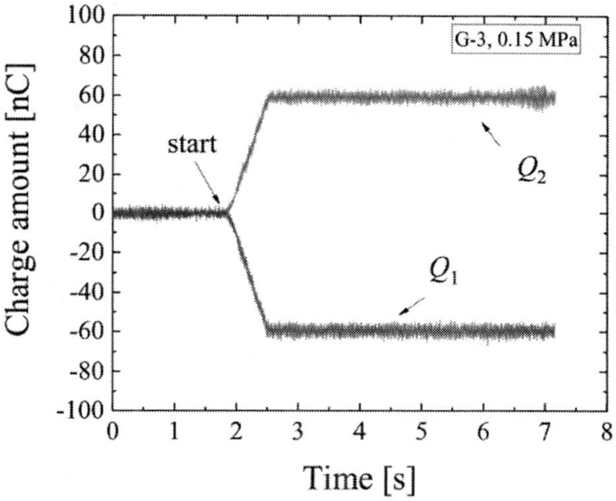


Fig. 2 Change of charge detected over time.

Figure 3 shows the charge to mass ratio of glass beads as a function of applied air. The range of applied air was between 0.1 MPa and 0.3 MPa in this study. When the applied air pressure was below 0.1 MPa, it was most likely not enough to send all of the powder through the pipeline. Each experiment was repeated five times, and the average values were reported. This graph does not contain an error bar, but these were about 10%. The results showed that the charge to mass ratio of both powders increased as the applied air increased. Although not shown in Figs. 2 and 3, the increase in air pressure also causes the slope of the graph to increase significantly. This can be explained by the fact that with increased pressure of applied air, the particles collide against the surface of the pipeline with more strength [9]. In order to investigate in more detail, the velocity of air at approximately 10 mm from the inside of the pipeline wall was measured using an anemometer (Kanomax, Anemomaster-6115). The approximate velocity of air flow was 9.5 m/s at 0.1 MPa, 12 m/s at 0.15 MPa, 15 m/s at 0.2 MPa, 20 m/s at 0.25 MPa, and 22 m/s at 0.3 MPa. It could be clearly seen that as the air pressure increased, the air velocity also increased. Another reason for this may be that because the air was applied in a spiral form, the particles may have collided more frequently against the pipeline when the air flow was increased. This exception will be investigated in detail in the future.

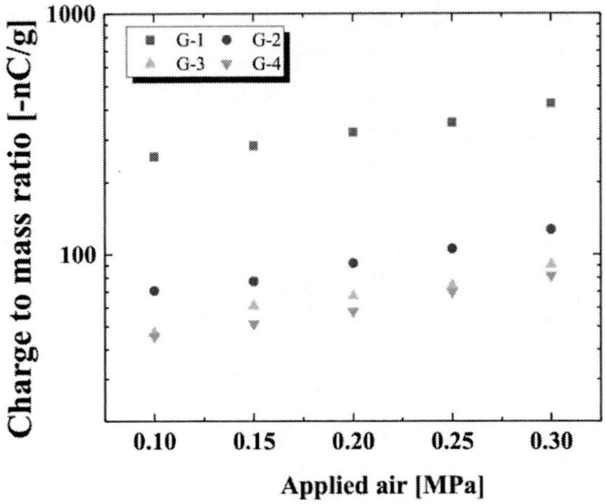


Fig. 3 Charge to mass ratio of glass beads as a function of applied air.

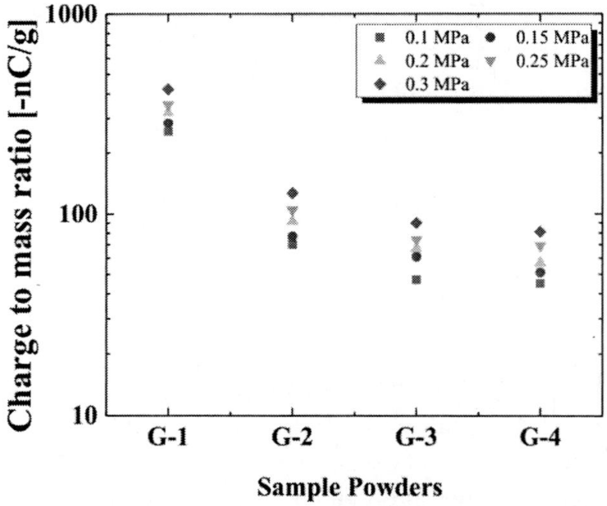


Fig. 4 Charge to mass ratio as a function of different glass bead sizes.

Figure 4 shows the charge to mass ratio as a function of different sizes of glass beads. The size distribution of each glass bead in this study was described in Table 1. The results showed that the charge to mass ratio of the glass beads increased as the powder size decreased. That is to say, smaller particles had a larger q than the larger ones. The total surface area of all of the smaller beads combined is greater than that of the larger ones, which results in a high charge to mass ratio development in the tribocharge. The value range of the q of the glass beads obtained in this study was about 40 nC/g to 100 nC/g. These values agreed with those obtained from the micronizing and pneumatic transfer process [10]. These values were high enough to generate incendiary electrostatic discharges during handling in industrial processes. This suggests that if glass beads are used with solvents in the cleaning process, they must be handled carefully, as incendiary discharges can ignite the solvents even though the glass beads themselves are nonflammable material.

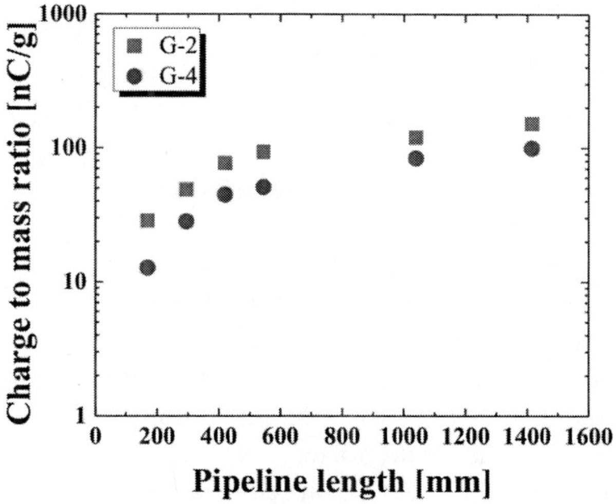


Fig. 5 Charge to mass ratio as a function of pipeline length.

Figure 5 shows the charge to mass ratio as a function of pipeline length. Six different pipeline lengths, *d*, were used: 170, 295, 420, 545, 1040, and 1415 mm. The glass beads used in this experiment were G-2 and G-4, and the air pressure was 0.35 MPa. Since the pipeline length changed with each test, it was difficult to attach the Faraday cup to the end of the pipeline. Therefore, in this study, only the value of $q_2$ was measured. As seen in Fig. 2, $q_2$ had the almost same value as $q_1$ but had an opposite sign. The length of the pipeline was measured starting from beginning of the powder port because the inside structure of the powder port was the same as that of the pipeline. Each experiment was repeated five times, and the average values were reported. The results showed that $q_2$ increased as the length of the pipeline increased for both sizes of glass beads. This can be explained by the fact that as the length increases, the time the glass beads come in contact with the pipeline also increases. The increase rate at the beginning section of the pipeline was especially large. It can also be seen through Fig. 5 that overall, G-2 had a larger $q_2$ than G-4 because of the total surface area. This confirms the reliability of the data shown in Fig. 4.

In addition, both beads seemed to reach saturation when the pipeline length was 1415 mm. It was compared to the maximum charge to mass ratio, $q_c$ [nC/g] calculated from the well-known equation shown below [11]:

$q_c = |\sigma_{max}|/(D \cdot \rho)$, --------------------------(2)

where $\sigma_{max}$ [kg/m$^3$] is the maximum surface charge density (27 μC/m$^2$); *D* [μm] is the particle diameter; and $\rho$ is the particle density ($1.5 \times 10^3$ kg/m$^3$).

In this study, the particle sizes of G-2 and G-4 were 131 μm and 225 μm, respectively, as shown in Table 1. The calculated maximum values $q_c$ and experimental maximum values $q_t$ are compared as shown in Table 2. The calculated and experimental values for both G-2 and G-4 agreed well with each other. Therefore, it can be seen that both G-2 and G-4 reached saturation.

**Table 2**. Calculated maximum values and experimental maximum values of the charge to mass ratio.

| | $q_c$ [nC/g] | $q_t$ [nC/g] |
|---|---|---|
| G-2 | 137 | 151 |
| G-4 | 80 | 99 |

## Conclusion

In this paper, a novel tribocharging system that utilizes the Faraday cup was presented and the electrostatic charge behavior of glass beads was experimentally investigated. The results were as follows:
1) For all of the glass beads, the charge clearly increased as the glass bead particles passed through the pipeline. The charge amount of the glass beads and the charge amount obtained from the pipeline had almost the same trend and absolute values. This result confirmed the reliability of the test and showed that there was no disturbance by external noise.
2) All of the glass beads were negatively charged when stainless steel metal was used as the contact surface. The value range of the charge to mass ratio of the glass beads obtained in this study was about 40 nC/g to 100 nC/g.
3) The charge amount of the glass beads increased as the applied air increased. It could also be seen that as the air pressure increased, the air velocity also increased.
4) The charge amount of the glass beads increased as the powder size decreased due to the total surface area of the beads.
5) The charge amount of the glass beads (G-2, G-4) increased as the length of the pipeline. The increase rate at the beginning section of the pipeline was especially large.

***Acknowledgments***
The authors gratefully acknowledge Mr. Naoto Nogera for his assistance during the test.

***References***

1. F. Riedewald (2012): "A fire in a secondary pharmaceutical powder transfer operation," *Process safety progress*, 31, 4, pp. 390-392.
2. W. Kaialy W, "A review of factors affecting electrostatic charging of pharmaceuticals and adhesive mixtures for inhalation," *Int.J.Pharm*. Vol.503, pp.262-276.
3. N. Gibson (1997), "Static electricity — an industrial hazard under control," *J. Electrostatics,* Vol. 40-41, pp. 21-30
4. K. Choi, K. Choi, K. Nishimura (2015), "Experimental study on the influence of the nitrogen concentration in the air on the minimum ignition energies of combustible powders due to electrostatic discharges," *J. of Loss Prevention in the Process Industries.* Vol.34.pp.163-166.
5. K.S. Choi, H.Sakasai, and K. Nishimura (2015), "Experimental study on ignitability of pure aluminum powders due to electrostatic discharges and Nitrogen's effect," *J. of Loss Prevention in the Process Industries.* Vol.35.pp.232-235.

6. K.S. Choi, H. Sakasai, and K. Nishimura (2015), "Minimum ignition energies of pure magnesium powders due to electrostatic discharges and Nitrogen's effect," *J. of Loss Prevention in the Process Industries.* Vol.41.pp.144-146.
7. N. Masui, and Y. Murata (1981), "charging tendency of the flowing powder through a metal tube (II), " *The five Annual meeting of Institute of Electrostatic Japan,* pp. 244-247.
8. S. Watano, S. Saito, and T. Suzuki (2002), "Feedback control of powder electrostatic charge in a pneumatic conveying process," *J. Soc. Powder,Japan*, Vol. 39, pp. 496-502.
9. P. Boschung, M. Glor (1980), "Methods for investigating the electrostatic behavior of powders, J. of elelctrostatic," Vol. 8, pp. 205-219.
10. IEC50404(2004), "Electrostatics- Code of practice for the avoidance of hazards due to static electricity," p. 66.
11. M.Glor (1988), "Electrostatic hazards in powder handling," *Research studies press LTD.*, p.26.

# 509266 Spray-Drying of a Layered Double Hydroxide Nanosuspension

**Abstract:**

Over the past few years, layered double hydroxide (LDH) has attracted particular interests in both academia and industry. LDH is 2D anionic clay of brucite-like layered structure with various organic or inorganic anions located between layers. The diverse composition of LDH, and its associated physical and chemical properties open up new possibilities in various applications. LDHs are successfully applied as precursors in synthesis of supported catalysts, sorbents, carriers delivering bioactive molecules, matrices containing active pharmaceutical ingredients for control release, electrochemical biosensors, etc. LDHs are mainly synthesized in a form of nanosuspensions. The application of nano-sized sorbents or catalysts in the liquid-phase processes requires a special technique to separate them from the final product. An uncontrollable agglomeration of solid nanoparticles makes their industrial application impossible in a dry phase. Moreover, the toxic effects of nanoparticles on human health and ecosystem have been discussed in the literature.

Utilization of larger micron-size agglomerates of nanoparticles is one of the alternative ways to safely handle nanoparticles in the dry phase on the industrial scale. Spray drying is one of the well-known techniques that have been widely applied for production of microspheres from nano-sized particles. Using this technique, it is possible to control the size distributions as well as the morphology of spray-dried agglomerates. However, there are only a limited number of publications related to spray drying of LDH nanosuspensions. Moreover, spray-dried LDH agglomerates of rather wide size distribution, low surface area and irregular shape were reported in the literature.

In this study, we explore the effect of process parameters, such as spraying air flow rate, drying air temperature, slurry concentration and slurry feed rate on the size distribution, morphology, specific surface area and pore volume of LDH agglomerates fabricated by spray drying of nanosuspensions.

The round-shaped agglomerates with a diameter of more than 2 and less than 15 micrometers were obtained by spray drying of suspensions of nanoparticles with median diameter of 171 nm without addition of dispersant and binder. Our results confirmed that the spraying air flow rate has the dominant effect on the agglomerate median size and the width of agglomerate size distribution as well as on the agglomerate specific surface area. The small agglomerates with narrow size distribution were collected during drying at the high flow rate of spraying air. Our results also indicate that lowering the temperature of drying air and using the dilute nanosuspensions promote generation of small agglomerates.

The specific surface area of spray-dried LDH agglomerates is in the range of 49 to 101 $m^2/g$. The agglomerates of high specific surface area were obtained at low flow rate of spraying air, high slurry feed rate of concentrated slurry and using low temperature drying air. These results also confirm that drying at low drying rate yields to formation of agglomerates of high surface area. Similar trends were observed in terms of the variation of total pore volume with drying process parameters.

Thus, by adjusting the spray drying conditions, it is possible to produce agglomerates of LDH nanoparticles satisfying the product requirements.

# 513829 Impact of Spheroidization of $UO_2$ Powders on the Filling of Press Molds

**Abstract:**

In industry, the use of powders is very common. This underlines the interest in understanding the flow of powders in process lines (hopper, silos, transportâ€¦) and finding the link between powder physical characteristics and its rheological properties. This study deals with the flowing properties of a UO2 powder in a powder metallurgy process for nuclear fuel fabrication. A reproducible filling of the press molds is required to avoid mechanical grinding of the sintered pellets.

A powder is characterized by its particle size distribution but also by the morphology of particles. Morphology has a major impact on the cohesion and then on the flowability of powders [1]. UO2 powders are generally made up agglomerates which are composed of aggregates, themselves formed by crystallites (Figure 1). Agglomerates are primarily governing the powder flowing behavior. Their characteristics sometimes do not allow them to meet the expectations of the process, especially for the filling of the press molds.

For this reason, in a powder manufacturing process, a spheroidization operation (Figure 2) could precede the shaping of pellets to improve the filling properties.

In this study we observe the influence of particles morphology on the rheological properties of UO2 powders. Whereas there is no standard and universal method to characterize the flow of powders [2], there is direct (shear cells) and indirect methods (dynamic flowability measurements like permeability). We use both methods: the indirect methods based on qualitative measures of powder cohesion, the direct methods to determine the powder cohesion for a specific consolidation state through the Freeman shear cell [3]. While a shear cell measurement permits to determine the powder flow in quasi-static system, a permeability measurement shows the flow index in dynamic system. We identify the link between theses measurements and the quality of the filling.

References:
[1] Xiaowei Fu and others, «Effect of Particle Shape and Size on Flow Properties of Lactose Powders», Particuology, Advances in Characterization and Modeling of Particulate Processes, Apr. 2012.
[2] P. Juliano, B. Muhunthan, and G. V. Barbosa-Cánovas, « Flow and shear descriptors of preconsolidated food powders », J. Food Eng., vol. 72, no 2, p. 157-166, Jan. 2006.
[3] J. Schwedes and D. Schulze, « Measurement of flow properties of bulk solids », Powder Technol., vol. 61, no 1, p. 59-68, Apr. 1990.

# 508203 Modelling a Twin Screw Granulator Using the Discrete Element Method


**Abstract:**

Wet granulation is a process used to create larger stable agglomerates (granules) from fine powders. This has many desirable outcomes such as improving flowability, compactibility, and homogeneity. Granulation is commonly employed in the food, pharmaceutical, detergent, and fertilizer industries, but despite its wide adoption it is often inefficiently operated, with high recycle ratios in continuous processes and high rejection rates in batch processes. Wet granulation is the most common type of granulation and, in pharmaceutical applications, it is a critical step in tablet manufacturing that affects the uniformity and compactibility of the final dosage form.

Traditionally batch granulation was the favoured granulation process method in the pharmaceutical industry due to the challenges faced from adopting a continuous approach, such as cost or inability to monitor product quality reliably. Since the introduction of Quality by Design (QbD) and Process Analytical Technology (PAT) by the FDA in 2003 there has been a move towards new manufacturing techniques such as continuous processing.

Although tremendous efforts have been made to gain scientific insight into the granulating process, a fundamental understanding of wet granulation is still lacking due to the complexity of the mechanisms involved. With twin screw granulation becoming a popularly employed method of wet granulation, an in-depth understanding of particle enlargement in the granulating process is necessary in order to improve the quality of the final product without the need for large-scale Design-of-Experiment studies.

Despite the extensive experimental research carried out for twin screw granulators in recent years, there has been little computational work carried out. This paper employs the Discrete Element Method (DEM) to study a 25 mm diameter, GEA ConsiGmaâ„¢ 1 twin screw granulator with a typical 60Â° forward configuration for kneading elements. The DEM simulations were conducted using the commercial code EDEM with a DEM contact model developed for cohesive solids implemented through an API. The contact model is based on an elasto-plastic contact with adhesion and uses hysteretic loading and unloading paths to model the elastic-plastic contact deformation. In these simulations, the adhesion is used to capture the effect of the binder liquid without the complication of modelling the liquid directly. The adhesion parameter is a function of the plastic contact overlap. The model has previously been shown to be able to predict the stress-history-dependent behaviour depicted by a flow function of the material. In this study the effect of particle size, gap tolerance and cohesion on the residence time distribution and local solid fraction at different elements in the granulator is assessed. The level of cohesion used is related back to uniaxial test results which can be used to measure the flowability of a cohesive granular solid.

# 508216 Hybrid Multiscale Modelling of a Twin Screw Granulator


**Abstract:**

Wet granulation is a process used to create larger stable agglomerates (granules) from fine powders. This has many desirable outcomes such as improving flowability, compactibility, and homogeneity. Granulation is commonly employed in the food, pharmaceutical, detergent, and fertilizer industries, but despite its wide adoption it is often inefficiently operated, with high recycle ratios in continuous processes and high rejection rates in batch processes. Wet granulation is the most common type of granulation and, in pharmaceutical applications, it is a critical step in tablet manufacturing that affects the uniformity and compactibility of the final dosage form.

Modelling processes such as wet granulation is not straightforward, but may be achieved using techniques such as Population Balance Modelling (PBM) and Discrete Element Method (DEM). PBM groups particles into different classes based on their properties and uses empirically derived differential equations to describe key processes such as aggregation or breakage. However, although PBM tracks the size and number of particles over time, it cannot account for the distribution of the particles and the effect of this local variability.

By contrast, DEM simulations track individual particles, and as such, phenomena such as segregation or liquid migration can be studied. DEM simulations also consider the effect that the geometry and its dynamics have on the particles' behaviour, which is something that PBM cannot account for directly. However, DEM simulations of fine materials can be computationally expensive and this often results in reduced numbers of particles being used to make the simulation computationally tractable. This is usually achieved through increased, unrealistic particle sizes or the simulation of isolated local zones without considering adjacent zones which may affect the behaviour.

Traditionally, large-scale Design-of-Experiment (DoE) studies would be needed to estimate key rate parameters for use in PBM modelling of granulation and this can be both costly and time-consuming. DEM simulations provide an alternative method to estimate these parameters. By coupling the two simulation methods, one can utilise the key benefits of each method to better simulate complex processes.

This paper presents a hybrid approach to modelling twin screw granulation via DEM–PBM coupled simulations to study a 25 mm diameter, GEA ConsiGma™ 1 twin screw granulator with a typical 60° forward configuration for kneading elements. The DEM simulation was conducted using the commercial code EDEM with a DEM contact model developed for cohesive solids implemented through an API. The PBM simulation is carried out with the commercial code gFORMULATE which is built on top on the advanced modelling platform gPROMS. A bi-directional coupling is implemented with information shared between the two simulation tools at specified intervals. Preliminary results for this coupled approach are presented and discussed.

# 509577 Blocking Rules for Discharging Granular Materials from a Flat Bottom Hopper

**Abstract:**

Hoppers and silos are widely used in storing granular materials in various industries, such as agricultural, chemical, food and pharmaceutical industries. Three key aspects need to be considered when hoppers are used to store powders and grains: 1) stress analysis; 2) hopper blockage ; 3) mass flow rate. A hopper must be designed strong enough to sustain the stress generated from the bulk materials during storage and discharge. On the other hand, a hopper must be designed properly so that the stored bulk materials can be discharged, i.e. without blockage and formation of ratholes and arches. Furthermore, the mass flow rate needs to be predicted and controlled. In this study, a combined experimental and numerical study was performed to determine the blocking rule for discharging granular materials from a flat bottom hopper. Granular materials of various sizes were considered and the minimal orifice size through which the bulk materials can be discharged was determined experimentally using a Flodex tester and numerically using DEM. The dependency of the minimal

orifice size on particle size was explored. An semi-empirical model was also developed. Using this model, the critical orifice size at which the flow will stop, i.e. blockage takes place, can be determined from the mass flow rate data. It was found that excellent agreement between the numerical results and experimental measurements was obtained. Both experimental and numerical results showed that, even for spherical particles, layered dead zones could be formed during the discharge from the flat bottom hopper. In addition, mono-layered dead zones were developed when large particles were used. Furthermore, the developed blocking rules were also verified experimentally and numerically.

# 508310 Characterization of Mesoscopic Structure in Cohesive Powder, Neat or Blended, By X-Ray Computed Tomography and Prediction By the Discrete Element Method

**Abstract:**

Understanding and predicting structural characteristics of cohesive powder systems is a technical challenge in many powder technology processes and applications. The discrete element method (DEM) [1] is a promising approach to model the structure and flow behavior of cohesive granular systems. Advances in experimental techniques such as X-ray computed tomography (XRCT) have now enabled characterization of fine cohesive powder systems with sub-micron resolution. Combined with standard bulk powder characterization methods (e.g. density), this technique allows for mesoscopic structural characteristics of cohesive powder systems to be compared with DEM simulation. In this study, DEM and XRCT are used to characterize the mesoscale structure, under controlled consolidation conditions, of two moderately cohesive 'model' (roughly spherical) powders with distinct material properties. The DEM simulations are based on a constitutive elasto-plastic model [2] that makes use of the JKR theory [3] to account for cohesive interparticle forces. Density and void structure obtained from DEM simulation are compared to characterization results obtained from XRCT image analysis. XRCT and DEM simulation approach is applied to examine structures of binary mixtures of the two model powders. The ability of DEM simulation and the contact model to accurately predict structural characteristics of single component and binary granular systems is discussed.

References

[1] P. A. Cundall and Otto D.L. Strack. A discrete numerical model for granular assemblies. Geotechnique, 29(1):47–65, 1979.

[2] C. Thornton and Z. Ning. A theoretical model for the stick/bounce behavior of adhesive, elastic-plastic spheres. Powder technology, 99(2):154– 162, 1998.

[3] K.L. Johnson, K. Kendall, and A.D. Roberts. Surface energy and the contact of elastic solids. In Proceedings of the Royal Society of London A: Mathematical, Physical and Engineering Sciences, volume 324, pages 301–313. The Royal Society, 1971.

# 511638 Numerical Research of Hydrodynamics in Gas-Solid Micro Fluidized Beds


**Abstract:**

Due to the advantages of less energy consumption, more safety and good operability, the miniaturization of fluidized bed, or named micro fluidized bed, has been extensively studied by a lot of researchers. The Micro Fluidized Bed Reaction Analyzer (MFBRA) was adopted by the Institute of Process Engineering (IPE), Chinese Academy of Sciences (CAS), to study the gas-solid reaction kinetics under isothermal differential condition[1, 2]. In the MFBRA, the hydrodynamics of micro fluidized beds play a vital role on the heat and mass transfer and which is a key to the accurate measurement of gas-solid reaction kinetics. In order to optimize and design the MFBRA, numerous experiments have been conducted to study the hydrodynamics of micro fluidized beds[3, 4]. Unfortunately, those works mostly focused on the bed pressure drop and minimum fluidization velocity, nevertheless, the detail information, such as local solid volume fraction and particle velocity profile in micro fluidized beds are reported rarely. It is because the existing experimental techniques, such as, optical fiber and capacitance probes are hard to measure those local parameters without disturbing the flow field.

The objective of this work is to study the hydrodynamics of micro fluidized beds and to investigate the isothermal characteristics during kinetics testing of graphite powder combustion in MFBRA. In this study, Discrete Particle Model (DPM) and Two-Fluid Model (TFM) are selected to investigate micro fluidized beds, respectively. The DPM is carried out by the commercial software EDEM-Fluent Coupling simulations. And for TFM, we used the Eulerian multiphase model in the ANSYS Fluent. From our numerical simulations, the detailed hydrodynamic characteristics of micro fluidized beds are explored systematically by comparing different column sizes and superficial gas velocity. Meanwhile, it is found that both DPM and TFM have the ability to model the MFBRA under practical operation condition with high superficial gas velocity. In the term of isothermal characteristics in micro fluidized beds, the temperature field distribution as well as the heat transfer between heat-carrier and graphite powder are explored by the combination of simulations and experiments.


Reference:


1. Xu, G., *Micro fluidized bed reaction kinetics analyzer. Supported by instru-ment development project of Chinese Academy of Sciences, China, Project numberY2005014. (in Chinese).* 2005.
2. Yu, J., et al., *Isothermal differential characteristics of gasâ€"solid reaction in micro-fluidized bed reactor.* Fuel, 2013. **103**: p. 29-36.
3. Liu, X., G. Xu, and S. Gao, *Micro fluidized beds: Wall effect and operability.* Chemical Engineering Journal, 2008. **137**(2): p. 302-307.
4. Guo, Q.j., Y. Xu, and X. Yue, *Fluidization Characteristics in Micro-Fluidized Beds of Various Inner Diameters.* Chemical Engineering & Technology, 2009. **32**(12): p. 1992-1999.

# 507312 An Improved Bubble-Based Drag ModelÂ for Accurate Coarse-Grid Two-Fluid Modeling of Geldart a Powder Bubbling Fluidization

**Abstract:**

An improved bubble-based energy minimization multi-scale (EMMS) drag model was formulated in which the original bubble-based EMMS drag model at macroscale bed level was extended to microscale sub-grid level for accurate coarse-grid two-fluid modeling (TFM) of Geldart A powder bubbling fluidization. At both the macroscale and microscale levels, the physically heterogeneous structures were modeled by three pseudo phases, i.e., bubble phase, emulsion phase, and inter-phase, where conservation and stability condition equations were applied. Through such scale and phase decompositions, the improved bubble-based EMMS drag model was formulated through a two-step macro-to-micro scheme, where the obtained drag coefficient is related to phase volume fractions, velocities, and accelerations within each coarse grid. Modeling results show that TFM combined with the improved bubble-based EMMS drag model can accurately reproduce the heterogeneous structures for a laboratory-scale Geldart A powder bubbling fluidized bed. At relatively coarse grid, the predicted axial and radial distributions of solid concentration agree very well with experimental measurements. This indicates that our proposed improved bubble-based EMMS drag model incorporated TFM is promising for accurate and economical simulation of engineering-scale Geldart A powder bubbling fluidization.

# 507860 Numerical Simulation on Fine Particle Transport Behaviour in Electrostatic Precipitators

**Abstract:**

Electrostatic precipitators (ESPs) are a major air pollution control device to remove particles from treated gas. Extensive research has been done on several aspects such as electrostatics, fluid dynamics, charging mechanism, and particle dynamics. The overall particle-collection efficiency of a modern industrial wire-plate ESP exceeds 99%. However, the removal efficiency of submicron particles by conventional ESPs is as low as 70â€"80% owing to the low charge carried by fine particles and the increase in mobility with decreasing size. Submicron particles possess a high superficial area, thus favouring toxic heavy metals and other pollutants. Therefore, fine particles are collected with much lower efficiency and further research is required to improve this aspect.

In this work, a numerical study on the charging and transport behaviour of fine particles has been carried out based on wire-plate electrostatic precipitators (ESPs) with multiple wire electrodes. The effect of the applied wire voltage, inlet height, and wire distance on particle charging and transport behaviour, and the influence of the precipitator structure on particle trapping are analysed in detail. The 2-D numerical model involves electric field, space-charge density, gas flow, and particle trajectories. The fluid-flow velocity is higher than 1 m/s in this study and the EHD flow has a negligible. A Lagrangian approach for particle transport is utilized with the trajectories of individual particles tracked by solving their equation of motion. The electric field and gas flow are calculated using the finite-volume method. The flow simulation is done using the commercial software FLUENT with other parts coded as user-defined functions. The following main results can be drawn:

(1) It was observed that the Y-displacement of the particles exhibited a universal increase for all particle sizes, especially for larger particles, at low injection positions ($Y < 25$ mm). The larger the particle size, the higher the Y-displacement of particles because larger particle sizes improve both the particle charging level and electric field strength. No significant improvements were found for submicron particles.

(2) Increasing the spacing of the discharge electrode within a certain range and therefore increasing the electric field, it was observed that the particle-trapping effect improved and the outlet level exhibited a universal increase for all particle sizes, especially for large particles.

(3) The ESP models M1 and M2 were used to investigate their influence on the outlet height. Except for particles with a diameter of 2 μm, all other particle sizes showed an increased outlet height in M2 compared to that in M0. In addition, the different discharge-electrode arrangements, especially in M2, exhibited significant effects on the charging and transport behaviour of particles with a diameter of 1 μm.

(4) Particles with an inlet height of 5 and 10 mm were studied in M1 and M2 precipitators. The changed discharge-electrode arrangement improves particle trapping, especially in M2. This is because of the electron trajectories being closer to the discharge electrodes (M2), giving the particles more charge and leading to a greater electric field.

# 507879 Lattice Boltzmann Simulations of Porous Particulate Flows

**Abstract:**

The comprehensive understanding of the particle-fluid interaction is very important for fluid-particles system, and thus many studies have been conducted for the particle-fluid system[1-3] of solid impermeable particles. In the real industrial fields such as the catalyst process in chemical engineering[4], the particles usually possess the porous structure which will have effects on the fluid-particles interaction[5-6]. However, there has been little open literature discussing the particle-fluid system of porous permeable particles. Therefore, the objective of the present study is to investigate the effect of the porosity and permeability on the porous particulate flows which are formulated by the volume-averaged governing equations[7] in terms of intrinsic phase average velocity. The lattice Boltzmann equation model (LBE)[8] is used to solve the governing equations. Firstly, we validate the LBE model by the numerical results existing in the literature for the flow around and through one porous particle. Then, the porous particles settling against gravity in a quiescent fluid, i.e., the DKT process[9] for porous particles, is investigated. The effects of the porosity and Darcy number on the porous particulate flows are studied numerically in details. Also, we present the simulation results of the fluidization of 512 porous particles. Our results demonstrate the significance of the porous structure of particles for the particles dynamics behaviors.

**Reference**

[1] Ladd, A.J.C., 1994. Numerical simulations of particulate suspensions via a discretized Boltzmann equation Part I. Theoretical foundation. J. Fluid Mech., 271, 285-310.

[2] Hill, R.J., Koch, D.L., Ladd, A.J.C., 2001. The first effects of fluid inertia on flows in ordered and random arrays of spheres. J. Fluid Mech. 448, 213-241.

[3] Van der Hoef, M.A., Beetstra, R., Kuipers, J.A.M., 2005. Lattice Boltzmann simulations of low-Reynolds-number flow past mono- and bidisperse arrays of spheres: results for the permeability and drag force. J. Fluid Mech. 528, 233-254.

[4] Zhao, Y.F., Li, H., Ye, M., Liu, Z.M., 2013. 3D numerical simulation of a large scale MTO fluidized bed reactor. Ind. Eng. Chem. Res. 52, 11354-11364.

[5] Noymer, P.D., Glicksman, L.R., Devendran, A., 1998. Drag on a permeable cylinder in steady flow at moderate Reynolds numbers. Chem. Eng. Sci. 53, 2859-2869.

[6] Zhu, Q.Y., Chen, Y.Q., Yu, H.Z., 2014. Numerical simulation of the flow around and through a hygroscopic porous circular cylinder. Comput. Fluids 92, 188-198.

[7] Wang, L., Wang, L.-P., Guo, Z.L., Mi, J.C., 2015. Volume-averaged macroscopic equation for fluid flow in moving porous media. Int. J. Heat Mass Trans. 82, 357-368.

[8] Chen, S.Y., Doolen, G.D., 1998. Lattice Boltzmann method for fluid flows. Annu. Rev. Fluid Mech. 30, 329-364.

[9] Fortes, A.F., Joseph, D.D., Lundgren, T.S., 1987. Nonlinear mechanics of fluidization of beds of spherical particles. J. Fluid Mech. 177, 467-483.

# 508370 Modeling of a Novel Multi-Particle Collision Model for Gas-Solid FlowsÂ

**Abstract:**

Fluidized beds are widely applied in the chemical, petrochemical, metallurgical, environmental and energy industries due to their favourable mass and heat transfer characteristics. In past few decades computational fluid dynamics has proven to predict hydrodynamics of fluidized beds. DPM (Euler-Lagrange) [1] and TFM (Euler-Euler) [2] are two most common modelling approaches for gas-solid flow in fluidized beds. In TFM the solid phase itself is treated as a continuum phase interpenetrating with the gas phase. Since the concept of individual particles has disappeared, the effects of particle-particle interactions in the TFM has to be incorporated in a statistically averaged way, via the solid phase pressure and viscosity. TFM is preferred model to study large scale fluidized beds but model suffer from high numerical diffusion. And the KTGF theory in TFM assume elastic collisions and is limited only to spherical particles [2]. In DPM, the gas phase is considered as the continuous phase whereas the particle trajectories are obtained by integrating Newtonâ€™s equations of motion. Particle-particle and particle-wall interactions are explicitly taken into account using various physical models such as the soft sphere model. DPM proves to be accurate modelling of gas-particle flow. However, lagrangian models quickly become computationally too expensive because of the large number of particle interactions. Therefore coarse graining of discrete particles is an emerging model that can handle large number of particles effectively. The new coarse-grained model developments are of vital importance for the advancement of large-scale modelling of gas-particle flows that are flexible to extended for polydispersity, and non-spherical particles. In this study a novel lagrangian multi-particle collision (MPC) method is developed for predicting dense particulate-fluid flow.

The MPC solver is developed in OpenFoam [5], where the fluid phase is treated as a continuum and the solids are modelled by lagrangian particles. The MPC model grew out of the multiphase particle-in-cell (MP-PIC) model [3], that employs a fixed Eulerian grid, and particles are traced in a lagrangian fashion, including a parcel approach where each simulated particle represents a large collection of real particles. Each parcel follows Newtonâ€™s equations of motion, where inter-phase momentum transfer is accounted through mapping and interpolating quantities back and forth between parcel locations and the eulerian grid. Particle collisions in MP-PIC are not fully resolved but are derived from particles stresses that are theoretically calculated from Harris and Crighton [4] stress closure. In the MPC model, we implemented a functional form for the particle stress tensor where the coefficients are obtained computationally by means of DPM simulations. The idea is to extend the model from spherical particles to non-spherical particles, for which theoretical particles closures are currently not available in the literature. The viscous stress and random forces are typically ignored in MP-PIC model due to their insignificant effect [3]. Importantly, we incorporated viscous stress and random forces representing scattering of particles by collisions, which are obtained from DPM simulations by measuring the actual variance. Details of model development and validation is presented in this study.

Results from MPC model are compared with MPPICFoam and, with DPM simulations from DEM-LIGGGHT [6], and MFiX [7] code respectively. Pseuduo-2D fluidized beds of dimension 0.01x0.15x0.45 m is considered for comparison with experiments [8]. Comparison is made on the basic of pressure drop, void fraction, bubble dynamics and solids motion. We observed that MPC is in good agreement with DPM, and experiments, when compared to MPPICFoam. Unrealistic over-packing of solids is predicted in MPPIC due to limitation of Harris and Crighton [4] model and the velocity field updates, whereas in MPC these limitations overcomes and no over packing of particles is predicted. Solid circulations are in good agreement with DPM results, where solid moving above in the center and downward near to the wall. Time-average porosity plots revels the path of bubble motion and is in close agreement with results of DPM. In conclusion a novel multi-particle collision model is developed and validated to study gas-solid fluidized beds. The model is efficient to handle large number of particles. In future study, we plan to extend the model for dense flow of non-spherical particles.

**Acknowledgments**

The authors thank the European Research Council for its financial support under its consolidator grant scheme, contract no. 615906 (NonSphereFlow).

**References**

1. N.A. Patankar, D.D. Joseph, 2001. Modeling and numerical simulation of particulate flows by the Eulerian Lagrangian approach, International Journal of Multiphase Flow, Volume 27, Issue 10, October 2001, Pages 1659-1684.
2. Gidaspow, D., 1994. Multiphase flow and fluidization continuum and kinetic theory description, Academic Press, Boston.
3. Andrews, M.J., O'Rourke, P.J., 1996. The multiphase particle-in-cell (MP-PIC) method for dense particulate ows, Int. J. Multiphase Flow 22, 379-402.
4. Harris, S.E., Crighton, D.G., 1988. Solutions, solitary waves and voidage disturbances in gas-fluidized beds, J. Fluid Mech. 195, 437.
5. OpenFOAM, 2015. The Open Source CFD Toolbox, Ver. 3.0.0. ESI Group, Paris, France.
6. https://www.cfdem.com/liggghtsr-open-source-discrete-element-method-particle-simulation-code
7. https://mfix.netl.doe.gov/mfix/
8. M. Goldschmidt, 2001. PhD thesis, Hydrodynamic Modelling of Fluidised Bed Spray Granulation.

# 508490 Dpm Analysis of Large Fluidized Catalytic Cracking (FCC) Reactors


**Abstract:**

DPM Analysis of Large Fluidized Catalytic Cracking (FCC) Reactors

Abstract:

Computational Fluid Dynamic (CFD) modeling is a powerful tool for equipment design and development and is widely used at UOP, particularly for the reactor and regenerator design of FCC process unit. Through CFD modeling, a technical solution was developed for a large capacity FCC process unit purposing an elevated amount of propylene production.

A key challenge in FCC process design is how to optimize the size of the reactor vessel, which is a function of the number and size of reactor cyclones (catalyst separationdisengagers). UOP developed a new cyclone cluster configuration via CFD study of the reactor internals, including UOP's VSS™ riser termination device and reactor cyclones, which reduce the reactor size significantly compare to the conventional design, while maintaining effective catalyst / vapor separation and mechanical reliability. The objective of the current CFD study is to evaluate the catalyst separation efficiency and performance of the new riser termination device technology, which accommodate more cyclones in a smaller reactor vessel.

To better understand the flow pattern, flow distribution and the gas residence time inside the UOP VSS™ , the analysis was broken down into two sections. Part 1 focused on single gas phase only and in part 2 the particles behavior is studied using the Discrete Phase Model (DPM) in Fluent.

The single gas flow analysis results showed that the flow in the VSS chamber section has the same rotational pattern as in conventional FCC/VSS units, where the catalyst will separate from the gas products. Also the gas flow distribution at each chamber arm is uniform, which is critical to catalyst separation efficiency.

The DPM analysis confirmed an equal catalyst particle distribution through all the cyclones and a high efficiency separation in the riser termination device. Having this information from CFD allows for the new design to move forward.

# 509623 Catalytic Reactor Design Using Multiphase CFD

**Abstract:**

Catalytic Reactor Design Usng Multiphase Cfddimitri Gidaspow the 2011 Book , "Computational Gas-Solids Flows and Reacting Systems ", Edited By S. Pannala, M. Syamlal and Tom O'Brien ,Engineering Science Reference, Shows the Progress We Have Made in Using Multiphase CFD in Modeling Fluidized Bed Chemical Reactors and Other Multiphase Systems, Such As the Pyroplastic Flow Emitted from Volcanoes. in This Presentation the Design of Two Types of Catalytic Reactors Will be Described: A Fischer_ Tropsch Slurry Bubble Column Reactora Polymerization Reactor the CFD Design of a New Slurry Bubble Column Reactor Was Already Described By Gidaspow ,Et Al in Chemical Engineering Science 134,784-799 (2015).in This Presentation the Design Will be Extended to Obtain 100, 000 Barrels of Diesel per Day, Twice That of the Largest Commercial Reactors. the Production of Synthesis Gas Using Oxygen or Carbon Dioxide in Catalytic Fluidized Beds Needed for the Fischer-Tropsch Reactor Will Also be Discussed Briefly. for the Polymerization Reactor Our CFD Code Described in " Computational Techniques " By D. Gidaspow and V. Jiradilok , Nova Science, 2009 Was Modified for Polymerization Reactors. Such Reactors and Their CFD Modeling Are Described in a Chapter in Pannala's Et Al Book from Professor Rodney Fox's Group. They Point out That the Fine Particle ( 80 microns ) Catalyst Radial Distribution, Is a Remaining Issue the Computations to be Presented Will Show That the Catalyst Will Have a High Concentration at the Reactor Walls for Various Operating Conditions. This Leads to a High Rate of Reaction at the Reactor Walls and May Result in Melting of the Polymers and Sheet Formation , Forcing Reactor Shut-Down. in Some of Our Simulations There Is Large Bubble and Slug Formation Similar to Those Mentioned in Fox's Chapter. Acknowledgement : I Thank Dr. Mayank Kashyap of SABIC for a Valuable Discussion and His Papers " Application of Multiphase Flow CFD in the Gas Phase Polymerization Process " Fluidization XV. May 24,2016 and a Related Paper Presented at Doe NETL, July 2016

# 509627 Simulation of Large Particle Turbulent Fluidization in Riser Reactors By Coarse Grain DEM-CFD

**Abstract:**

The complex two-phase flow in the riser is an important element of many industrial circulating fluidized bed reactors. In some cases, e.g. gas-solid polymerization, relatively large particles are transported upward in a turbulent regime, in which the high gas velocity results mostly from heat removal requirements. Strong solids acceleration, radial and axial velocity profiles for both the gas and solid, intense clustering and inhomogeneities make it impossible to reliably predict the riser head or local bed density profile. Detailed numerical simulations proved capable of showing some of the peculiarities of the flow, but limitations in the scale, lack of grid independence, need for special drag corrections and sub-grid models still hinder confident use of such tools in design and scale-up, particularly in industrial applications characterized by particle size and Reynolds numbers that are not well covered by the available scientific literature.

In the present contribution, coupled Discrete Element Method-Computational Fluid Dynamics (DEM-CFD) simulations are carried out using a coarse graining technique for the solid particles. According to this method, small number of particles are grouped into a computational particle (*parcel*) and represented by it as a spherical "lump," to reduce the computational cost of the simulations and allow lab-scale systems to be represented in full detail. The investigation focuses on the influence of the coars-graining degree, i.e. the number of real particles represented by a *parcel*, on the simulated flow hydrodynamics both in terms of its advantages (a quantification of the computational savings) and disadvantage, primarily the real capability of the method to resolve the behaviour observed and produce results in line with experimental findings. The gas and solid motion are studied, with the aim to characterize the clustering phenomena observed during the up-flow and its effect on the pressure drop. Assuming prescribed gas and solid inlet velocity fields, the potential tendency to show core-annulus flow, the intensity of solids clustering and the local bed density within the riser are examined at different number of particles per *parcel*.

# 509640 Numerical Study of Particles Shape Effects on Solid-Liquid Fluidizations

**Abstract:**

Numerical study of particles shape effects on solid-liquid fluidizations

Esmaeil Abbaszadeh Molaei, Zongyan Zhou*

Laboratory for Simulation and Modelling of Particulate Systems, Department of Chemical Engineering, Monash University, Clayton, VIC 3800, Australia

Solid-liquid fluidizations have been observed in many industries due to their high heat and mass transfer rate, high operation flexibilities, and reduced back mixing of fluid and solid phases. In most of these multiphase operations, particle properties such as size, density, and shape may experience significant changes during the process because of attrition, coalescence, comminution or chemical reactions. The difference in particle properties can result in significant segregation problem in liquid fluidized beds. To gain an understanding of solid-liquid fluidisation behaviour, most existing studies, either experimentally or numerically, have focused on solid flow behaviour with mono-sized particles or binary and ternary mixtures with different particle sizes or densities. In spite of substantial studies on the hydrodynamics of liquid fluidized beds, the effects of particles shape on the solid-liquid flow behaviour are still less reported. Except for two physical experiments, the previous studies considered particles as spheres, and few studies have been remarked on liquid fluidization of non-spherical particles. However, particles are generally irregular in most of the processes such as ore beneficiations using liquid-solid fluidized beds separator. Particle shape can cause different flow behaviour in liquid fluidizations rather than that observed for ideal spherical particles. Nonetheless, the answers to some fundamental questions such as how and why particle shape affects flow phenomena are not clear, and hence still poorly understood.

Therefore, in this work, a combination of computational fluid dynamics (CFD) and discrete element model (DEM) [CFD-DEM] is used to examine the effects of particle shape on the flow characteristics in liquid fluidizations. In the simulations, ellipsoidal particles are used as they can represent a wide range of particle shapes from oblate to prolate particles. Different aspect ratios are employed, and results are analysed mainly in terms of particle flow patterns, pressure drop, and bed expansion. In addition, force and particle orientation analysis is also conducted to examine the effect of particle shape and explain the occurrence of entrainment phenomenon.

The results show that ellipsoidal particles intend to enter to freeboard region and entrainment phenomenon occurs, and consequently, the interface between the bed and freeboard cannot be observed clearly. In addition, the pressure gradient variation with bed height is uniform for spheres; however, this uniformity is replaced with â€œdistorted parentheses-shapeâ€ profile for disc-shape and elongated particles. Finally, bed expansion analyses show that spherical particles have the lowest bed expansion. On the other hand, disc-shape particles intend to expand more significantly than spherical and elongated particles. This causes the entrainment phenomenon which has been found to be due to the effects of particles shape on particles terminal velocity.

* Corresponding author. Tel: 61 3 9905 0846; email: zongyan.zhou@monash.edu.

# 514318 CFD Investigation on Gas-Solids Flow and Heat Transfer in Two Fluidized Catalyst Cooler

**Abstract:**

Our previously experimental work showed that the internal circulation of solids in an annular catalyst cooler (ACC) had an ability to enhance the heat transfer between bed and wall, comparing with the base catalyst cooler (BCC). The current work simulated in detailed hydrodynamics of gas-solids flow and heat transfer to advance our understanding of the ACC by using the two-fluid model. The effective particle diameter $d_p^{âŽ}$ was determined to 170 Î¼m and the specularity coefficient Ï† was 0.3 in the 3D simulations. The catalyst cooler can be divided to four distinct regions including draft tube region, annulus region, bottom region and particle diffluence region according to the distributions of solids volume fraction and solids velocity. The effect of gas-solids flow on bed-to-wall heat transfer was compared in the different regions of the BCC and ACC. And the effects of operational gas velocity and bed structure were discussed. The results showed that the proper selection of the ratio of gas velocities and the position of the heat transfer tube were needed to form a stable internal solids circulation in the ACC. The ACC had a combined hydrodynamic feature of up-flow and down-flow catalyst coolers with bigger solids volume fraction and smaller particle resident time, which are beneficial for improving the heat transfer between solids and wall.

# Hydrodynamics and mixing characteristics of a new-type particle mixer

**Mengxi Liu, Chunxi Lu, Zhenliang Meng**

State Key Laboratory of Heavy Oil Processing, China University of Petroleum, Beijing 102249, PR China

E-mail: mengxiliu@sina.com (M. Liu)

**Abstract:**

In order to further enhance the particle mixing in the internally circulating fluidized bed, especially along the radial direction, a new-type particle mixer with slots on the draft tube was proposed in the present work. The hydrodynamics of the mixer was investigated by using a multi-scale computational fluid dynamics with the EMMS drag. The prediction in terms of cross-sectionally averaged solid holdup and solid velocity were compared to experimental data for validation. The results show that in draft tube a low pressure region can be observed near the slots and in annulus the pressure in the vicinity of the slots seems larger than that in the region between the adjacent slots, which induces horizontal particle immigration and mixing. Nearly 25.4% to 29.7% of gas in the draft tube bypasses and flows through the slots into the annulus. The ratio of the mass flux of the tracer gas bypassing into the annulus to that of total tracer gas is approximately equal to 0. With increase of the superficial gas velocity in the draft tube from 0.3 m/s to 0.45 m/s, the total solid mass flux increases from 4.87 kg/s to 10.86 kg/s, while the mass flux through slots decreases from 2.69 kg/s to 0.52 kg/s.

## 1. Introduction

The characteristics of particle flow and mixing have a considerable influence on the mass and heat transfer, as well as the product quality. According to the previous study, the mixing degree of particles along the axial and radial direction is quite different [1-5]. Mostoufi and Chaouki [4] discovered that the diffusivities varied from $3.3\times10^{-3}$ $m^2/s$ to $5.6\times10^{-2}$ $m^2/s$ along the axial direction and $2.6\times10^{-4}$ $m^2/s$ to $1.5\times10^{-3}$ $m^2/s$ along the radial direction. Berruti et al. [6] also revealed that the particle mixing rate for radial direction was an order of magnitude smaller than for the axial direction. Therefore, the weak particle flow along the radial direction greatly limits the mixing ability of a fluidized bed. Internally circulating fluidized bed (ICFB) has been widely recognized as an efficient fluidized bed reactor, in which orderly and macroscopic particle circulation occurs. Compared with the conventional fluidized bed, ICFB offers multiple particle circulation between the draft tube and the annulus, which leads to relatively strong particle convection, diffusion and shear, and thereby improves the particle mixing performance along the radial direction [7]. In order to further enhance the particle mixing for some processes in which a uniform particle mixing is required, a new-type particle mixer with slots on the draft tube was proposed in the present work.

With increase of the computational power, the computational fluid dynamics (CFD) provides an additional tool to predict particle flow and mixing in the local region, which is difficult with the experimental methods. In particular, the gas-solid flow in a gas-solid fluidized bed can be accurately described by the energy minimization multi-scale (EMMS) model [8]. In this paper, the hydrodynamics of the mixer was investigated by using a multi-scale CFD based on the EMMS drag model [9, 10].

## 2. Experimental apparatus, Materials and Operating conditions

The structure of the particle mixer is shown in Fig.1. The mixer has a height of 5 m and an inner diameter of 286 mm. A draft tube with a height of 1 m and an inner diameter of 215 mm is coaxially mounted into the column. The four rectangular slots of 80×40 mm are symmetrically opened on the draft tube at a height of $h$=860 mm, in order to strengthen the particle radial flow. A gas distributor plate and a ring distributor are installed below the draft tube and the annulus respectively. The height of the gap between the gas distributor plate and the draft tube is 60 mm. For the sake of the convenience, the mixer is divided into six zones: the pre-mixing zone, feed influence zone, draft tube zone, gas-solid separation zone, annulus zone and the bottom zone. A perforated plate is mounted at the bottom of the pre-mixing zone, which is connected with a central downcomer, extending into the draft tube. Particles are first mixed in the pre-mixing zone and then the pre-mixed particles flow downward through the central downcomer and into the feed influence zone. In the feed influence zone, the pre-mixed particles are mixed with the particles moving upward in the draft tube, flow upward and circulate between the draft tube and the annulus. In the annulus, particles move downward to the bottom. Then a part of particles flow out of the mixer and the other part flow horizontally into the draft tube and recirculate.

The solid particles used in the experiment were FCC catalyst ($\rho_p$=1498 kg/m$^3$), with an average diameter of 76 μm. Air ($\rho_g$=1.225 kg/m$^3$) was introduced into the mixer to fluidize the particles. The superficial gas velocity ranged from 0.3 to 0.45 m/s in the draft tube (based on the cross-sectional area of the draft tube) and remained at 0.05m/s in the annulus (based on the cross-sectional area of the annulus). The mass flux of cold particles remained at 13.8 kg/m$^2$s. As we only investigated the influence of slots on the flow and mixing of particles in this paper, no hot particles were induced into the mixer.

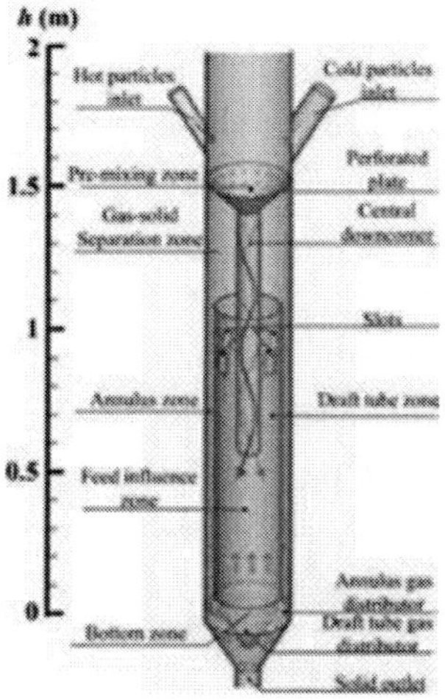


Fig. 1 Schematic diagram of the new-type particle mixer

## 3. CFD model and simulation method

The hydrodynamics of the mixer was performed using the Eulerian- Eulerian of Ansys Fluent

®13.0. The governing equations [11] include a set of mass and momentum equations for each phase, as shown in Table 1.

A 3-D mesh containing 580,000 cells was generated by using Gambit®2.4. The no-slip boundary was assigned for the gas phase and the partial-slip boundary condition developed by Johnson and Jackson [12] was used for the solid phase. In order to ensure a steady-state operation of the mixer, simulations were first performed for 30 s, and then the time-averaged data was analyzed for another period of 10 s. Other parameters settings in the model are shown in Table 2.

Table 1 Governing equations

Gas-phase Continuity equation $\frac{\partial}{\partial t}(\varepsilon_g \rho_g) + \nabla \cdot (\varepsilon_g \rho_g \boldsymbol{u}_g) = 0$

Solid-phase Continuity equation $\frac{\partial}{\partial t}(\varepsilon_s \rho_s) + \nabla \cdot (\varepsilon_s \rho_s \boldsymbol{u}_s) = 0$

Gas-phase momentum equation $\frac{\partial}{\partial t}(\varepsilon_g \rho_g \boldsymbol{u}_g) + \nabla \cdot (\varepsilon_g \rho_g \boldsymbol{u}_g \boldsymbol{u}_g) = -\varepsilon_g \nabla p + \nabla \cdot \boldsymbol{\tau}_g + \varepsilon_g \rho_g \boldsymbol{g} - \beta(\boldsymbol{u}_g - \boldsymbol{u}_s)$

Solid-phase momentum equation $\frac{\partial}{\partial t}(\varepsilon_s \rho_s \boldsymbol{u}_s) + \nabla \cdot (\varepsilon_s \rho_s \boldsymbol{u}_s \boldsymbol{u}_s) = -\varepsilon_s \nabla p - \nabla p_s + \nabla \cdot \boldsymbol{\tau}_s + \varepsilon_s \rho_s \boldsymbol{g} + \beta(\boldsymbol{u}_g - \boldsymbol{u}_s)$

Granular temperature equation

$$\frac{3}{2}[\frac{\partial}{\partial t}(\varepsilon_s \rho_s \Theta_s) + \nabla \cdot (\varepsilon_s \rho_s \Theta_s \boldsymbol{u}_s)] = (-p_s \boldsymbol{I} + \boldsymbol{\tau}_s) : \nabla \boldsymbol{u}_s + \nabla \cdot (\kappa_s \nabla \Theta_s) - \gamma_s - 3\beta\Theta_s$$

Gas-phase stress $\boldsymbol{\tau}_g = \varepsilon_g \mu_g [\nabla \boldsymbol{u}_g + (\nabla \boldsymbol{u}_g)^T] - \frac{2}{3}\varepsilon_g \mu_g (\nabla \cdot \boldsymbol{u}_g)\boldsymbol{I}$

Solid-phase stress $\boldsymbol{\tau}_s = \varepsilon_s \mu_s [\nabla \boldsymbol{u}_s + (\nabla \boldsymbol{u}_s)^T] + \varepsilon_s (\lambda_s - \frac{2}{3}\mu_s)(\nabla \cdot \boldsymbol{u}_k)\boldsymbol{I}$

Solid-phase pressure $p_s = \varepsilon_s \rho_s \Theta_s [1 + 2(1+e_s)\varepsilon_s g_0]$

Solid-phase shear viscosity $\mu_s = \mu_{s,col} + \mu_{s,kin} + \mu_{s,fr}$

Where $\mu_{s,col} = \frac{4}{5}\varepsilon_s \rho_s d_p g_0 (1+e_s)\sqrt{\Theta_s / \pi}$, $\mu_{s,kin} = \frac{10\rho_s d_p \sqrt{\Theta_s / \pi}}{96\varepsilon_s (1+e_s) g_0}[1 + \frac{4}{5} g_0 \varepsilon_s (1+e_s)]^2$, $\mu_{s,fr} = \frac{p_s \sin\phi}{2\sqrt{I_{2D}}}$

Radial distribution function $g_0 = [1 - (\varepsilon_s / \varepsilon_{s,max})^{1/3}]^{-1}$

Diffusion coefficient $\kappa_s = \frac{150\rho_s d_p \sqrt{\Theta_s / \pi}}{384(1+e_s) g_0}[1 + \frac{6}{5} g_0 \varepsilon_s (1+e_s)]^2 + 2\rho_s \varepsilon_s^2 d_p (1+e_s) g_0 \sqrt{\Theta_s / \pi}$

Collisional energy dissipation $\gamma_s = \frac{12(1-e_s^2)\varepsilon_s^2 \rho_s g_0 \Theta_s^{3/2}}{d_p \sqrt{\pi}}$

Heterogeneity index $H_d = \frac{\beta_e}{\beta}$, $\beta_e = \beta H_d$

Drag based homogeneous structure $$\beta = \begin{cases} \frac{3}{4} C_d \frac{\varepsilon_s \varepsilon_g \rho_g |\boldsymbol{u}_g - \boldsymbol{u}_s|}{d_p} \varepsilon_g^{-2.65}, \varepsilon_g > 0.8 \\ 150 \frac{\varepsilon_s^2}{\varepsilon_g} \frac{\mu_g}{d_p^2} + 1.75\varepsilon_s \frac{\rho_p}{d_p} |\boldsymbol{u}_g - \boldsymbol{u}_s|, \varepsilon_g \le 0.8 \end{cases}$$

Where $C_d = \begin{cases} 0.44 & \mathrm{Re}_s \ge 1000 \\ \frac{24}{\mathrm{Re}_s}(1 + 0.15\mathrm{Re}_s^{0.687}), & \mathrm{Re}_s < 1000 \end{cases}$, $\mathrm{Re}_s = \frac{\varepsilon_g \rho_g d_p |\boldsymbol{u}_g - \boldsymbol{u}_s|}{\mu_g}$

Table 2 Simulation Parameters

| | |
|---|---|
| Gas-solid model | Eulerian-Eulerian |
| Solids Pressure | Lun et al. |
| Particle-wall specularity coefficient | 0.5 |
| Gas-wall friction coefficient | No-slip |
| Pressure-Velocity Coupling | Simple |
| Gradient discretization | Least squares cell based |
| Momentum discretization | Second-order upwind |
| Volume fraction discretization | Quick |
| Max. number of iterations per time step | 40 |
| Pressure outlet (atm) | 1 |
| Inventory (kg) | 56 |
| Incipient solid holdup | 0.6 |

## 4. Results and discussion

### 4.1 Model validation

Fig. 2 gives the distribution of cross-sectionally averaged solid holdup and particle velocity in the mixer. As shown in Fig. 2(a), the solid holdup is larger in the bottom zone ($h$=0.04 m) and due to the accelerated upward motion of particles, the solid holdup decreases with increase of the axial height (when $h \leq 0.16$ m). The cross-sectionally averaged solid holdup is nearly constant in the draft tube and gas-solid separation zone of $0.4 \leq h \leq 1.25$ m. The cross-sectionally averaged particle velocity is shown in Fig. 2(b). The particle velocity is smaller in the bottom zone and keeps almost invariable in the draft tube and the gas-solid separation zone. Compared with the experimental data, the particle velocity seems slightly higher, while the solid holdup in the simulation agrees reasonably with the experimental results.

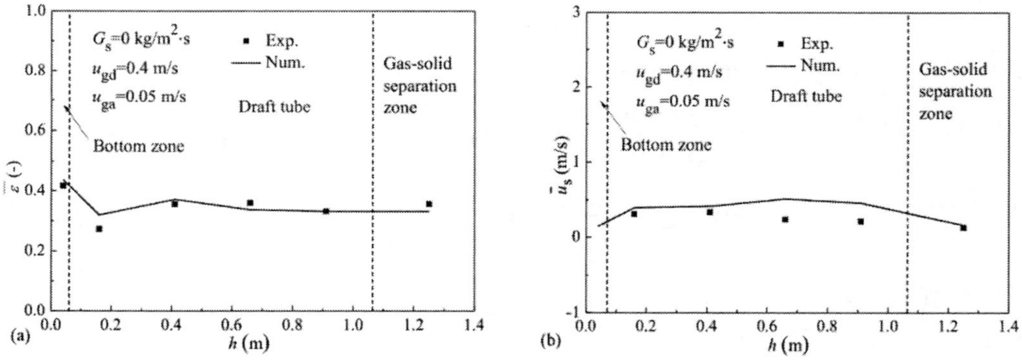


Fig. 2 Distribution of cross-sectionally averaged solid holdup and particle velocity as a function of axial heights

### 4.2 Pressure distribution near the slots

Fig. 3 shows the time-averaged contours of static pressure on the cross-section of $h$=0.9 m, the cross-section on the middle of the slots. It is seen that in the draft tube a low pressure region can be observed near the slots, smaller than the pressure in the region between the adjacent slots. This pressure drop forced particles inside the draft tube to move towards the slots and pass through them, which significantly facilitates the mixing of the particles along the radial direction. In the annulus the pressure in the vicinity of the slots seems larger than that in the region between the adjacent slots, inducing horizontal particle immigration and mixing.

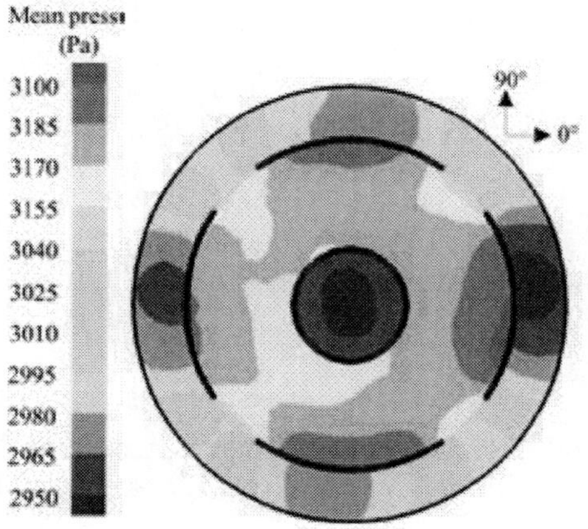


Fig. 3 Time-averaged contours of static pressure at the cross-section of the slots ($u_{gd}$ =0.4 m/s, $G_s$=13.8 kg/m$^2$s, $h$=0.9 m)

4.3 Gas bypassing

In the present work, the gas bypassing behavior was investigated by CFD simulation, and a tracer gas having identical physical properties with air was defined. After the mixer running stably, the tracer gas was introduced into the mixer from the gas distributor plate to replace the original gas. A cross section ($h$=0.06 m) on the exit of the annulus was selected and the mass flux of the tracer gas through the cross section was registered. The results are shown in Fig. 4. $f_B$ refers to the ratio of the mass flux of the tracer gas through the cross section ($h$=0.06 m) to that of total tracer gas. It is seen that $f_B$ varies between -0.05 and 0.05 with a frequency of 10 Hz, indicating an unstable gas bypassing from the draft tube to annulus. This suggests that the gas bypassing into the annulus was finally brought out of the annulus by the circulating particles. The average $f_B$ is approximately equal to 0, much smaller than that in a conventional ICFB (usually within 15%) [13], because massive particle circulation in the mixer makes it difficult to bypass from the draft tube to annulus.

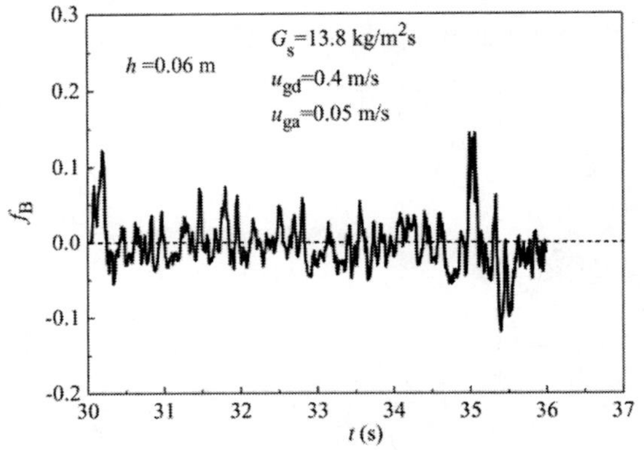


Fig. 4 Gas bypassing from draft tube to annulus in the bottom zone

As discussed previously, in the mixer there exists a pressure gradient from the draft tube to the vicinity of the slots, and then the annulus. Such a pressure difference leads to horizontal flows of particles and gas from the draft tube to the slots and then to the annulus. In the mixer the upside and downside of the slots is located at $h$=0.94 m and 0.86 m, respectively. Mass flux of the gas bypassing through the slots was obtained by monitoring the gas flux through the lower cross section ($h$=0.7 m) and the upper cross section ($h$=1 m) in the draft tube. Fig. 5 shows the gas mass flux under different operating conditions. The total gas mass flux in the draft tube increases with increasing gas velocity in the draft tube. With increase of the gas velocity from 0.3 to 0.45 m/s, the mass flux of the gas bypassing through the slots increases from 0.157 to 0.221 kg/m$^2$s. But the mass flux of bypassing gas does not increase linearly with the gas velocity. It is also seen that the mass ratio of the gas through the slots to that in the draft tube drops from 29.7% to 25.4%.

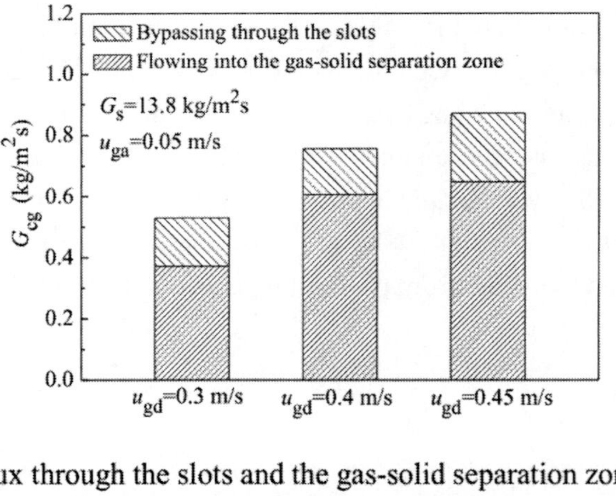


Fig. 5 Gas mass flux through the slots and the gas-solid separation zone as a function of superficial gas velocity in the draft tube

4.4 Particle mass flux

Fig. 6 shows the particle mass flux in the mixer. It is seen that the mass flux increases with increase of the superficial gas velocity in the draft tube. The mass flux through the slots accounts for about 50% of the total mass flux when $u_{gd}$=0.3 m/s. Because the axial particle velocity rose with increase of the superficial gas velocity from 0.3 m/s to 0.45 m/s in the draft tube, making it difficult to change the direction of the particles. Meanwhile, the radial particle velocity above the draft tube increases. Consequently, with the increase of the superficial gas velocity in the draft tube , the mass flux through the slots considerably decreased, from about 55.3% of the total mass flux (about 2.69 kg/s) at $u_{gd}$=0.3 m/s to only 4.8% (about 0.52 kg/s) at $u_{gd}$=0.45 m/s. Meanwhile, the total particle mass flux in the mixer increased from 4.87 to 10.86 kg/s as $u_{gd}$ increased from 0.3 to 0.45 m/s.

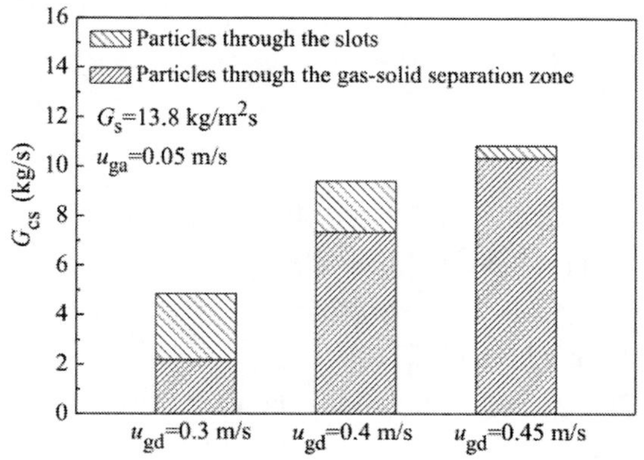


Fig. 6 Particle mass flux through the slots and the gas-solid separation zone as a function of superficial gas velocity in the draft tube

## 5. Conclusions

The characteristics of particle flow and mixing have a considerable influence on the mass and heat transfer, as well as the product quality. In order to further enhance the particle mixing in the ICFB, especially along the radial direction, a new-type particle mixer with slots on the draft tube was proposed in the present work. The hydrodynamics of the mixer was investigated by using a multi-scale computational fluid dynamics with the EMMS drag. The prediction in terms of cross-sectionally averaged solid holdup and solid velocity were compared to experimental data for validation. The results show that in draft tube a low pressure region can be observed near the slots and in annulus the pressure in the vicinity of the slots seems larger than that in the region between the adjacent slots, which induces horizontal particle immigration and mixing. Nearly 25.4% to 29.7% of gas in the draft tube bypasses and flows through the slots into the annulus. The ratio of the mass flux of the tracer gas bypassing into the annulus to that of total tracer gas is approximately equal to 0. With increasing the superficial gas velocity in the draft tube from 0.3 m/s to 0.45 m/s, the total solid mass flux increases from 4.87 kg/s to 10.86 kg/s, while the mass flux through slots decreases from 2.69 kg/s to 0.52 kg/s.

## Notation

| | | | |
|---|---|---|---|
| $G_s$ | solid mass flux (kg/m$^2$ s) | $G_{cg}$ | gas mass flux in the mixer (kg/m$^2$ s) |
| $G_{cs}$ | solid mass flux in the mixer (kg/s) | $h$ | height (m) |

| | | | |
|---|---|---|---|
| $p$ | pressure (Pa) | $R$ | radius of the reactor (mm) |
| $r$ | radial position of measurement points (mm) | $t$ | time (s) |
| $u_{ga}$ | superficial gas velocity in the annulus (m/s) | $u_{gd}$ | superficial gas velocity in the draft tube (m/s) |
| $\bar{u}$ | cross-sectionally averaged axial particle velocity (m/s) | | |

*Greek*

| | | | |
|---|---|---|---|
| $\varepsilon_g$ | voidage | $\varepsilon_s$ | solid holdup |
| $\bar{\varepsilon}_s$ | cross-sectionally averaged solid holdup | $\rho$ | density (kg/m$^3$) |

**References**


[1] F. Niklasson, H. Thunman, F. Johnsson, B. Leckner, Estimation of solids mixing in a fluidized-bed combustor, Ind Eng Chem Res, 41 (2002) 4663-4673.

[2] L.H. Shen, M.Y. Zhang, Effect of particle size on solids mixing in bubbling fluidized beds, Powder Technol, 97 (1998) 170-177.

[3] L.H. Shen, M.Y. Zhang, Y.Q. Xu, Solids Mixing in Fluidized-Beds, Powder Technol, 84 (1995) 207-212.

[4] N. Mostoufi, J. Chaouki, Local solid mixing in gas-solid fluidized beds, Powder Technol, 114 (2001) 23-31.

[5] D.Y. Liu, X.P. Chen, Lateral solids dispersion coefficient in large-scale fluidized beds, Combust Flame, 157 (2010) 2116-2124.

[6] F. Berruti, D.S. Scott, E. Rhodes, Measuring and Modeling Lateral Solid Mixing in a 3-Dimensional Batch Gas Solid Fluidized-Bed Reactor, Can J Chem Eng, 64 (1986) 48-56.

[7] M.M. Fang, K. Luo, S.L. Yang, K. Zhang, J.R. Fan, Computational Fluid Dynamics-Discrete Element Method Investigation of Solid Mixing Characteristics in an Internally Circulating Fluidized Bed, Ind Eng Chem Res, 52 (2013) 7556-7568.

[8] J. Li, M. Kwauk, Particle-fluid Two-phase Flow: the Energy-minimization Multi-scale Method, Metallurgical Industry Press, Beijing, 1994.

[9] Z.S. Shi, W. Wang, J.H. Li, A bubble-based EMMS model for gas-solid bubbling fluidization, Chem Eng Sci, 66 (2011) 5541-5555.

[10] K. Hong, Z.S. Shi, W. Wang, J.H. Li, A structure-dependent multi-fluid model (SFM) for heterogeneous gas-solid flow, Chem Eng Sci, 99 (2013) 191-202.

[11] D. Gidaspow, Multiphase flow and fluidization: Continuum and Kinetic Theory Descriptions, Academic Press, Boston, 1994.

[12] P.C. Johnson, R. Jackson, Frictional–collisional constitutive relations for granular materials, with application to plane shearing, J. Fluid Mech. 176 (1987) 67–93.

[13] B.H. Song, Y.T. Kim, S.D. Kim, Circulation of solids and gas bypassing in an internally circulating fluidized bed with a draft tube, Chem Eng J, 68 (1997) 115-122.

# 514498 Power Spectral Density Analysis of Pressure Signal in 18 m Circulation Fluidized Bed Riser

**Abstract:**

Studies on the gas-solids flow characteristics using Power Spectral Density based on the pressure data from different axial locations in a riser of diameter 80 mm and height 18 m. Results show that the pressure signal fluctuation is mainly affected by gas-solids interaction, which included gas-solids friction, particle-particle collision and particle-particle aggregation. With the increasing solids circulating rate, frequency of particle interactions including particle-particle collision and particle-particle arrogation become strong and the pressure fluctuation also increased. It can be seen that the pressure signal fluctuation intensity decreased gradually with the increasing axial position of the riser. The low frequency but high power phenomenon nearly disappears in the fully developed zone near the top of the riser. Based on the results, it can be concluded that the gas-solids interaction is stronger at the bottom of riser. On the other hand, the gas-solid interaction is more homogeneous, and the pressure fluctuation is relatively weak in the fully developed zone.

# 514514 Characteristics of Pressure Fluctuations in Particle-Transport Inclined Pipe of a Circulating Fluidized Bed


**Abstract:**

In a large circulating fluidized bed device, the characteristics of pressure fluctuation in a particle-transport inclined pipe under different solid circulating flux were investigated based on measuring the dynamic pressure of gas-solid two-phase flow by using FCC equilibrium catalyst as the particles. The experimental results showed that with increase of particle mass flux, the flow patterns of FCC particles above and below the butterfly valve in the inclined pipe both successively changed, and the fluctuation amplitude of dynamic pressure increased. The standard deviation of the dynamic pressure below the butterfly valve possessed a linear relation with the particle mass flux, which could be used to characterize the real-time dynamic pressure in the inclined pipe. By wavelet analysis, it is indicated that there exists a main frequency caused by inclined pipe flow structures in the pressure fluctuations above and below the butterfly valve, besides, there is a secondary frequency in the pressure fluctuation below the butterfly valve caused by gas turbulence and gas-solid interactions. With the increase of particle mass flux, the wavelet energy ratio of the secondary frequency gradually decreased. The fluctuation pressure caused by discharge in the inclined pipe is a low frequency pressure and the excitation source of inclined pipe vibration.

# 508323 Dynamic Modeling of Attrition and Reactions in Circulating Fluidized Bed Reactors

**Abstract:**

Fluidized bed reactors are widely applied in chemical processes due to the excellent gas-solid contact. A problem that occurs in fluidized beds is the mechanical stress that particles are exposed to during the fluidization, which causes particle attrition. Additionally, fluidized beds are often operated in batch mode, which will cause ever-changing operating conditions with time. This makes fluidized bed modeling especially challenging, because the distributed particle properties, like size, wetness or reaction state, have to be tracked over the whole course of the simulation.

At the Institute of Solids Process Engineering at Hamburg University of Technology, novel process simulation software named DYSSOL is developed, particularly tailored for solids processes. The software is based on the flowsheeting concept. This means that chemical unit operations, like for example chemical reactors or separators, are put into process modules, which are interconnected into a process flowsheet and then mathematically solved. To make dynamic simulations available, new dynamic models had to be proposed for the involved equipment. The models themselves are based on empirical and semi-empirical correlations for hydrodynamics and kinetic particle reaction models are integrated.

The dynamic flowsheeting software was then used to simulate a system of interconnected fluidized bed reactors, which are used for Chemical Looping Combustion of solid fuels. This process is especially interesting, because heterogeneous reactions take place. Further, the process is a reactor-regenerator system causing the reactive solid to be oxidized and reduced and, hence changing its conversion state. Attrition plays a major role for the cost effective operation of the process.

In a small lab scale bubbling bed reactor, the jet and bubbling attrition rate constants for the solid oxygen carrier were experimentally determined under ambient conditions. In a small stand-alone cyclone the cyclone attrition was investigated. Furthermore, the reaction kinetics of the solid material with fuel gases was experimentally determined. All this information was included into the flowsheet simulation of a pilot scale circulating fluidized bed reactor, which is actually operated in our institute to deliver validation data.

The dynamic flowsheet simulation of the pilot plant, which was enhanced with kinetic and attrition information of smaller scale facilities, was conducted over 1000 seconds. A fuel load change was simulated and the chemical conversion was tracked. Over the whole simulation, the attrition and small particles leaving the reactor system was also covered. When comparing the simulation with experiments it could be shown that the gas conversion was sufficiently calculated and the conversion state of the oxygen carrier tracked correctly. The attrition modeling, however, with its parameters measured had to be adapted, so that it fitted the outlet particle flows measured under reaction conditions.

The financial support of DFG (Deutsche Forschungsgemeinschaft) within the priority program SPP 1679: â€œDynamic simulation of interconnected solids processesâ€ is gratefully acknowledged; grant number HA 6935/2-1.

# 514599 Analysis of FCC Cyclone Fault Diagnosis Technology Based on Particles Information

**Abstract:**

Cyclones are key equipment in fluid catalytic cracking unit (FCCU), which are used for separating the catalyst particles from oil gas or flue gas. Hence, it is very necessary to carry out condition monitoring and fault diagnosis for the running cyclone. The properties of particles vary at different positions in the particle circulation loop, such as size distribution, morphology and element composition, etc. In consequence, the particles contain much information about operation. Once the cyclone operates abnormally, the information of run-off catalyst particles appear abnormal too, and there exists a corresponding relationship between them. So, taking the catalyst particles as the carrier of fault information, this paper put forward cyclone fault diagnosis technology by extracting catalyst particle information, such as particle packing density, particle size distribution, particle micrographs etc, to build the relationship model between the cyclone fault and catalyst particle information.

# 514600 The Multi-Hole Throttling Effect and the Erosion Characteristics of the High Pressure Natural Gas

**Abstract:**

Based on Euler-Lagrange multi-phase flow numerical simulation method, the single hole and multi-hole throttling process of the high pressure natural gas was studied by using the RNG *k-Îµ* turbulent model, the Sutherland Viscosity Law compressible fluid viscosity model, and the advanced erosion model numerical simulation. The calculated mass flow variation law in the throttle choke is comparable with the theoretical calculations, which verified the adopted models and calculating methods. And then the throttling effect and the erosion characteristic of the high pressure natural gas was studied to check its specific law with different throttle holes by using the advanced erosion model. The results showed that, high pressure natural gas can reduce pressure and expansion after flowing through the single hole and multi-hole throttle choke, in which lead a high-speed jet region came up near its outlet. But the maximum jet velocity of multi-hole throttle choke was significantly lower than that of the single hole. At the same time the multi-hole uniform jet may reduce the vortex entrainment and the wall erosion damage, in which can be easier to mix the cold core and achieve stable pressure recovery at the role of the multi-jet mixing effect. Therefore, the multi-hole throttling can effectively restrain the erosion damage and the gas hydrate formation probability.

# 514601 Numerical Simulation of Flow Field In a Gas Pipe Distributor of the FCCU Regenerator


**Abstract:**

The flow field in the pipe distributor used in regenerator of FCCU (Fluid Catalytic Cracking Unit), was analyzed based on the numerical simulation. Firstly, the distributing gas performance of the pipe distributor was investigated. The result shows that the pressure increases along the branch pipe of the pipe distributor and the axial flow rate decreases in the branch pipe, which leads to the nozzle flow rate increasing gradually. Furthermore, an obvious bias flow was found in the inlet of branch pipe, which brings some nozzle flow rate less than the designed flow rate and causes an uneven gas distribution in the fluidized bed as well. Meanwhile, the momentum equation with variable mass flow was applied to analyze the flow in the pipe distributor. The flow type in the branch pipe belongs to the "momentum exchange controlling model". The pressure in the branch pipe always increases from the inlet to the end, and the pressure distribution is not consequentially uniform. The uneven gas distribution of the pipe distributor results in fluidized gas non-uniformity in the bed and erosion in the branch pipe and nozzle. Finally, according to the pressure characteristics in fluidized bed system, relationship between the pressure drop of branch pipe and pressure drop of nozzles, as well as of the critical nozzle pressure drop and the designed nozzle pressure drop, were discussed. The gas distributor performance could be optimized by means of the structural improvement of branch pipe.

# 514602 Full-Loop Simulation of Gas-solids Flow in Circulating Fluidized Bed of FCC System

**Abstract:**

Circulating fluidized beds are widely used in petrochemical, coal, metallurgy, and other industrial fields. A typical circulating fluidized bed (CFB) includes the riser, gas-solid separator, downer and solids flow control device. To obtain detailed information and get better understand of gas-solids flow in CFB, computational fluid dynamics (CFD) had been applied extensively in many systems such as circulating fluidized bed combustion (CFBC) systems and fluid catalytic cracking (FCC) systems. In most studies, the riser, the core part in CFB system, was isolated as a single simulation object by applying proper boundary conditions. However, there are some limitations of riser-only simulation, without consideration of dynamics in solids circulation rate and effects of other part of CFB system. Therefore, the full-loop simulation is a more promising way to describe complex gas-solids interaction in CFB by treating the whole CFB system as a simulation object. Nowadays, most of current full-loop simulations were focused on CFBC systems, there was little work about FCC system, with different type of particles and recycle mechanism from CFBC system.

In this work, both full-loop and riser-only simulations were carried out by TFM approach to investigate a pilot-scale gas-solid circulating fluidized bed of FCC system which has been set up in our group. The simulation condition was chosen according to experiments with different superficial gas velocity, valve opening and fluidizing air flow rate. Detailed comparison between full-loop and riser-only numerical simulations has been conducted with respect to the hydrodynamics inside the riser, which showed full-loop simulation had advantages on predicting dynamics of solids circulation rate and the gas-solids flow behavior. Distribution of pressure, solids holdup and velocity in system was investigated which was helpful to understand the hydrodynamics in CFB. The effects of operating condition were then studied, which revealed the system stability mechanism.

# 506758 Immobilization of Sulfur-Oxidizing Bacterium, *Thioalkalivibrio Sp.* D301 on Magnetic Nanaoparticles and Biodesulfurization

**Abstract:**

There is an urgent need to develop desulfurization methods for halo-alkaline wastewater. *Thialkalivibrio sp.* D301, an obligatory chemolithoautotrophic, haloalkaliphilic, sulfur-oxidizing bacterium, was used in the removal of sulfide in wastewater. We developed a process to immobilize D301 cells by using superparamagnetic $Fe_3O_4$ nanoparticles (NPs) under halo-alkaline conditions. $Fe_3O_4$ NPs were synthesized by coprecipitation of ferric and ferrous salts in the presence of $N_2$ gas. TEM image shows the spherical shape of $Fe_3O_4$ NPs with an average diameter of 10 to 15 nm. The particles had superparamagnetic properties. The saturation magnetization (Î´$_s$) of the $Fe_3O_4$ NPs was 73.985 emu g$^{\text{âˆ'1}}$ with zero remanence and coercivity. Analysis of zeta potentials and hydraulic diameters showed that the $Fe_3O_4$ NPs and bacterial cells interact mainly through the nano-size effect. The NPs-coated cells have the same sulfur oxidization activity as that of free cells. Experiment results showed that the immobilized cells still retained >90% of their sulfur oxidization activity after six times of reuse. The cells coated with $Fe_3O_4$ NPs did not interact through electrostatic attraction. The mechanism of NPs adsorption mainly involved the nano-size effect. Magnetic immobilization is a convenient technique for HA-SOB immobilization under halo-alkaline conditions and is a promising technology for large-scale industrial applications.

# 507944 Simulation of Bulk Solids and Granular Systems By Using Combined Discrete Element Models

**Abstract:**

Granular materials can be found in many industrial processes from chemical to energy, agriculture, mining, pharmacy and many other fields. They are almost always opaque, making the acquisition of experimental data below the surface difficult. DEM (Discrete Element Method) simulations are an effective way to study the physics of granular materials because they can simulate the micro-dynamic behavior of granular material on an individual particle scale and the motion of every particle can be traced. Since DEM was first considered, many modified discrete element models have been developed. However, most existing DEM simulation are just for spherical particles or a specific type of non-spherical particles. In this work, a combination of several types of discrete element models including spherical models, super-ellipsoids, multi-sphere models (with no interaction and deformation between sub-spheres), and bonded-spheres models (with interaction and deformation between sub-spheres) were used to simulate the flow of granular materials with many different shapes. Several simulation cases including the flow of particles in tumblers, blenders, and other devices were performed to validate the modeling approaches. The simulation results show that it is practical and efficient to combine different discrete element models together for simulating both spherical and non-spherical particles in bulk solids and granular systems.

# 515703 Material Optimization of Perovskite Films using High Throughput Synthesis and Multi-Dimensional Analysis

**Abstract:**

Over the past years, Perovskites received special attention in the field of renewable energy as they are one of the prime candidates for energy storage and conversion materials. Perovskites act as the active light harvesting layer of a solar cell and provide excellent performance.[1] A common synthesis strategy is through the liquid phase under inert atmosphere.[2,3] However, this comes with the challenge of a huge parameter space that in principle would have to be first synthesized, then characterized and finally optimized based on the therefrom deduced knowledge.

To overcome these limitations, we follow a fully automated approach, which allows automated high throughput (HT) synthesis and HT characterization of perovskites with high accuracy. As an efficient route to explore the parameter space, we use design of experiments (DoE) to find optimized settings for process parameters such as synthesis and annealing time and temperature, but also material aspects like the molar ratio of the reactants.

For all experiments, a Chemspeed© technologies (CST) swing XL platform with a liquid handling tool and a heatable shaking rack was used. The robotic platform is enclosed within an Mbraun© glovebox with inert atmosphere (Nitrogen). After precursor preparation, deposition was realized by a liquid handling tool on glass substrates inserted in a customized aluminum 24 well plate, which is then transferred onto a hot plate for film formation.

In a first step, after annealing at different temperatures, a characterization cascade was developed and applied that contains the subsequent elimination of samples that exhibit undesired properties and side products. Fast absorption analysis, photoluminescence (PL) measurements and finally X-ray diffraction (XRD) were performed to identify optimum chemistry and process parameters. Although this is a very reliable procedure, it is still comparatively demanding in terms of the time effort. Therefore a more efficient, purely optical approach by which the experimental parameter space is only explored in terms of rapid emission analysis was developed in a second step. Noteworthy, both strategies led to the same optimum conditions, however, at significantly reduced time effort.

In conclusion, our work shows how by the combination of automation, DoE and efficient characterization, optimum conditions for new materials like perovskites are rapidly identified. This opens the door towards faster material development which is important whenever new perovskite structures, e.g. lead free materials for industrial applications, are strived for.

### Literature

1. Green, M.A.; Ho-Bailie, A.; Snaith, H.J., "The emergence of perovskite solar cells", Nature Photonics 2014, 8, (7), 506-514.
2. Baikie, T.; Fang, Y.; Kadro, J.M.; Schreyer, M.; Wei, F.; Mhaisalkar, S.G.; Graetzel, M.; White, T.J., "Synthesis and crystal chemistry of the hybrid perovskite (CH 3NH3)PbI3 for solid-state sensitised solar cell applications", J. Mater. Chem. A 2013,1 , 5628-5641.
3. Yan, K.; Long, M.; Zhang, T.; Wei, Z.; Chen, H.; Yang, S.; Xu, J., "Hybrid Halide Perovskite Solar Cell Precursors: Colloidal Chemistry and Coordination Engineering behind Device Processing for High Efficiency", J. Am. Chem. Soc. 2015, 137, (13), 4460–4468.

# 509509 Solid Liquid Separation Via Particle Flow Instability

**Abstract:**

A new separation strategy using novel particle-flow-instability physics is successfully developed for a *difficult-to-separate suspension* in which there is some combination of a small density difference between solid and liquid, high viscosity, and small-sized particles. The method we propose here requires no dilution for medium, high or extremely high viscosity slurries and can produce effective solidâ€"liquid separation with low capital and operational inputs. The results have relevance for many biotechnological and/or chemical processes.

# 509543 Milling and Grindability Assessment of Pharmaceutical Materials

**Abstract:**

Milling is one of the most important unit operations in the pharmaceutical industries. Most of the Active Pharmaceutical Ingredients (APIs) and excipients are produced by crystallisation, and grown to a sufficiently large particle size suitable for isolation and drying. Therefore most of the materials have to be milled to desirable particle sizes. Milling is influenced by a combined effect of milling dynamics (input power, mill geometry, etc) and material properties. Considerable research has been carried out on analysing the particle behaviour in various mills. Nevertheless, the prediction of the milling rate and milled product properties (e.g. product size distribution) is still very challenging.

In this work, the milling behaviour of several pharmaceutical materials in the PicoPlex mill (pin mill unit of the Picoline, Hosokawa Alpine) has been studied. A combination of experiments and Distinct Element Method (DEM) simulations is used to explore the dynamic behaviours of the materials in the mill. The results are then used to validate a novel approach of grindability assessment of pharmaceutical materials using the Scirocco disperser of the Malvern Mastersizer 2000 (Bonakdar et al. 2016). Based on this method the ease of grindability of the materials in the mill can be predicted.

Bonakdar, T., Ali, M., Dogbe, S., Ghadiri, M., and Tinke, A. (2016). A method for grindability testing using the scirocco disperser. *International Journal of Pharmaceutics,* 501: 65-74

# 507857 Bi-Directional Thermal Control of Twin Screw Granulation Process Via a Specialised Annular Heat Pipe

**Abstract:**

The continuous process of twin screw granulation forms a hotbed in pharmaceutical research as it is expected to succeed the current batch production of granules. In essence, the scale up of the material output is a function of production time, instead of equipment size (for batch granulators). This way, a â€œjust-in-time productionâ€ of the desired amount of material can be performed according to specific needs. In this work, a twin screw granulator was utilised in the granulation of both the placebo and active pharmaceutical formulations. Subsequently, the same equipment was used in granulating ceramic raw material powders for tile pressing. During processing, the spatial constraint in the barrel and screw compartments caused friction, resulting in excess heat generation. For the thermally-sensitive pharmaceutical formulations, this must be controlled to avoid denaturation of the granule products. A common solution is by using flowing cool water, but a good control has been proven to be difficult to achieve as different granulation zones generate different heat levels (kneading blocks > conveyer blocks). Heat pipe is a passive and highly effective thermal conductor and stabiliser. A carefully-designed annular heat pipe, filled with a small amount of methanol was successfully employed to stabilise the temperature in the barrel space, cooling the granulating material before it reached its critical temperature. For the ceramic powders, the methanol filling of the heat pipe was replaced by water and it was then heated up to provide an in-situ drying operation as the powders granulating. The in-situ drying in this instance, reduces energy consumption for the downstream drying processing as the product granules contain less moisture to be removed. The annular heat pipe, with its bi-directional operation of heating and cooling is a technically viable temperature controller for twin screw granulators. Compared to the established active cooling methods, the heat pipe provides a passive alternative, with only a discreet amount of fluid required to last its operational lifetime, thereby providing an economic thermal control solution.

# Mixing Grains with Different Elongation in a Rotating Drums

*Claudia Piacenza[1,2], Marco Marconati[1], Colin Hare[1], Andrea C. Santomaso[2], Marco Ramaioli[1,*]*

[1]*Department of Chemical and Process Engineering, Faculty of Engineering and Physical Sciences, University of Surrey, Guildford GU2 7XH, UK*
[2] *APT Lab, Department of Industrial Engineering, University of Padova, Padova, Italy*

[*] *Email: m.ramaioli@surrey.ac.uk*

**Abstract**

Mixing grains with different properties is a remarkably challenging process, relevant to many industrial applications. Rotating drums have been used extensively as model systems to study granular media flow and mixing and segregation. Numerous studies considered the mixing of grains with different sizes, but only few studies considered shape and elongation, which have been already identified as important characteristics affecting packing and inducing segregation. In this contribution, the mixing of binary mixtures of grains having the same volume, but different elongations was studied experimentally. The mixing dynamics of a layered granular medium was characterized in a rotating drum, highlighting the impact of the grain shape, of the drum angular speed and of the drum filling ratio. A mixed or segregated state is reached very rapidly, but experiments are continued to verify that the state reached is a steady state. The experiments demonstrated that below a critical difference of elongation, the grains can be mixed effectively. Conversely, when grain elongations are very different, a central core is formed, rich in the more elongated grains. In this case only limited mixing can be achieved. These results can guide the formulation of mixtures of grains with different elongations.

## 1. Introduction

Mixing is a process operation aimed at obtaining a homogeneous system starting from two or more components. On the contrary, when there is a lack of homogeneity within the mixture segregation occurs. Segregation phenomena represent a relevant problem in the case of mixing solid particles having different physical properties such as particle size, density, shape or surface roughness. Differently from liquids and gases, solid particles are not thermal activated system [1] and so in order to obtain a certain degree of mixing they need to be mechanically activated. Typically three mixing mechanisms are largely recognized:

- dispersion, analogous to diffusion in fluids, it involves single particles that randomly move due to collision between each other;
- convection, it involves a portion of the bulk material, with a given direction and velocity;
- shear mixing: it is peculiar to granular solids and it is a combination of the two previous mechanisms.

Powder mixers are relatively simple machines of low capital cost. They are classified into two types: tumbling mixers and agitated mixers. Tumbling mixers are the simplest. A totally enclosed vessel is rotated about an axis causing the particles within the mixer to tumble over each other on the mixture surface. Tumbling vessels provide low-shear environments. Convective mixers instead are characterized by the action of impellers or paddles to move the powders around and generate a well-mixed product. Convective mixers provide more shear into the mixture and can be used with cohesive materials.

Rotating drums belong to the tumbling mixers. They are extensively used in the chemical and process industries. With respect to mixing processes, rotating drums are often used to study and understand mixing and segregation phenomena; as a result, many experimental and numerical studies were developed on granular behavior in rotating drums [1-3]. Different granular flow regimes can occur in the drum, each one with its specific flow behavior. In particular, different flow regimes can be assessed by using the dimensionless Froude number, $Fr = \omega^2 R/g$ [4], where $\omega$ [$s^{-1}$] is the drum rotation speed, $R$ [m] the drum radius and $g$ [$m/s^2$] the gravitational acceleration. With $10^{-4}<Fr<10^{-2}$ and filling level $\varphi>0.1$ rolling regime takes place. This motion is characterized by a uniform, steady flow of a particle layer on the surface (cascading or active layer) while the larger part of the bed (plug flow region) is moved upwards by solid rotation with the rotational speed of the wall. A uniform and good mixing is achievable with this type of motion, but when particle size distribution is very broad segregation can appear. The rolling bed is preferred, since is the simplest regime providing favorable conditions for heat transfer and high quality of the product even when mass flow rates are large.

When particles to be mixed have different properties such as different size, different density or different shape segregation typically occurs and particles tend to separate. [5, 6]. Difference in size is by far the most important cause of particle segregation. Particle size distribution influences also strongly bulk properties such as packing fraction and bulk modulus [7].

Different size segregation mechanisms can be identified:

- trajectory mechanism: according to the Stokes' law the velocity of fine particles is lower than that of big particles. This can cause segregation when small particles move through air;
- percolation mechanism: if a bulk material is disturbed, a rearrangement in the particle packing occurs and for fine particles it is easier to fall down in the inter-particle voids;
- inertial mechanism: in pouring and filling operations bigger particles tend to flow selectively to the bottom of the pile since they have a higher momentum;
- elutriation mechanism: in presence of air the very fine particles can remain in suspension while larger ones tends to settled. This can occur in charging storage vessels or hoppers.

A homogeneous binary granular mixture, with differences in size and density, in a two-dimensional rotating drum typically gives radial segregation. Considering a rotating drum operating in the rolling regime filled with particles of different sizes, bigger particles mainly tend to the edge of the system, while smaller particles accumulate to the centre of the drum. The phenomenon is fast and the rotational speed of the drum seems not to affect the phenomenon; the size ratio of the particles is, instead, the parameter that mainly affects size segregation phenomena. Most of the experimental investigations about segregation in the rotating drum involved spherical or nearly-spherical particles; the experimental studies of the flow of non-spherical particles are, in fact, rather limited [6,8]. The shape of the particles, however, strongly influences the motion in the system. In particular, the static and dynamic angles of repose of non-spherical particles are found to be very different from the ones measured for spherical particles suggesting that differences in particle mobility can affect the segregation process [9].

## 2. Materials and methods

### 2.1 Particles

Both spherical and non-spherical particles were used for the experiments. Their size (diameter $d$ for spheres, major and minor axis for ellipsoids $a$ and $b$ respectively) , aspect ratio $AR$, projected area $A_p$, volume $V$, and true density $\rho$ are reported in Table 1.

**Table 1.** Physical properties of the particles

| Type | $d$ [mm] | $a$ [mm] | $b$ [mm] | $AR$ [-] | $A_p$ [$mm^2$] | $V$ [$mm^3$] | $\rho$ [$g/cm^3$] |
|---|---|---|---|---|---|---|---|
| Spheres S | 8 | - | - | 1 | 50.3 | 268 | 1.09 |
| Ellipsoids E1 | - | 10.48 | 6.98 | 1.5 | 57.5 | 267 | 0.96 |
| Ellipsoids E2 | - | 18.44 | 5.26 | 3.5 | 76.2 | 267 | 0.98 |

Spherical particles with size 8 mm were used for the experiments. The ellipsoids were 3D printed with *AR* equal to 1.5 and 3.5. They were printed with the Ultimaker 3D printer and made of polylactic acid (PLA). The software used to realize the 3D design of the particles used in this work is FreeCAD, which is an open-source parametric 3D modeller allowing to obtain STL files.

The true density of all the particles used for the experiments was measured by gas pycnometry (Multivolume Pycnometer 1305, Micromeritics). The measured density of the spheres was very close to the typical density of the nylon 66 (1.14 g/cm$^3$). The PLA ellipsoids had a density that was lower than the typical density of the PLA (1.24 g/cm$^3$). This is because the particles were printed with a set infill density of the 20%, so they where partially empty inside.

### 2.2 Tumbling mixer

The quasi-2D drum was made of two transparent polycarbonate disks with a diameter of 30 cm and a thickness of 6 mm; between them a wooden spacer was placed, with the same external diameter, an internal diameter of 25 cm and a thickness of about 9.1 mm. The spacer allowed to create an internal volume of finite thickness separating the two polycarbonate disks.

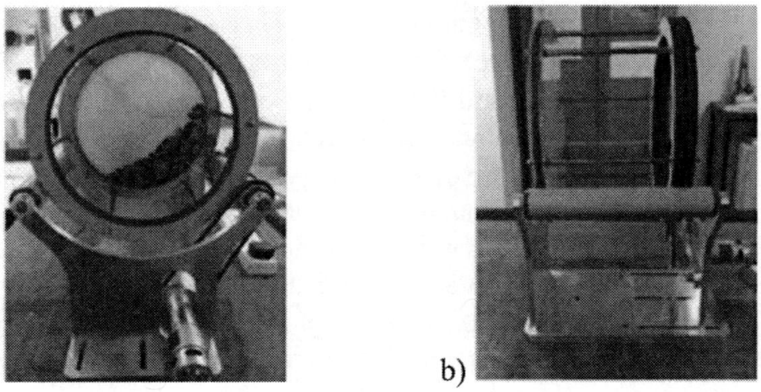

a) b)

**Figure 1.** a) Front view and b) side view of the rotating drum (the drum is on the left side of the rigid structure).

The internal lateral wall of the volume was covered with sandpaper in order to provide friction and avoid undesired slip of the particles. To keep the drum in vertical position and allow its rotation, three additional wooden spacers were used, at a distance of about 16.3 cm. This rigid structure could rotate at variable rotational speed thanks to a motor connected to a voltage supplier. Two rollers mounted on a rigid support made the rotation of the drum possible; a pulley transmitted the motion to the driving roller, while the second roller was passive. The rotational speed could be changed by varying the motor voltage. The structure could be removed from the rollers and by unscrewing the bolts, the drum could be easily opened so that to load/unload the particles.

### 2.3 Image acquisition

Frontal pictures of the drum were taken by a color camera and analyzed by the image processing software ImageJ [10]. To avoid the reflection of the camera and other objects on the front transparent disk, a squared LED panel was used which allowed to have a dominant diffused light illuminating the drum. The LED had a power of 8 W and a color temperature of 4000 K; it was placed behind the drum and opposite to the camera. A circular sheet of paper, which was attached to the drum between it and the LED, diffused the light properly. The camera was a Basler ace camera – model acA2000-165uc, with a maximum resolution of 2040 × 1086 pixels. The focus and the diaphragm of the camera could be adjusted manually.

## 2.4 Experimental plan

The attention was focused on mixtures of particles with different shapes, since segregation phenomena due to size had been widely investigated in the past. Three mixtures were considered, with two possible initial top-bottom segregated configurations. Filling level and rotational speed where changed also as detailed in Table 2.

**Table 2.** Experimental tests performed with differently shaped particles.

| *AR* 1 [-] | *AR* 2 [-] | *ω* [rpm] | *φ* [-] | Init. Config. |
|---|---|---|---|---|
| 1 | 1.5 | 4.22 | 25% | E1/S |
| 1 | 1.5 | 4.22 | 25% | S/E1 |
| 1 | 1.5 | 4.22 | 47% | E1/S |
| 1 | 1.5 | 4.22 | 47% | S/E1 |
| 1 | 1.5 | 11.51 | 25% | E1/S |
| 1 | 1.5 | 11.51 | 47% | E1/S |
| 1 | 3.5 | 4.22 | 25% | E2/S |
| 1 | 3.5 | 4.22 | 25% | S/E2 |
| 1 | 3.5 | 4.22 | 47% | E2/S |
| 1 | 3.5 | 11.51 | 25% | E2/S |
| 1 | 3.5 | 11.51 | 25% | S/E2 |
| 1 | 3.5 | 11.51 | 47% | E2/S |
| 1.5 | 3.5 | 4.22 | 25% | E2/E1 |
| 1.5 | 3.5 | 4.22 | 25% | E1/E2 |
| 1.5 | 3.5 | 4.22 | 47% | E2/E1 |
| 1.5 | 3.5 | 4.22 | 47% | E1/E2 |

*AR*1 and *AR*2 indicate the aspect ratios of the particles in the binary mixture. It is reminded that the spheres (*AR*=1) and the two types of ellipsoids (*AR*=1.5 and *AR*=3.5) had the same volume. The E1/S initial configuration indicates a segregated configuration with short ellipsoids above spheres (this was assumed as basis configuration); the S/E1 configuration is inverted with spheres above short ellipsoids. A similar convention is applied to the experiments considering mixtures of ellipsoids. All the experiments were performed with binary mixtures with a composition of 50% and 50% on a number (and hence also volume) basis. The filling level $\varphi$ of the 25% was obtained with a total number of particles equal to 200; with 320 total particles the filling level of the drum was equal to 47%.

## 2.5 Image analysis procedure

Image analysis techniques were used to identify the particles on the basis of their different shape. The procedure was automated using customized ImageJ macros. Contrast and brightness of the images were first adjusted. Color images were then set to gray scale and segmented by applying a threshold. The operation was carried out twice to separate light and dark particles with different thresholds. Watershed segmentation was then used to separate touching objects. This last operation was more difficult for the particles at the border of the drum; in some case, in fact, the ellipsoids were erroneously identified as spheres. Also, the presence of the border of the drum complicated the watershed operation.

The centre of mass of each particle ($x_{m,i}$ and $y_{m,i}$) was estimated and the radial distance ($d_i$) from the centre of the drum ($x_c$, $y_c$) calculated as:

$$d_i=[(x_c-x_{m,i})^2+(y_c-y_{m,i})^2]^{0.5}$$

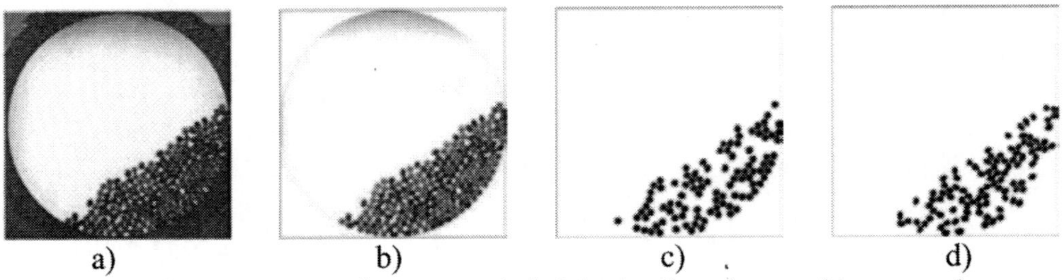


**Figure 2.** Example of image processing steps. a) Original colour image; b) grayscale conversion and circle cut from the original picture; segmentation applied to c) light particles and d) dark particles.

The mean value of distance ($d_m$) for the spheres and for the ellipsoids was then calculated as:

$$d_m = \Sigma\, d_i/N$$

Only for very elongated ellipsoid ($AR$=3.5) the watershed operation failed resulting in over-segmentation errors. Single particles could not be identified unequivocally in this case; only one center of mass for all the elongated ellipsoid was estimated and the average distance ($d_m$) of the particles from the drum center calculated.

## 3. Results and discussion

Figure 3 shows two typical experiments, carried out at the lowest rotational speed and filling ratio, with ellipsoids ($AR$=1.5 at left and $AR$=3.5 at right) on top and spheres on bottom in the initial and final configuration. It is evident even by visual inspection that the final state presented some level of radial segregation and that a different final segregation intensity was reached in the two systems.

**Figure 3.** Drum at a) t=0 and b) t=30 min. On the left the initial configuration is E1/S (using ellipsoids with $AR$=1.5) and on the right the initial configuration is E2/S ($AR$=3.5). In both experiments $\omega$=4.22 rpm and $\varphi$= 25%.

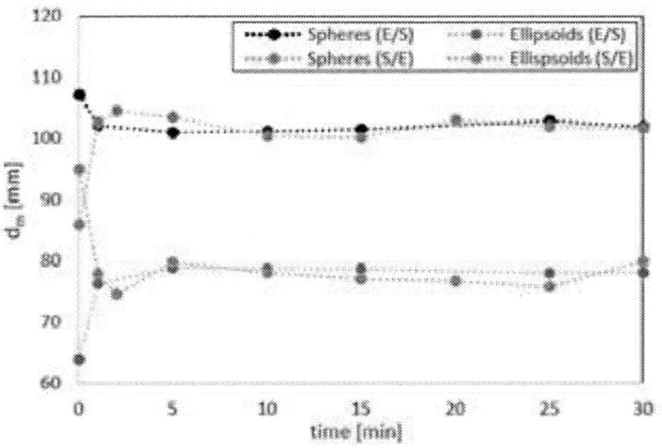


**Figure 4.** Evolution of the mean radial distance vs. time for the mixture of spheres and ellipsoids with $AR$=3.5. The initial configuration E2/S is compared to the initial configuration S/E2.

A more quantitative analysis was obtained by plotting the trend of the mean radial distance ($d_m$) of the two species at different time instants. As a typical example of the system evolution, the case of E2/S

is shown in Figure 4. The $d_m$ reached a steady value after a short time. Dedicated experiments on a shorter time scale (2 minutes, not shown here) revealed that steady state value for $d_m$ was reached after 30s for mixtures with ellipsoids E1 ($AR$=1.5) and after 60s for mixtures with E2 ($AR$=3.5).
At steady state, the radial distance was in both cases smaller for the ellipsoids, meaning that they mainly remained concentrated on top or at the center of the granular bed while the spheres mainly concentrated at the periphery of the drum (close to the drum wall). This tendency was much stronger for ellipsoids E2. So a slower segregation dynamics compensated by a larger segregation intensity was observed with the more elongated particles.
As Figure 4 shows, inverting the initial configuration from E/S to S/E did not affect the dynamics and the final result in all the experiments performed with these particles. Initial configuration resulted to have no effect on the segregation propensity of the mixtures analyzed.

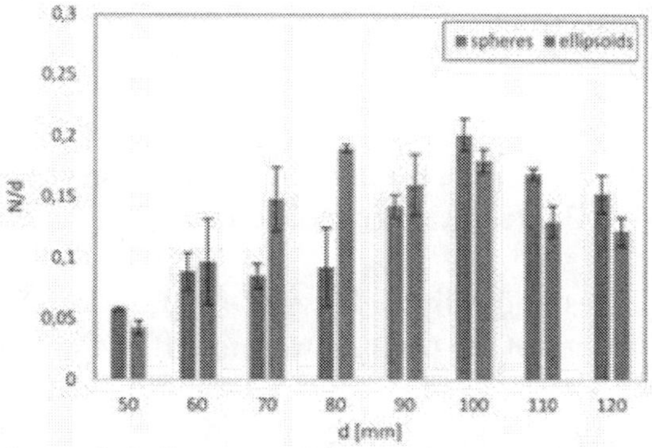


**Figure 5.** Distribution of the radial distance of spheres and ellipsoids with AR=1.5, obtained by averaging the results of three repetitions. Other conditions are t=30 min, $\omega$=4.22 rpm, $\varphi$= 25%.

A more in-depth analysis can be carried out considering the distribution of the radial distance of the spheres and ellipsoids separately. The radial distance from the centre of the drum ($d$) was divided in classes of 10 mm and the number of each type of particles ($N$) in each class was computed. To obtain the distribution, $N$ was normalized on the radial distance as $N/d$ and plotted versus $d$. In the case of completely homogeneous mixtures and assuming that the medium takes the shape of a circular sector, $N/d$ should be constant and equal for the two types of particles in each class. The peak of the distribution of the ellipsoids was found at lower distance than the peak of the distribution of the spheres. Moreover, at lower distance, the number of ellipsoids was higher than the number of spheres, especially for distances between 70 and 90 mm; at higher distance (closer to the drum wall), instead, the number of spheres was higher than the number of ellipsoids, thus confirming the previous analysis based on $d_m$.

Also, binary mixtures made of the two types of ellipsoids E1 and E2 ($AR$=1.5 and $AR$=3.5 respectively) were taken into account at the same filling levels of the previous experimental tests and at the lower rotational speed. Stable segregation was again observed with intermediate results with respect to the two previous cases (S+E1 and S+E2). So as a general comment on these experiments, stable segregation was always observed when mixing particles having same volume but different $AR$. The intensity of segregation increased with increasing the difference in $AR$ between the particles. In particular particles with larger $AR$ tended to accumulate at the center of the granular bed, while particle with lower $AR$ accumulated at the periphery. Segregation occurred in the surface active layer where the more elongated particles, because of their reduced propensity to flow (particle rolling was prevented by their elongated shape), were systematically overtaken by the less elongated (and faster) particles. In this way a spontaneous separation of particles occurred in the active layer, with those having a lower $AR$ arriving first at the drum wall, while those having an higher $AR$ remaining trapped in the core.

### 3.1 Effect of filling level

The comparison of the behavior of the mixtures at the two filling levels can be carried out by normalizing the mean radial distance $d_m$ of the two types of particle on the mean distance of all the particles $d_c$, calculated as $(d_{m,\ spheres}+d_{m,\ ellips.})/2$. The trends of $d_m/d_c$ are shown in Figure 6

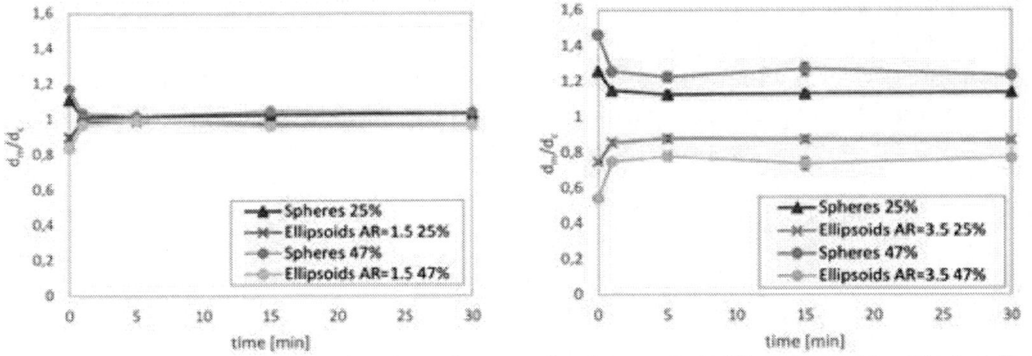


**Figure 6.** Trend of $d_m/d_c$ vs. time for the mixture of spheres and ellipsoids with $AR$=1.5 (left) and AR=3.5 (right) at two different filling levels.

In the case of mixture of spheres and ellipsoids with $AR$=1.5, the curves at different filling levels overlap, showing that the behaviour of the mixture was not affected by the filling level of the drum. In the case of the other two mixtures (S+E2 and E1+E2, not shown), the curves did not overlap , especially for the case S+E2 shown in Figure 6 (right). An explanation can be found in terms of bed geometry. When increasing the filling level (below 50%) the length of the surface active layer increased as well. This gave more chances for particles to separate, in particular when the difference in $AR$ was large enough (S+E2 and E1+E2). However for the system S+E1 where the difference in $AR$ was the smallest, the difference in active layer length in the two cases was not sufficient to create observable differences in the segregation pattern.

### 3.2 Effect of rotational speed

In addition to experiments carried out at 4.2 rpm, tests were repeated at 11.5 rpm with spheres and ellipsoids (25% filling level). Figure 7 shows that rotational speed did not affect the segregation patterns, at least in the range of the analyzed speeds.

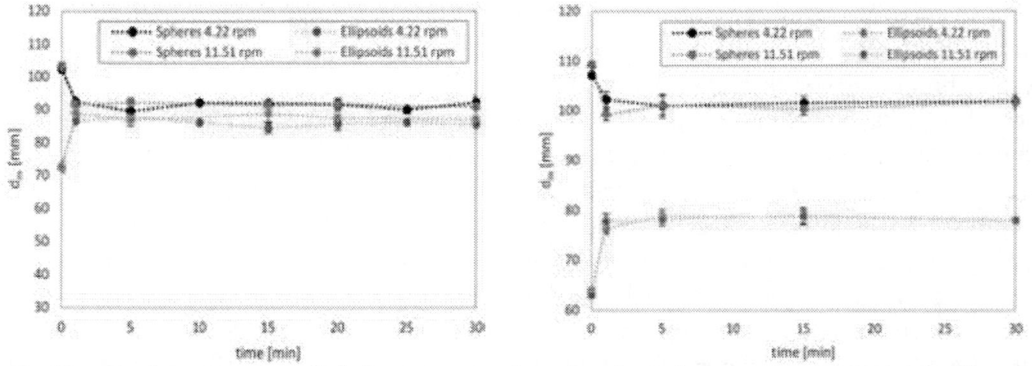


**Figure 7.** Trend of the mean radial distance vs. time for the mixture of spheres and ellipsoids with $AR$=1.5 (left) and $AR$=3.5 (right) at two different rotational speeds.

## Conclusions

From the experimental tests described in the previous paragraphs, it was found that binary mixtures of particles with different shape but with the same volume do not mix uniformly and segregation was always observed. In particular, the particles with lower aspect ratio tended to stay closer to the drum walls, while the more elongated particles were found preferentially at the centre of the granular bed. The most significant case was represented by the mixture of spheres and ellipsoids with $AR$=3.5; for this mixture, in fact, the segregation was more evident and the non-spherical particles clearly accumulated in the centre of the granular bed. The same final state was obtained regardless of whether the longer particles were initially located to the top or the the bottom of the mixture. A proportionality between $AR$ difference and segregation intensity was observed. Some variables such as rotational speed and initial bed configuration were found to not affect the segregation dynamics and intensity. Only filling level (in addition to $AR$) was found to affect segregation intensity. A mechanistic explanation was given in terms of differences in particle flowability occurring in the surface active layer (confirmed also by an analysis of the angles of repose in the drum). In this specific case therefore percolation and inertial effects (typical of size segregation) were not the main driving mechanisms for the observed segregation (the particle indeed had the same volume and similar mass so that sterical and inertial phenomena were minimized). Segregation mainly occurred because of differences in particle velocity in the active layer due to a reduced mobility of particles with larger $AR$.

## References


1. GDR MiDi (2004), "On dense granular flows", The European Physical Journal E, 14, 4, 341–365.
2. Jarray A., Magnanimo V., Ramaioli M., Luding S. (2017) Scaling of wet granular flows in a rotating drum, EPJ Web of Conferences, 140,03078.
3. Volpato S., Canu, P., Santomaso A.C. (2017), "Simulation of free surface granular flows in tumblers", Advanced Powder Technology 28(3), 1028-1037.
4. Mellmann J. (2001), "The transverse motion of solids in rotating cylinders – forms of motion and transition behaviour", Powder Technology, 118, pp 251-270.
5. Windows-Yule C.R.K., Scheper B.J., van der Horn A.J., Hainsworth N., Saunders J., Parker D.J., Thornton A.R. (2016), "Understanding and exploiting competing segregation mechanisms in horizontally rotated granular media", New Journal of Physics, 18, 2, 023013, 2016.
6. Santomaso, A.C., Olivi, M., Canu, P. (2004), "Mechanisms of mixing of granular materials in drum mixers under rolling regime", Chemical Engineering Science, 59 (16), 3269-3280.
7. Kumar N., Magnanimo V., Ramaioli M., Luding S. (2015), "Tuning the bulk properties of bidisperse granular mixtures by small amount of fines", Powder Technology, 293, 94-112.
8. Ramaioli M., Pournin L., Liebling Th. M., (2005), "Numerical and experimental investigation of alignment and segregation of vibrated granular media composed of rods and spheres", Proceedings of the 5th International Conference on Micromechanics of Granular Media, 2, 1359-1363.
9. Santomaso, A.C., Canu, P. (2007), "Single particles properties vs. bulk flowability", Proceedings of 5th International Conference for Conveying and Handling of Particulate Solids.
10. Schneider, C. A., Rasband, W. S., Eliceiri, K. W. (2012), "NIH Image to ImageJ: 25 years of image analysis", Nature methods 9(7): 671-675.

# 507441 Generation of Particles with a Special Morphology By Desublimation of Copper Phthalocyanine

**Abstract:**

The morphology of submicron particles becomes more and more important concerning new functional materials. For example the special electric properties of particles can be used in electronic devices [1]. Moreover the electric, optical or catalytic properties of particles can be optimised by adjusting the morphology of the particles. This may imply a risk for the health but makes it possible to realize specific product characteristics. The morphology of the particles should therefore be considered e.g. for filter tests and for the developing and testing of sensor systems.

To explore the characterization of the particle morphology the DFG-AiF cluster project "multi-parametric characterization of particle-based functional materials using innovative online measurement systems" (MPaC) includes the development of online sensor systems for the size and structure of the particles as well as the development of generation methods for particles with a special morphology. The production of spherical particles can be achieved with different commercial aerosol generators. Several generation techniques for special particles are described in literature [2] [3] [4]. To test the sensor systems with particles of different shapes a new type of aerosol generator was developed. The aim for the new generator is to produce the particles in the gaseous phase and provide the particles directly for sensor or filter testing.

The developed generator uses the desublimation of copper phthalocyanine (organic pigment) to structure the surface of titanium dioxide. The bulk material copper phthalocyanine is dispersed into a nitrogen gas stream and transported to a tube furnace. Here the bulk material sublimates and the gas stream carries the material to a laval nozzle where the adiabatic expansion takes place. This leads to a rapid decrease of the temperature behind the laval nozzle and an oversaturation in the gaseous phase. In addition to the decrease of the temperature core particles of titanium dioxide were initiated behind the laval nozzle. The copper phthalocyanine desublimates on the core particles and causes a structured surface. The morphology of the generated particles is characterized by SEM. The particles show a punctual coating of the titanium dioxide particles with copper phthalocyanine. The size of the titanium dioxide particles is about 500 nm and the punctual coating of copper phthalocyanine measures a few nanometres.

The generation method and the design of the generator will be presented and the morphology of the generated particles will be described.

[1] K. Deppert, M. H. Magnusson, S. B. Carlsson, B. Gustafson, T. Junno, . L. Montelius, B. J. Ohlsson, L. Samuelson und L. E. Wernersson, „Future electronic devices by aerosol techniques," *Journal of Aerosol Science,* Nr. 31, 2000.

[2] J. Raula, A. Kuivanen, A. Lähde, H. Jiang, M. Antopolsky, J. Kansikas und E. . I. Kauppinen, „Synthesis of image-leucine nanoparticles via physical vapor deposition at varying saturation conditions," Journal of Aerosol Science, Nr. 38, 2007.

[3] M. Heurlin, M. H. Martin, D. Lindgren, M. Ek, R. L. Wallenberg, K. Deppert und L. Samuelson, „Continuous gas-phase synthesis of nanowires with tunable properties," Nature, Nr. 492, 2012.

[4] K. Nielsch, K. Deppert, M. H. Magnusson, G. Dumpich, E. Kruis und H. Fissan, „Formation of ultrafine particles from powders of compounds with incongruent sublimation behavior," Journal of Aerosol Science, Nr. 28, 1997.

# 516184 Rubbery Milling of Seed Endosperms for Improved Sustainability by Natural Functionality Preservation

**Abstract:**

Increasing demand for functionalised goods addressing individual needs requires a profound understanding of process-induced alterations. Development of models and process-structure-property relationships are essential to control the resulting quality and properties of processed food. A wide range of food systems contain solid or semi-solid particulates as major structure building blocks. Generation and tailoring of such particulates strongly depends on their size distribution, shape, morphology, phase behavior (amorphous/crystalline), interfacial properties and related inter-particle force interactions.

A novel processing technology was developed widening the frame of possible alterations, aiming towards tailoring such particle properties generating novel food structures or preserving techno-functional characteristics of the raw material. The technology employs targeted adjustments of the initial moisture content in combination with milling temperature. The resulting transition from the glassy to the rubbery state generates different material properties and processing behaviour. The technology is highly versatile, owing to the universality of moisture induced state transitions.

Seeds from locus bean gum or guar gum were wetted with excess water in order to adjust a rubbery state of the seed endosperm material. In this state milling within a novel type of impact mill with superimposed drying and sifting steps causes more gentle material disintegration compared to conventional dry milling. This preserves the high molecular weight fraction of e.g. galactomannans (micro-scale structure) from degrada-tion and allows at the same time the adjustment of a micro-porous morphology (macro-scale structure) improving wetting, water uptake and dispersing characteristics of the resulting thickener powders. Such structure preservation /modification increased the thickening characteristics extraordinarily as will be demonstrated.

From this prominent rubbery milling example we have entered in related systematic work addressing a wider range of seed endosperms with various â€œnature-synthesizedâ€ func-tionnalities to be further addressed.

/1/ Pollard, M., Eder, B., Fischer, P., and Windhab, E. J. (2010) "Characterization of galacto-mannans isolated from legume endosperms of Caesalpinioideae and Faboideae subfamilies by multidetection aqueous SEC." *Carbohydrate Polymers*, 79(1), 70-84.

/2/ Illmann, S., Stoob, M., Sieber, C., Konz, A.K. and Windhab, E.J. (2012) "New Thickening Quality of Galactomannan Polysaccharides by Tailored Impact Milling" *XVIth Int. Congress on Rheology*, Lisbon, Portugal, August 5-10

# 503422 Particle Technology Education at Purdue University

**Abstract:**

Degree programs and coursework at Purdue University focused on particle technology will be presented. Additional information may be found at: https://engineering.purdue.edu/CP3

# 506453 Teaching Particle Technology in Portugal â€“ University of Coimbra


**Abstract:**

Education in Particle Technology started in Portugal in the 80â€™s (last years of the decade). The Chemical Engineering degree of the University of Coimbra was the first engineering degree, in the country, to include this topic in the curriculum. Particle Technology was recognized by us, at that time, essential for the education of chemical engineers, in parallel to education in the traditional areas of Chemical Engineering which were still focused on monophasic systems.

Since then, Particle Technology has always been present in the Chemical Engineering curriculum of the University of Coimbra, in different configurations. During the first years particle characterization was combined with design of operations involving particulate systems (grinding and pneumatic and hydraulic transport). From 2005 to 2010, with the curriculum reformulation in relation with the Bologna Process, the two topics started to be addressed in different courses: one course on Materials Characterization and the process topic in a course entitled Structuring and Stabilization Processes, dealing with colloids and emulsions, grinding, agglomeration and spray drying and, for the first time, an integrated modelling strategy for particulate processes based on the population balance formulation was introduced in this course. The two courses were compulsory, at this time.

From the school year 2011/2012 till present, the configuration of this area in the Chemical Engineering degree of the University of Coimbra changed again, even if not deeply (see https://apps.uc.pt/courses/EN/course/361). The course on materials characterization, which includes general notions on particle characterization is kept as an autonomous course included in the 1$^{st}$ cycle of the Chemical Engineering degree of the University of Coimbra. A second course on design of processes involving particulate systems now exists in the 4$^{th}$ year of the MSc degree (1$^{st}$ year of the 2$^{nd}$ cycle), entitled Disperse Systems Technologies, which is an optional course, and which deals with all types of disperse systems, i.e gas, liquid or solids as disperse phase. This course starts with a review of the mechanisms/interactions occurring at the interface with the disperse phase, goes on, again, with a chapter on population balances modelling, and proceeds with strategies for disperse systems stabilization, linking this subject with formulation concepts, ending with the design of some processes involving particulate systems, such as colloids and emulsions, crystallization, grinding, agglomeration and spray drying. This course is also offered to other engineering degrees, such as Materials Engineering and Biomedical Engineering.

Also at the PhD level a course on Disperse Systems Technologies can also be offered during the formal lectures period of the PhD program on Chemical Engineering (or others), depending on the students background and interests.

Since the introduction of the Particle Technology topic in the curriculum of the Chemical Engineering degree of the University of Coimbra, the feedback from employers has been very positive, and the competences in this field acquired by our students have been quite valued by the job market. Till today, the Chemical Engineering degree of the University of Coimbra is still the one, in Portugal, with a stronger emphasis on this topic in the curriculum. In our curriculum, the topic of Particle Technology interacts directly with another course on Product Design, also in the 4$^{th}$ year of the MSc degree and, of course, also with the Project Design course (5$^{th}$ year of the MSc degree), besides other more traditional courses in a Chemical Engineering degree such as in Reaction Engineering and with courses related to environmental technologies.

At present, we have started developing multimedia tools for the teaching/learning of this topic, especially in the field of particle characterization, to be made available on-line, as is the case of the interactive video on nanoparticles characterization (http://labvirtual.eq.uc.pt/siteJoomla/index.php?option=com_content&task=view&id=277&Itemid=470, in Portuguese). Also, the European Federation of Chemical Engineering (EFCE) is considering investing in this type of tools for the field of particle technology, as the result of a collaboration between the Working Groups of Characterization of Particulate Systems and Education.



**Acknowledgement**

The authors would like to acknowledge the financial support of Pest/C/EQB/UI0102/2013 from FCT/MCTES (PIDDAC), co-financed by the European Regional Development Fund (ERDF) through the program COMPETE (POFC).

# 506629 Highlights of Particle Technology Teaching in Singapore

**Abstract:**

Our approach to education in particle technology is to incorporate elements of our research in this field, relevant industrial case studies and apply a Technology Enhanced Learning (TEL) approach for implementing pedagogical interventions to enhance the learning experiences of students.

For supercritical $CO_2$ processing, case studies in extraction and drying are used to teach different chapters in the curriculum such as:

- Particle characterization
    - The bulk density, size, shape of real industry samples are given to students for analysis using software such as ImageJ
- Flow through packed bed
    - The pressure drop and scale-up of packed bed design for supercritical $CO_2$ flow through packed bed are used as a worked example and assignment problem to help students appreciate the role of density and viscosity in the pressure drop in a packed bed
- Drying
    - Supercritical $CO_2$ as an alternative drying media as compared to hot air or superheated steam. Case studies and examples provided

For gasification and chemical looping combustion applications, case studies in the treatment of wastes and carbon capture are similarly used as part of the syllabus:

- Particle characterization
    - The effect of polydispersity on various fluidization characteristics (entrainment, transport dis-engaging height, clusters, segregation, etc)
    - The effect of agglomeration
- Reliability of correlations and simulations
    - Correlations in literature give predictions orders-of-magnitude apart
- Scale-up
    - Rule-of-thumbs in scaling up the various regimes

A design software has been developed within the paradigm of Technology Enhanced Learning (TEL) to facilitate learning via a repeated practice approach. The software generates detailed solution steps to typical engineering design problems encountered within this module. Students can compare the complete solutions generated with their own hand calculations, and thereby apply a repeated practice approach towards their learning of engineering design calculations. The software has been successful in enhancing the performance of students in the final examination.

# 507116 Particle Technology Course at the University of Salerno

**Abstract:**

The Course of Particle technology at the University of Salerno is a 6 EC, that is a 60 hours course including lectures, as well as classroom and laboratory practical. The course dedicated to Master student in Chemical Engineering. It a compulsory course within the â€œinnovative process and nanotechnologies curriculumâ€ can be optioned by any mechanical or chemical engineering student at the master level.

The course includes basic particle size and shape characterization concepts, fundamentals on interparticle forces, fundamentals of powder mechanics and silo discharge, fundamentals on-fluid particle interactions and fluidization.

The main learning outcome is the ability of the student to understand the phenomena occurring in processes involving particulate solids and to carry out basic design operation on storage and fluidized bed units.

# 507382 Professional Master of Engineering Degree in Particle Technology

**Abstract:**

Particles are ubiquitous in the process industries yet nearly absent in the undergraduate education of chemical engineers. This poster descibes a professional Master of Engineering degree program at University of Delaware (the MEPT) that closes that gap by exclusively focusing on particulate matter, its processing, and its incorporation into products. The program can be completed in a single year although we welcome already-employed students who wish to take it part time. Now in our fourth year, we will describe our program and our experiences including guiding distance learners successfully through the program.

In its 1-year realization, a core curriculum of 18 credits is augmented by 6 credits of electives which the student may use to tailor the program to their specific interests. The program is capped by a 10-week, 6-credit summer internship. The 12 core credits taken in the fall teach the fundamentals of particle transport, particle properties & characterization, particle formation and transformation kinetics and the mathematics of particulate systems. The 6 core credits taken in the spring teach the student to apply the fundamentals to both (a) equipment design (in a unit operations course) and (b) product and process design (in a product engineering course).

The presenters of this poster are the program directors, both with long industrial careers preceding this venture into education. In addition, we have developed a strong network of industrial partners who share a wealth of additional experience and expertise with our students both as adjunct professors and as internship mentors. We have successfully placed international students in internships at US companies via the curriculum practical training program (CPT) who have gone on to practice in industry after graduation under the optional practical training program (OPT).

# 507780 Recent Developments in Particle Technology at the Universidad Tecnica Federico Santa Maria

**Abstract:**

The Department of Mechanical Engineering of the Universidad Técnica Federico Santa María formed in 2003 a Research Center for Transporting Granular Materials (CITRAM) to support educational activities and experimental research, as an extension for those ME students taking the regular course on Bulk Solids Handling and those developing a thesis in this area.

Over the years, many ME students have studied, investigated and modeled different topics including: pneumatic conveying of solid particles of different nature, size and density; identification of flow patterns developed in the pipeline, minimum conveying velocity and pressure drop; abrasive wear using a novel rubber-wheel tester; segregation of ores in stockpiles and silos; flow of granular materials in hoppers and chutes; heat transfer of granular materials; among others. The aim of this poster is to present and share the results of this work.

# 508520 Using Perusall to Enhance Student Learning of Particle Technology at Graz University of Technology

**Abstract:**

Perusall is a digital platform that allows students to collaboratively annotate textbooks, papers, and other similar classroom reading assignments. In order to enhance student learning, Perusall was introduced to help teach Particle Technology at Graz University of Technology (TU Graz, Austria). At Graz University of Technology, two Particle Technology courses exist in sequence; the first course is focused on fundamentals topics, whereas the second course is focused more on applications and separations topics. Approximately 60-75 students are registered for each course. For both courses, Perusall was used to assign reading materials, which were primarily journal articles that complemented the course topics. Although these assignments were non-compulsory due to national requirements, students were asked to use Perusall to read and annotate the course readings. Student participation and input regarding this digital approach to learning was recorded. This poster will highlight how Perusall was implemented in both Particle Technology courses as a tool to enhance student learning.

# 508524 A Graduate Course in Fluidization and Gas-Solid Flow Systems

**Abstract:**

The objective of this course is: a) to provide students with a fundamental understanding and design of Fluidization and Fluid- Particle flow systems, b) to provide students with an understanding of Computational Fluid Dynamics (CFD) and the kinetic theory approach to fluid-particle flows c) to provide students with the knowledge required to solve flow, heat and mass transfer problems in processes which involve particles and d) to provide students with sufficient knowledge to read and understand technical articles in areas related to fluid-particle processes. The course is designed for Master and PhD students in Chemical, Mechanical and aerospace engineering. The course includes: Fluidization concepts and regimes including bubbling, turbulent and circulating beds; Particle characterization, diagnostics and size reduction and agglomeration/granulation; Concept and design of Standpipes, non- mechanical valves, gas distributors, cyclones and hoppers; CFD approach to fluid-particle flow systems and fluidization including kinetic theory and dispersed phase modeling; Computational transport phenomena of fluid-particle systems including method of moments; and examples from energy, chemical and pharmaceutical industries. In addition the students will visit PSRI to see large-scale fluidized beds in operation.

The students are required to finish a team project. The teams consist of 3-4 students working on a project which will count as their final exam. The projects will primarily involve background research (about 80%) but some of the investigation (about 20%) should involve taking an additional step from what is currently known. This additional step can take the form of a research proposal, new theory, new insight based on current information, new calculations, new simulation and/or new data. The results of the projects will be communicated to the class and to the Course Instructors via a written report of 10-20 pages, and an oral presentation involving the professors and all of the group members.

# 508742 Education of Fluid-Solid Multiphase Flow at Department of Mechanical Engineering, Osaka University

**Abstract:**

In mechanical engineering field, education curriculum in undergraduate course concentrates on the basic subjects of mechanical engineering such as mathematics, mechanics, material mechanics, fluid mechanics, thermodynamics, control engineering, design, etc. Concerning the course relating to powder technology, we have a class of "Fluid-Solid Multiphase Flow" in the curriculum of graduate school of engineering.

The contents of "fluid-solid two-phase flow" is as follows;

1 Introduction: examples of multiphase flow

2 Definitions and fundamental concepts

3 Particle response time and Stokes number

4 Fluid force acting on particle

5 Fluid force acting on particle in dense system

6 Particle-wall and particle-particle interaction

7 Heat and mass transfer

8 Continuous phase flow equations

9 Modeling of fluid-solid multiphase flow

10 Discrete particle modeling and simulation

11 Numerical simulation method for continuous phase flow

This course includes basic concepts of powder technology, dynamics of particles in fluid flow, derivation of basic equations for locally-volume-averaged fluid flow, discrete particle modeling including DEM (Discrete Element Method), and basics of numerical analysis and modeling.

# 509420 Bulk Solids Handling Education at the KSU Bulk Solids Innovation Center

**Abstract:**

Kansas State University Bulk Solids Innovation Center (BSIC) was started in 2015 with university and industry partnership and is located in Salina, Kansas. This center is dedicated to bulk solids handling research, education and outreach and equipped with world class pneumatic conveying system (both dense and dilute) full scale test lab and other material testing equipment. The vision of this center is to become valued resources for companies who produce or handle bulk solids during their operations. Currently, this center offers short courses in various subject lines related to bulk solids handling such as storage and flow, pneumatic conveying systems, discrete element method, dust explosion and control and mixing and segregation. Duration for these short courses varies from two to three and half days and offered as live course with class room teaching and practical training. The fundamental course on bulk solids handling is offered as distance education online format. This center is also in the process of starting two year online (distance education) master degree program named professional master of technology specialized in bulk solids handling and one year graduate certificate program (online) in bulk solids handling.

# 508861 Dynamic Simulation of Interconnected Solids Processes


**Abstract:**

1. Introduction

A novel flowsheet simulation system, aimed at the dynamic modelling of complex production processes mainly in the area of solids processing, is being developed within the Priority Program SPP 1679 “Dynamic simulation of interconnected solids processes (DYNSIM-FP)” of the German Research Foundation (DFG) since July 2013 [1]. The Priority Program consists of 29 separate projects from different German universities and is thematically divided into four main groups:

- A: Development and implementation of new dynamic models of individual apparatuses and process substeps;

- B: Calculation of material parameters and development of models to describe properties of solids materials;

- C: Development of advanced algorithms and methods for dynamic simulation of complex process structures;

- Z: Development of a novel framework for the flowsheet simulation of dynamic processes.

This flowsheet simulation framework, named Dyssol (acronym for the DYnamic Simulation of SOLids processes), is presented in this contribution. The main objective of the system is to combine models, methods and algorithms, developed within the whole SPP 1679 program, into a single modelling framework that allows performing dynamic modelling of complex process structures involving solid, liquid and gaseous phases as well as their mixtures.

2. Calculation algorithm and applied methods

The Dyssol simulation system implements a sequential-modular approach [2] (SMA), where each unit is relatively independent from others and can be computed separately by its own numerical method or calculation algorithm using an individual simulation time step. Obtained results after computation of each unit are transferred between units according to the flowsheet structure. This approach leads to a relative independence between models and provides high flexibility of the simulation system itself.

The SMA implies sequential computation of models in a certain order, so it is necessary to know all inlets of a model to start calculation. Flowsheets, which consist solely of serially connected modules, satisfy this condition by default. On the other hand, applying of SMA is complicated if there are recycle streams in a process structure. In this case a flowsheet must be additionally converted into sequential form by tearing recycle streams [3]. Tearing here means search for recycle streams and setting some initial values into them.

For the dynamic simulation a modified waveform relaxation method (WRM) [4] is used. The whole simulation time is divided into smaller non-constant time intervals, on which convergence can be reached faster. Units are solved separately on each time window using some initial guess for the solution. Calculations are repeated on the current time interval until the convergence is reached. Then, the system proceeds to the calculations on the next time window. To initialize each next time window extrapolated values from the previous time intervals are used. The number of WRM iterations, needed to reach convergence, depends on the difference between the extrapolated and actual results. A set of extrapolation algorithms has been implemented in Dyssol to initialize the first iteration on each time interval.

## 3. Parallelization methods

Dynamic flowsheet simulation is a computationally demanding process. Sequential-modular algorithm of the Dyssol makes it relatively easy to implement different parallelization mechanisms [4] that can significantly reduce calculation time. Several strategies are implemented in the simulation system Dyssol:

- Parallelization of basic functions of the simulation system, such as matrix transformation and mixing of material streams.

- Parallel calculation of the model itself. Feasibility of this approach depends on the particular model, and the specific implementation is the task of a model's developer.

- Simultaneous calculation of individual time points for steady-state models. If the outlet parameters of model are independent of its internal state, and are determined only by the input parameters, model state at different time points can be computed in parallel.

- Parallel calculation of models, which are located on independent branches of a flowsheet and thereby are not directly dependent from each other.

## 4. Handling of multidimensional distributed parameters

One of the main challenges during the simulation of solids processes is related to the dispersity of granular materials: the solid phase can be distributed along several interdependent properties, such as size, shape, moisture content, density, etc. forming a multidimensional set of distributed parameters. On the other hand, each operation on the material flow, such as separation, mixing, agglomeration, etc. may require a change in one or more distributed parameters. In this case it may be necessary to calculate all dimensions from a multidimensional set even if the value of a single parameter had been changed. Therefore, it should be considered that during the simulation interdependency between all parameters should not be lost. This means that applying the same methods for calculation of solids as for liquid-gaseous systems is impossible or can lead to their incorrect handling.

All parameters of the granular materials are represented in the simulation system Dyssol in the discretized form. The entire interval, on which the parameter is determined, is represented as a set of shorter intervals. To describe the distribution each such interval is associated with the number or mass of particles whose parameters fall within this interval. The discretization scheme must be fine enough that the numerical errors were small compared to the accuracy of the used models.

To represent interdependency between several distributed parameters multidimensional matrixes [4, 5] are used, where the number of dimensions is equal to the number of parameters. Each cell of such matrix contains the mass fraction of the solid material with certain combination of parameters, whereas the total sum of all entities of the matrix is equal to one.

For storing of such discretized multidimensional distributed parameters a sparse data format is used. The multidimensional matrix is represented by a flat tree data structure, where each level of this tree is one of the dimensions. If there are some classes, which are empty (contain zeroes), there is no need to store data on lower levels. This approach provides an opportunity to reduce memory consumption and calculation time. For more efficient usage of this structure dimensions are arranged in such a way, that the more zeroes a dimension contains, the higher it is placed in the structure.

For correct handling of solids an approach with transformation matrices is used [4, 5]. Each cell of a transformation matrix describes the mass fraction of material that passes from one class with a particular combination of parameters to some other class with another combination of parameters. This concept allows preserving information about all parameters, even those which are not considered or not changed in the current model. Thus, instead of an explicit calculation of the model's parameters transformation matrix can be generated and applied to obtain output stream in each unit.

When using this approach, the model of each particular unit is slightly complicated because not the final distribution of parameters is being calculated, but the laws of their transformation. However, this concept ensures that all information about the distributed parameters of granular materials will be retained during the computation and properly considered.

Acknowledgement

We gratefully acknowledge the financial support of the German Research Foundation (DFG) within the priority program SPP 1679 "Dynamic simulation of interconnected solids processes DYNSIM-FP".

References

[1] www.dynsim-fp.de (2017).

[2] Hillestad, M., Hertzberg, T.: Convergence and stability of the sequential modular approach to dynamic process simulation. Comput. chem. Engng. 12, 1988, 407-414.

[3] Towler, G., Sinnott, R.: Chemical engineering design: principles, practice, and economics of plant and process design. Second edition. Butterworth-Heinemann Press, 2013.

[4] Dosta, M.: Hamburg-Harburg, Techn. Univ., Diss., 2012.

[5] Pogodda, M.: Hamburg-Harburg, Techn. Univ., Diss., 2007.

# 509589 Simulation of Powder Segregation Phenomena during Filling and Emptying of an Industrial Storage Hopper

**Abstract:**

This presentation describes a numerical simulation of segregation of a powder mixture when a storage bin is used in a powder process. A continuum framework tracks composition changes inside and exiting the bin during its dynamic operation, including filling and emptying. The simulation treats particle segregation on moving free surfaces and includes velocity profiles within the powder during discharge. The simulation combines historical treatments of powder flow (mass and funnel flow) from the 1960â€™s with the recent developments in powder segregation modeling from Northwestern University. This simulation was programmed using gPROMS and therefore it can be connected to other unit operations within the gSOLIDS library for an overall dynamic powder process simulation. We will show how this modeling approach is useful to industry in mitigating segregation risk with examples.

# 509662 Experimental Investigation of Single Droplet Drying Above Boiling Point

**Abstract:**

An experimental method has been developed to enable dried particle structure, properties and drying rate to be predicted based on droplet drying history, particularly focusing on effects driven by boiling including bubble nucleation, growth and collapse. The method involves a single droplet drying rig which will be used to investigate drying kinetics above boiling point and across a broad range of drying conditions, drop size and composition space. The drying rig was improved as compared to previous apparatuses reported in literature, in that it is designed to measure droplet drying rates and morphological changes at high temperatures and under conditions that more closely resemble those found in industrial applications. The aim is to conduct drying experiment for droplets as small as 300μm diameter, in an air stream which temperature may be as high as 350°C.

Drying kinetics and morphological evolution will be determined, particularly focusing on the second drying stage as well as inflation/deflation periods. Initially, the constructed drying rig is checked and evaluated by conducting experiments on droplets of pure water. Afterwards, experiments will be conducted on solutions of different materials such as Hydroxypropyl Methylcellulose (HPMC), sodium sulphate and sucrose. An assessment of morphologies, transformations and drying behaviour under isothermal drying would be completed at a wide space of process condition, initial water content, initial drop size and composition space. The investigation will look to link material properties, particle size, initial moisture content and drying history to morphology. Results will be compared with single droplet drying models reported in literature.

An initial investigation has focused on mapping the morphological drying behaviour of HPMC, sodium sulphate and sucrose. It has shown that the three materials show different behaviour; sucrose showed deflation and inflation cycles, sodium sulphate droplets supersaturated and crystallized while HPMC showed inflation and rupture. Differences in drying behaviour is due to difference in material and transport properties. Therefore, the next stage of this work is focused on conducting some physical properties measurement at different water concentrations and temperature. In addition, different materials will be dried on the rig and the drying behaviour will be determined. The experimental data from this work will be used to develop, a universal morphology regime map which will link material properties, particle size, initial moisture content and drying history to morphology.

# 508450 On the Development of an Innovative and Sustainable Spray Drying Technology

**Abstract:**

Spray drying is the main process used for producing powders from a liquid feed in food, pharmaceutical and chemical industries. The spray drying process has unique advantages, such as the ability to produce powders with a specific particle size and moisture content, handling heat sensitive products, and possibility of high volume application. In recent years, major focus has been given on understanding and modeling the spray drying process in order to facilitate the choice of operating conditions, which are product dependent, and to optimize energy usage. However, spray drying still remains a complex energy intensive process.

Spray drying is a convective drying method, which has low thermal efficiency. A liquid feed is atomized at the top of a drying chamber into a flow of hot air. For several products, this leads to a rapid reduction of surface moisture, which causes hindrance to moisture content transfer from the inner part of the particles. Consequently, a long drying time is required, which explain the generous height of industrial spray dryer chambers. Large volume of the chamber and long drying time make it difficult to control the morphology of the dry powder, and hence, reduce the quality of final product. Furthermore, a large volume of drying chamber leads to longer warm-up period, and makes the process complex, as a bigger hot air flow needs to be ensured.

Unlike the convective heating, electromagnetic drying is rapid and energy efficient, it promotes expulsion of inner water to the surface, and it allows selective heating of the materials depending on the local water content preventing damage to the surface and to the already dry area.

The goal of this project is to improve thermal efficiency of the spray drying process by combining electromagnetic and convective heating. Such an improvement might aid development of a more flexible and sustainable process for the production of functionalized powders starting from a liquid feed.

In this presentation, a new single-particle electromagnetically assisted drying-kinetics device is shown. It allows acoustic levitation of a single-particle and evaluation of the drying kinetics during combined convective drying and electromagnetic energy. In order to present the potential of this novel methodology, drying-kinetics and dynamic morphological changes of trial food particles in case of pure convective and electromagnetically assisted drying are compared.

Combination of electromagnetic energy and spray drying is a novel technology and it is expected to improve drying efficiency in food, pharmaceutical and chemical industries, and it could contribute to the development of innovative products in the powder form.

# 509611 Energy Efficient Spray Drying By the Use of Superheated Steam with a Focus on Process Evaluation and Powder Properties of Food Products

**Abstract:**

The major part of energy for spray drying is used to evaporate water, without the possibility for further use of the applied energy. A coupling of processes acting as energy source and energy sink is a way to overcome this situation. Prerequisite is, that the right amount of energy is available with the required temperature at a certain time. With conventional air spray drying it is not possible to reuse this energy as the dew point is around 40-45°C. Most of the related processes require energy input at higher temperatures. As an alternative, it is possible to use superheated steam instead of hot air for drying. In this case, the excess steam can be used for other processes at temperatures of about 100°C. Thus, the coupling with a pre-concentration step e.g. a multi stage thin film evaporator can reduce the energy requirement significantly.

Heat recovery of the heat of evaporation allows also the efficient drying of low concentrated feeds, which would be very energy intense in conventional air spray drying. This applies especially to products that are difficult to atomize due to high viscosity, like xanthan gum. These materials can be produced more cost efficient by superheated steam drying, provided that the condensed excess steam can be used elsewhere.

Particle size distributions and powder structures resulting from different atomizing techniques will be shown in comparison to air drying results. Bulk density i.e. particle porosity, powder flow and surface wetting of the produced powders will be discussed in detail. Key learnings of the study are related to process design: Due to the high product temperatures, it is essential to remove the dry product from the steam atmosphere as fast as possible. A special rotary valve solution was designed and characterized. It is equally important to keep the temperature impact on the feed stream at an acceptable level. This was achieved by defined cooling of the nozzle system. In a final step, the process was characterized by determining the residence time distribution of the powder in dependence of the particle size.

Pilot plant trials were done with dairy powders and a couple of other food powders and compared to air spray drying. The dairy applications were chosen because of the high production volume worldwide with the greatest potential for a reduction in energy consumption. Other products are interesting as well, because the process itself has an influence on particle properties. This results in an improved solubility of the powder in water based applications. The high temperature of the steam atmosphere has a sterilizing effect which gives the product a benefit In terms of product safety.

Advantages regarding the product quality are related to the absence of oxygen and a short residence time, which is caused by the smaller possible specific volume of the drier due to higher heat capacity of steam. Furthermore, the inert atmosphere is beneficial in terms of process safety, because dust combustion in such systems is not possible. Disadvantages are a more complex instrumentation and process control and a longer start up and shut down process. A big challenge in food production is the high product temperature. This is not the case for available applications in chemical and detergent industry as those products can withstand temperatures of 120 to 140°C. Therefore, in food applications, residence time for sensitive products needs to be as short as possible.

# 508583 Model Enhanced Prediction of Self-Heating in Detergent Spray Dryer Wall Build-up

**Abstract:**

In the spray drying of detergent powders, it is common for layers of powder to accumulate on the inner walls of the spray drying tower. At the high temperatures at which these towers operate these powder layers have the propensity to self-heat, whereby exothermic reactions occurring within the build-up cause an increase in temperature at the core of these layers. Significant self-heating in these systems can cause the powder to char, leaving burnt particles which compromise the quality of the finished product, while in extreme cases thermal runaway of these layers can occur. Limiting this threat requires a greater understanding of these systems, methods to characteristics these powder, and a means of modelling this behaviour. This work explores methods of characterising the self-heating reaction kinetics and thermal properties of these detergent powders, including a novel parameter estimation approach, and develops numerical models to predict the behaviour of self-heating detergent powder systems, such that a workflow can be developed to limit these problems for both current and new detergent powder formulations.

In characterising these detergent powders, the self-heating reaction kinetics were measured using basket heating methods, namely the steady state approach and cross-point temperature method. In these methods the temperature profiles, as measured by a number of thermocouples embedded in baskets of detergent powder, are used to determine zero-order self-heating reaction kinetics. These methods are well established and have been previously used to measure self-heating reaction kinetics for a range of powdered materials. As part of this investigation these basket approaches have been applied to a number of detergent powder formulations and the measured kinetics compared to those measured using a novel parameter estimation approach.

This parameter estimation approach is performed using a 2D transient model of an oven heated basket of powder developed in gPROMS ModelBuilder. This model captures the heat and mass transfer in an equi-cylindrical basket of self-heating detergent powder. The model predictions are compared to a series of experimental basket temperature profiles, measured by an array of thermocouples embedded in the basket of powder. The prediction error is minimised to determine the best-fit self-heating reaction kinetics, thermal conductivity, and specific heat capacity. The correlation of these parameters is explored, and to improve this approach, Differential Scanning Calorimetry (DSC) and the Ozawa-Flynn-Wall approach has been applied to determine the reaction activation energy of the powder prior to fitting. The advantage of this approach is that fewer experiments are required compared to the steady-state method, and this approach has been found to be less sensitive to thermocouple errors than the cross-point temperature method.

The reaction kinetics found using this novel parameter estimation approach compare well with those from the other basket heating techniques. Using the best-fit parameters the 2D model was successfully validated against critical ambient temperatures (temperature above which thermal runaway occurs) of oven heated baskets of powder, and the influence of different conditions on the self-heating behaviour of these baskets explored. This model and the parameter estimation approach provide a viable and more efficient means of characterising current and new detergent powder formulations. Adapting this model to represent self-heating in tower wall accumulations, the influence of tower temperatures, airflow patterns, layer thickness, and other process conditions on the critical ambient temperature and optimal tower operating temperature of different formulations has been explored.

# 507931 Use of Process Analytical Technology (PAT) for Monitoring and Optimizing Powder Mixing Processes


**Abstract:**

**Use of Process Analytical Technology (PAT) for monitoring and optimizing powder mixing processes**

The Process Analytical Technology (PAT) approach, which is well established in the pharmaceutical and chemical industry, was successfully applied on food manufacturing. PAT is about the IDENTIFICATION of critical process parameters in the manufacturing process as well as in-line sensors enabling real time process MONITORING and CONTROL for the continual improvement of quality and costs.

A food manufacturing process was analyzed in view of critical unit operations and process parameters. A strategy to assess and develop in-line sensors to monitor and control the critical unit operations was put in place. Sensors were used to make the invisible visible, which further improved our process understanding and enabled the optimization of the manufacturing process. The advantages and limitations of PAT will be discussed for powder mixing, which is one of the most important unit operations when producing powdery products. During mixing, the foundation of a compliant and consistent product is laid. But the homogeneity does not only effect product quality but quite often also the asset intensity of downstream unit operations. For a given installation and recipe, the homogeneity of the mixed mass is mainly influenced by the mixing time, the speed of the mixing devices as well as the filling degree of the mixer. Traditionally, the analysis of the powder mixer performance is quite time intensive, which makes the studies of parameter variations difficult. Several samples are taken after different mixing times, analyzed with off-line analytics and the variance is calculated to assess the mixing progress with time [1].

In his PhD thesis "Continuous dynamic mixing of cohesive powders" [2], Kehlenbeck has already shown the power of near-infrared (NIR) spectroscopy to access continuous and batch powder mixing processes [3]. An industrial application of the developed self-cleaning near-infrared probe (air flushing) was not possible as a contact between probe and product was required:

- In contrast to the model ingredients maize starch and calcium carbonate used in the framework of the PhD thesis, most food powders are moisture as well as temperature sensitive and tend to form layers in the mixer. After a short time of operation, a layering on the measuring window of a near-infrared spectrometer would occur, which cannot be removed by air flushing and would cause an incorrect measuring result.

- Furthermore, industrial mixers do not have openings allowing an immersion of near-infrared probes in the fluidized mass without drilling holes.

In the meantime, powerful near-infrared spectrometers for a contactless measurement in distances of up to 60 cm were commercially available. The assessment of this technology for process monitoring and optimization was successfully restarted in pilot and industrial scale.

The following topics will be covered:

- Process Analytical Technology in general

- Impact of the mixer filling degree on the mixing homogeneity

- Impact of the sample size on the measuring result

- Use of near-infrared spectroscopy for monitoring continuous and batchwise working powder mixers in pilot and industrial scale (time and trajectory plots)

- Comparison of off-line reference and in-line near-infrared measurements

- Until the World Congress on Particle Technology (WCPT8), it is expected that the impact of the particle size on the mixing result was also investigated and can be discussed.



[1] Kehlenbeck, V.; Sommer, K.: Different methods to determine the mixing performance of a batchwise working screw mixer; Powder handling & processing Vol. 15 (2003) No. 5, pp. 318-327

[2] V. Kehlenbeck: Continuous dynamic mixing of cohesive powders; PhD Thesis Technical University Munich 2016

[3] V. Kehlenbeck: Use of Near Infrared Spectroscopy for in- and off-line performance determination of continuous and batch powder mixers: opportunities & challenges; Procedia Food Science 1 (2011) pp. 2015 – 2022

# 505965 Establishment of a General Correlation Accounting for Impeller-to-Wall Distance in a Bladed Powder Mixer Operating in the Cataracting Regime

**Abstract:**

As for any unit operation, the search and establishment of correlations between dimensionless numbers is essential to equipment design, scale-up, scale-down, energy minimization and of course operation. In the case of powders, two main issues are arising quickly:

- Dimensional analysis is extremely tedious as powder characterization is still a work-in-progress. As a consequence, the correlations that have been derived are strongly powder-dependent.
- According to rotational speeds and powder flowability, typically free-flowing or cohesive, different mechanisms and flow regimes can take place: rolling, cascading, cataracting, fluidized. Correlations have then to be adapted for the case considered, the usual limit between rolling and cataracting corresponding to a value of the Froude number equal to 1.

Together with tumbler mixers, convective mixers are widely employed throughout the industry to mix powders of different nature. They are mostly designed through a set of blades mounted on a shaft that rotates in a certain range of speeds. In the cataracting regime, the powder is taken by the blade and is projected towards a direction that depends on its inclination. The shear that is enhanced by the blade, together with the projection effect, makes it a suitable mechanism to promote in the case of cohesive systems as a way to disrupt the particle-particle interactions that are counteracting the mobility of individual particles. As a consequence, impeller-to-wall distance in the case of single-impeller shafts, or impeller-to-impeller distance for multiple-impeller systems is an important geometrical parameter to account for in the establishment of a correlation. In the present work, we report an experimental study for which the influence of the impeller-to-wall distance has been considered in the establishment of a correlation between the Froude number and the Newton number in the case of a polyvalent bladed mixer of our conception. Several cohesive powders have been investigated, so as to give a more general projection to the correlation.

The mixer considered is a 4-blade system that is mounted on a shaft in a vessel of cylindrical shape. In addition to process variables such as fill level or rotational speed of the stirrer, the equipment itself has multiple options: blade angle, equipment angle, mixing volume, distance between wall and blades ... that can all be set initially at different values [1]. Removable disk-shaped walls can be placed at different positions inside the cylinder, either symmetrically or asymmetrically with respect to the blade system. This gives rise to different mixing volumes for a given filling ratio, that can be figured out by the impeller-to-wall distance L, or in dimensionless representation the ratio between blade radius R and L. Four different lactose cohesive powders (Flowlac 100, Granulac 140, Granulac 230, Tablettose 70) are studied and characterized through the help of a FT4 rheometer, as well as a tap-tap volumenometer and LASER-diffraction particle size analyzer. Stirrer's rotational speed is varied, as well as blade-to-wall distance, the filling ratio being the same for all the experiments conducted.

A torque-meter is placed along the shaft and allows recording the torque exerted by the powder, and therefore the power that is consumed at a given rotational speed. The Newton number Ne, that is the ratio between the torque and the product MgR, where M is the powder mass, is calculated and employed to derive correlations between Ne, Fr and R/L.

Results show that a single correlation of the form $Ne = k\,(R/L)^a\,Fr^b$ can be derived for all the powders studied as long as the mixer operates in the cataracting regime, coefficients a and b being invariants: $a = 1$ and $b=0.3$. This holds true for a free flowing powder that has been considered to complete the study, and suggests that the cataracting motion of the powder is governed by the blade system itself, regardless of the powder flowability, or at least at a much lesser level than in the rolling mode. This also gives a much precise idea of the blade's zone of influence that can serve as guidelines for multiple impeller system's location along the shaft.

[1] Legoix L., Gatumel C., Milhé M., Berthiaux H., Mizonov V. Powder flow dynamics in a horizontal convective blender: tracer experiments, Chemical Engineering Research and Design, 121 (2017), pp 1-21.

# 507901 X-Ray Particle Tracking in Vertical Bladed Mixers

**Abstract:**

The motion of individual granular particles in bladed industrial equipment such as vertical axis mixers, filter dryers, and high shear granulators is poorly understood despite being used in multiple industries including food and pharmaceuticals. Particle motion in this equipment is subject to competing mechanisms of mixing including diffusion, convection, and slip plane mixing. Visual observation of the particle motion is restricted to free surfaces or near the walls, whereas most velocimetric techniques are unable to capture the orientation of the particles as they move through the system.

In this study, experiments were performed in a transparent acrylic cylinder fitted with two 45$^{o}$ blades to simulate a vertical axis bladed mixer. Tagged particles were introduced at different radial and axial locations in the cylinder. X-ray particle tracking was employed to obtain the trajectory, orientation, and velocity profile of the tagged particles. Tracking more than one point on the tagged particles allowed for the measurement of particle orientation as they moved through the mixer. This X-ray imaging technique will be used in the future to obtain unique data to validate computational models describing granular mixing.

# 506951 Modeling Granular Material Mixing and Segregation Using a Finite Element Method and Advection-Diffusion-Segregation Equation Multi-Scale Model

**Abstract:**

A multi-scale model is presented for predicting the magnitude and rate of powder mixing and segregation in blenders. The model combines diffusion and percolation correlations from small-scale discrete element method (DEM) simulations with advective flow field information from large-scale finite element method simulations. The multi-scale model predictions closely match results from both large-scale DEM-only simulations and experiments, but they are much more computationally efficient than DEM-only simulations. Therefore, in contrast to DEM, the multi-scale model is well suited for predicting blending in industrial-scale systems.

# Particle Measurement in High-Temperature Gas based on Light Scattering

Li-feng LU[1,3], Xiao-lin WU [1,3], Zhong-li JI [1,2], Ming-xing WANG[1,3]

[1] Department of Chemical Engineering, China University of Petroleum, Beijing, P. R. China

[2] Department of Mechanical and Transportation Engineering, China University of Petroleum, Beijing, P. R. China

[3] Beijing Key Laboratory of Process Fluid Filtration and Separation, China University of Petroleum, Beijing, P. R. China

**Abstract:** In order to achieve the particle measurement at 650 °C, a single particle measurement method and device suitable for high-temperature gas were developed. The influence of gas temperature on optical sensor of optical particle counter (OPC) was investigated in detail based on geometric optics theory and heat transfer theory. The thermal expansion of aerosol tube along the airflow direction was caused by the heat conduction between the tube and the gas, resulting in the position of the light apertures to shift to obscure the partial light beam. The invar material can be adopted to eliminate the above mentioned disadvantages. According to the principle of gas state equation, the optical parameters (refractive index) of the gas at different temperatures change with the change of its state parameters. Based on the geometrical optics theory and the geometric mathematical relationships, the theoretical and experimental dynamic models of the optical measuring volume with the change of the gas temperature were established. The established models were then used to analyze the influence of gas temperature on the optical measuring volume. The results showed that the influence of the temperature on the optical system is negligible. Meanwhile, particle detection experiments were conducted at different temperature to achieve a good performance of OPC.

**Keywords:** High-temperature gas; Optical measuring volume; Dynamic model

## 1 Introduction

With the continuous improvement of environmental protection and the safety requirements for industrial production, the technology of high-temperature purification is drawing more and more attention, especially in the fields of clean coal power generation, coal gasification, petrochemical industry and metallurgy [1]. For example, nuclear power plants often produce a large number of low radioactive solid wastes, and the combustible wastes usually account for a considerable proportion. The incineration and pyrolysis are effective treatment techniques for the combustible waste, during which the temperature of flue gas is 600~800 °C. With the aim to purify the flue gas, the German Karlsruhe Nuclear Research Institute, the French nuclear power station and the Japanese nuclear power station have installed the silicon carbide ceramic filter tube, aluminum silicate fiber and alumina ceramic filter tube at gas temperatures above 650 °C, respectively, and investigated their properties properties [2-4]. These filters have a direct influence on the purification quality of gas and the particle concentration in the downstream. In an industrial cracker system, the particle concentration is required to reach 0.014 mg/m$^3$ after the filters; the Integrated Gasification Combined Cycle (IGCC) requires the particle size to be smaller than 1 um with a particle concentration of less than 2.2 mg/m$^3$ and in the range of 1~10 um with less than 5.8 mg/m$^3$, and to remove the dust size of larger than 10 um [5]. In a Shell Coal Gasification Process (SCGP), the particle concentration is required to be less than 20 mg/m$^3$ [6]. In order to detect the concentration and size of the particles in pipes in real time for calculating the separation efficiency of evaluating the purification performance of the filtration system, the particle detection technology is needed to be improved [7].

The widely-used particle detection devices in industry are mainly based on the principle of light scattering [8-10], and the particle size and concentration are obtained through an inversion of the scattering light intensity and particle size, which are suitable for the conditions of below 200 °C. The method of first cooling and then detecting can be adopted for the particle detection under higher-temperature conditions, but droplet condensation and sublimation will inevitably occur and result in an incorrect particle samples.

With the aim to achieve the particle detection at 650 °C, an optimization was conducted on the existing particle detector (OPC) in this paper. And based on geometric optics theory [11] and heat transfer theory, a detailed research was performed on the influence of gas temperature on OPC optical sensor, including the dynamic change of the aerosol tube and the optical measuring volume (OMV). Finally, the experiments were conducted on the on-line particles measurement of different temperatures.

## 2 Experimental setup and method

Figure 1 gives the experimental setup for detecting particles in high-temperature gas, mainly including the heating unit, optical detection unit and gas purification and processing unit.

The heating unit consists of the vibration particle feeder and the air heater. The gas brings the particle out of the feeder, and is heated up to 1000 °C by the heater, which is monitored quickly by a PID temperature control instrument with easy operation.

The optical detection unit is used to achieve the experimental measurement of the characteristics of the OPC optical system and the on-line detection of particles, including the optical sensor and the OPC control host. The optical sensor is shown in Figure 2, which is composed of an aerosol tube and optical system. Aiming to avoid the influence of high-temperature gas on optical system, the aerosol tube (shown in Figure 3) has been developed, and is in a distance of 15 mm to the optical lens. In addition, the insulation coating with lower thermal conductivity (-80-1700 °C) is adopted to treat the outer wall of the aerosol tube. A K-type micro thermocouple was bent and installed into the light aperture to monitor the gas temperature at the center of the aerosol tube in real time without affecting the airflow and optical paths. Figure 4 shows the optical system of the sensor, which is composed of an incident light path and an observed scattering light path. Both the paths are arranged in 90° with the same optical lens and elements [12-14], and the intersected area is the center of the aerosol tube to form an OMV [15]. The intersected area is enough small and allows particles to pass through it one by one [16, 17]. The OPC host consists of the light source system (wavelength: 300-700 nm) [18], the photoelectric detection system and the signal analysis and processing system. In the particle detection, particle is illuminated by the incident light when passing through the OMV, and the scattering light is collected and transmitted to the photoelectric detector to achieve the conversion of a pulse voltage signal from an optical signal. Finally, the tested particle size is calculated based on the calibration curve of particle size and voltage amplitude [14, 19].

The gas purification and processing unit is composed of a vacuum pump, a flow controller, a dust collector and a heat exchanger. The high-temperature gas form the aerosol tube is completely cooled during the heat exchanger, the flow rate of which is adjusted by the flow controller. In order to avoid the damage of particles on the pump, the dust is captured and collected by a filter in the dust collector.

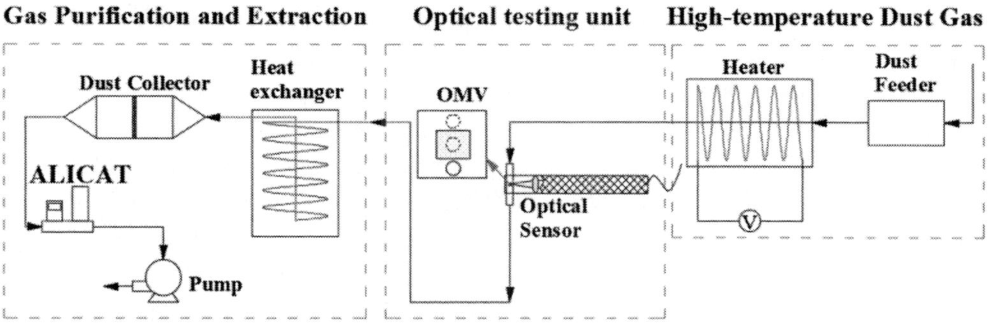


Figure 1. Experimental setup for detecting particles in high-temperature gas

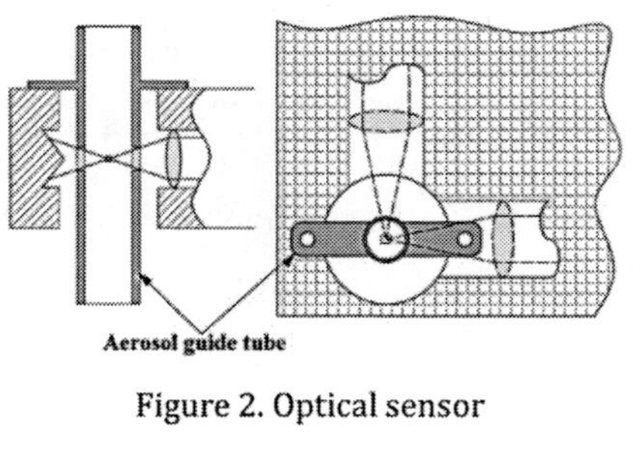


Figure 2. Optical sensor

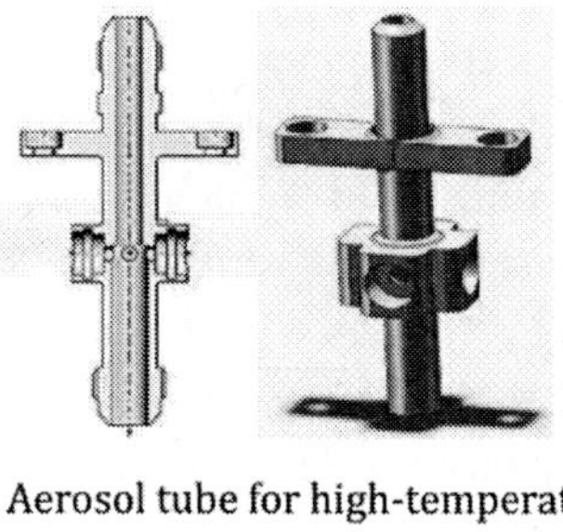

Figure 3. Aerosol tube for high-temperature gas

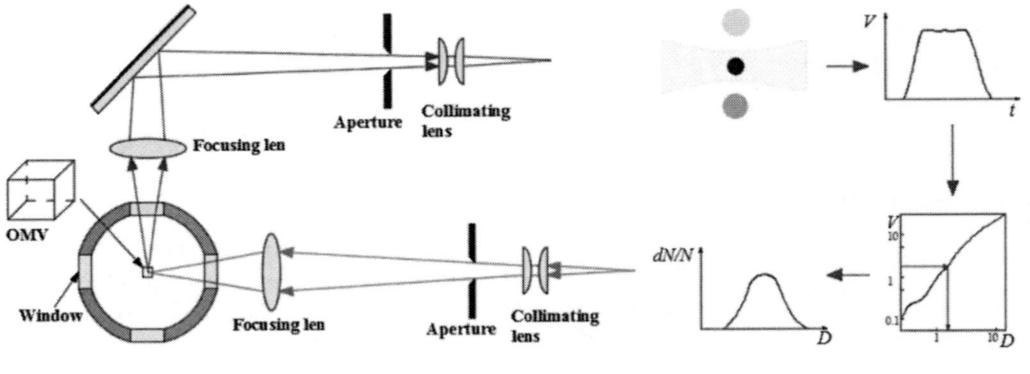


Figure 4. Optical system and principle of OPC

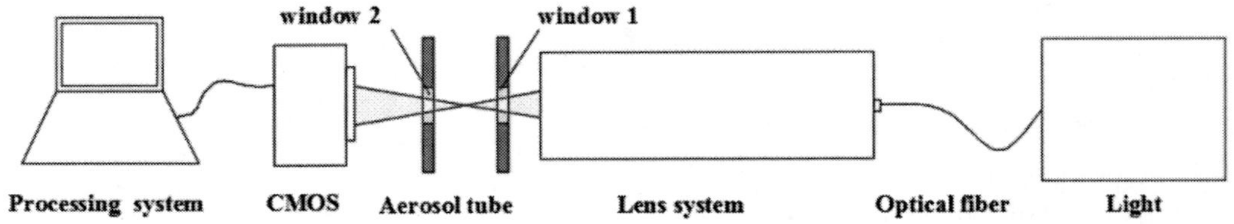


Figure 5. Evaluation and analysis device for light beam

## 3 Analysis and discussion

### 3.1 Influence of temperature on aerosol tube

The aerosol tube shown in Figure 3 has four apertures along the center axis, in which the glasses with a thickness of 2 mm were installed, as shown in Figure 4. When the gas flows in the tube, it can reduce the damage of high-temperature gas to the optical system elements. But the heat resistance of the tube has a direct influence on the imaging of the optical system at the center of the aerosol tube.

Figure 5 gives an evaluation and analysis device for light beam, including the light source system, the light path system, the aerosol tube and the light beam analyzer (CINOGY CMOS-1201, Germany), which can be equipped at the optical testing unit as shown Figure 1. For the gases at different temperatures, the light passes through the glasses in the aerosol tube and images on the CCD of CMOS-1201 to detect the spot of the light beam in real time.

After eliminating the background noise, the spot image at different temperatures can be acquired. When the aerosol tube is made of 304 stainless steel, the measurement results are shown in Figure 6. It can be found that the shape of the light spots is distorted and shows the imperfection of the imaging with an increase in gas temperature. That's due to thermal expansion along the airflow direction caused by the change of gas temperature, which results in the shift of light apertures position to obscure the partial light beam. A bad result must be acquired for smaller particle size and lower particle count efficiency because of the decreased incident light. Meanwhile, the shape of the OMV was changed to greatly reduce the performance of OPC detection by a failure of eliminating the error of edge detection in optical system.

With the aim to eliminate the above mentioned disadvantage, the invar material with a linear expansion coefficient of $1.6\times10^{-6}$ m/°C was adopted for high-temperature experiment. The results are shown in Figure 7. At different temperatures, no deformation was found in the shape of light spots, indicating that the material can meet the testing requirement for high temperature up to 650 °C with a good heat resistance.

Table 1. Comparison of the application conditions of the two materials

| Material | Light divergence angle /° | Gas temperature /°C |
|---|---|---|
| 304 (Stainless steel) | <10 | <200 |
| 4J36 (Invar) | <30 | <700 |

The application conditions of the two materials are

summarized in Table 1. Compared with the stainless steel, the invar material is of great significance for the development of high-temperature detection device.

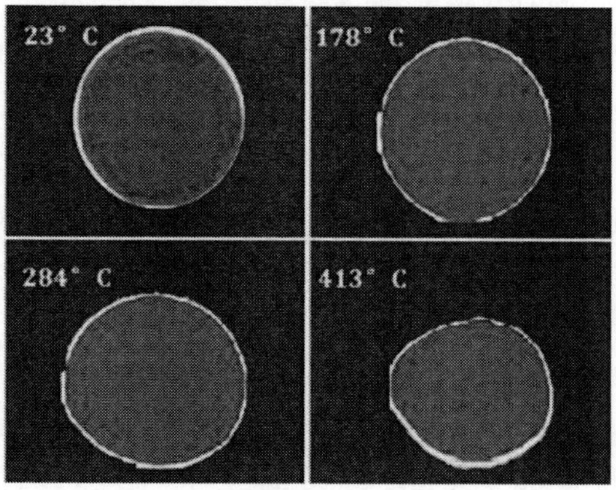

Figure 6. The spot images measured at four different temperatures of gas for 304 stainless steel.

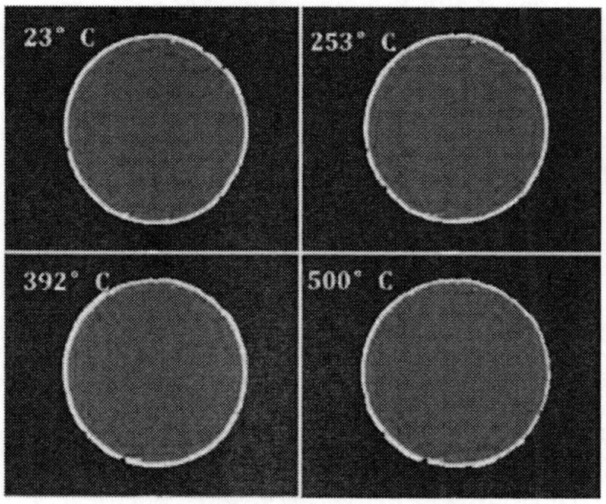

Figure 7. The spot images measured at four different temperature of gas for 4J36 (Invar).

### 3.2 Influence of temperature on optical system

In general the optical sensing system in the particle detection device is only applied at atmospheric temperature, but can also satisfy the temperature up to 150 °C. Under high-temperature condition, the gas volume is going to expand with the reduction in the gas density. The transmission characteristic of light beam could be changed, resulting from the decrease in the optical performance of the sensor. In order to investigate the influence of temperature on optical system, the specific causes are analyzed by theoretical method, and the extent of its influence is determined through experimental verification.

#### 3.2.1 Theoretical method

By calculating the refractive index of the gas at different temperatures, the change of the OMV with the temperature can be obtained.

**a) Calculation of gas parameters**

The gas medium is air, so its refractive index can be calculated by the formula proposed by Rueger at the working temperature, given by:

$$n_{ph} = \frac{273.15}{1013.25} \cdot P \cdot N_{sph} \cdot \left(a_0 + \frac{a_1}{T} + \frac{a_2}{T^2} + \frac{a_3}{T^3} + \frac{a_4}{T^4}\right) \times 10^{-6} + 1 \quad (1)$$

Where $P$ is the gas pressure; $a_0$, $a_1$, $a_2$, $a_3$ and $a_4$ are the correlation coefficient and constant; $N_{sph}$ is the gas refractive index at atmospheric temperature, given by:

$$N_{sph} = (n_{sph} - 1) \times 10^6 = 287.6155 + \frac{1.62887}{\lambda^2} + \frac{0.01360}{\lambda^4} \quad (2)$$

The wavelength range is 400-700 nm. The gas refractive indexes at different temperatures in the range of 25-650 °C are calculated by Eq. 1 and 2. The results are shown in Figure 8. It can be found that the refractive index of air decreases with the increase in the gas temperature. And it can be inferred that the light divergence angle at high temperature is going to increase compared with that at atmospheric temperature. Consequently, the position of the beam focus is offset, leading to larger optical measuring volume, as shown in Figure 9.

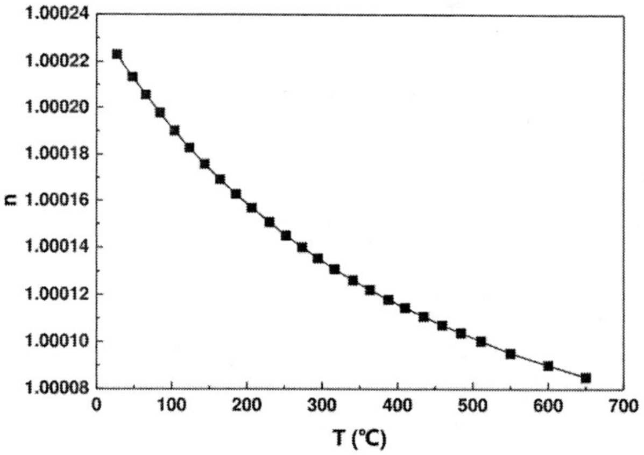

Figure 8. Gas refractive index at different temperature

In order to analyze the above unfavorable influence, a model of the OMV and gas temperature is needed to be built, mainly including the functions of the determination of the optical parameters and the calculation of the dynamic variable.

**b) Calculation of optical parameters**

On the basis of the Lorentz-Lorenz formula and actual gas state equation, the relationship between gas pressure and refractive index is given by:

$$\frac{n^2-1}{n_0^2-1}=\frac{\rho}{\rho_0}=\frac{PT_0Z_0}{P_0TZ} \tag{3}$$

Where $T_0$, $P_0$, $\rho_0$, $Z_0$ and $n_0$ are the gas temperature, pressure, density, compression factor and refractive index under atmospheric condition, respectively; $T$, $P$, $\rho$, $Z$ and $n$ are the gas temperature, pressure, density, compression factor and refractive index under high-temperature condition, respectively.

The light divergence angle is calculated by the refraction law shown below:

$$n_0^{'}\sin\alpha = n\sin\beta \tag{4}$$

Where $n_0'$ and $\alpha$ are the air refractive index and light divergence angle under atmospheric condition; $\beta$ is the light divergence angle under high-temperature condition.

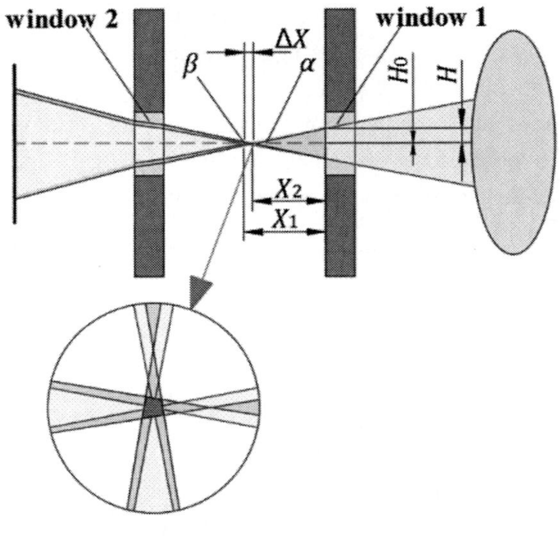


Figure 9. Schematic of the enlargement of the OMV

#### c) Calculation of dynamic variable

Based on the geometric optical theory and the mathematic relationship, the distance between the light focus position and the glass window 1 is triangular with the equivalent spot radius of the inner side of the glass window 1. According to $X_1$ and $X_2$ (the distance between the light focus position and the inner side of the glass window 1 under atmospheric condition and high-temperature condition), the theoretical model of the OMV with gas temperature was built. $X_1$ and $X_2$ are calculated by:

$$X1=\frac{H-H_0}{\tan\alpha} \tag{5}$$

$$X2=\frac{H-H_0}{\tan\beta} \tag{6}$$

The change of the light focus position ($\Delta X$=$X1$-$X2$) can be calculated by Eq. 5 and Eq. 6.

The change of the equivalent spot diameter of the OMV can be calculated based on the triangle relation, as shown below:

$$\Delta D = 2\Delta X \tan\beta \tag{7}$$

Eq. 6 and Eq. 7 are the formulas of the theoretical dynamic model of the OMV with the change of the gas pressure.

Using the model, the optical parameters at different gas temperatures in the range of 25 ~ 700°C were calculated with three light divergence angles (4.02°, 9° and 13°, tested under atmospheric condition, respectively). The results are presented in Figure 10. It can be found in Figure 10 that $\Delta X$ increases with the gas temperature with approximate two patterns. In addition, $\Delta D$ changes by a smaller margin (less than 1 um). However, $\Delta D$ and $\Delta X$ satisfy a trigonometric relation. It can be also noted that the enlargement rate of the OMV volume ($\delta$) has a small increase with the increase in gas temperature. Under the constant temperature, $\delta$ is going to increase for the larger light divergence angle, but it is still relatively smaller (<1%) compared with the OMV at atmospheric condition. Consequently, the influence of the gas temperature on the OMV can be ignored based on the above theoretical analysis.

### 3.2.2 Experimental method

In order to verify the reliability of theoretical analysis, an experimental dynamic model of the OMV with gas temperature variation was established based on the image analysis and the mathematical geometry relations, including the measurement of the change of light spot, the calculation of the geometric relation and the calculation of the relative variable.

By the refraction law, the changes of light spot on the inner and outer side of the glass window 2 under high-temperature condition ($\Delta D$ and $\Delta D'$) meet $\Delta D$=$\Delta D'$. Taking the light focus position as the central axis, a triangle relation was formed by the distances between the light -

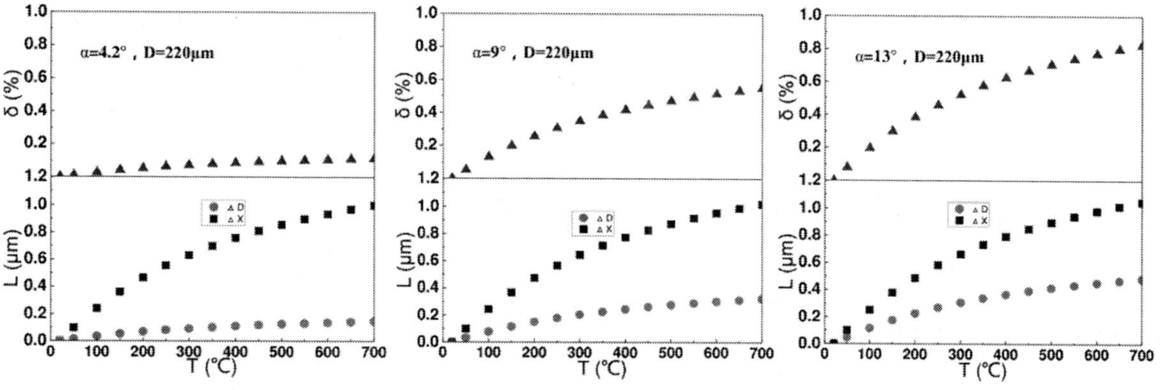


Figure 10. Results calculated based on theoretical dynamic mode with three light divergence angles (4.02°, 9° and 13°)

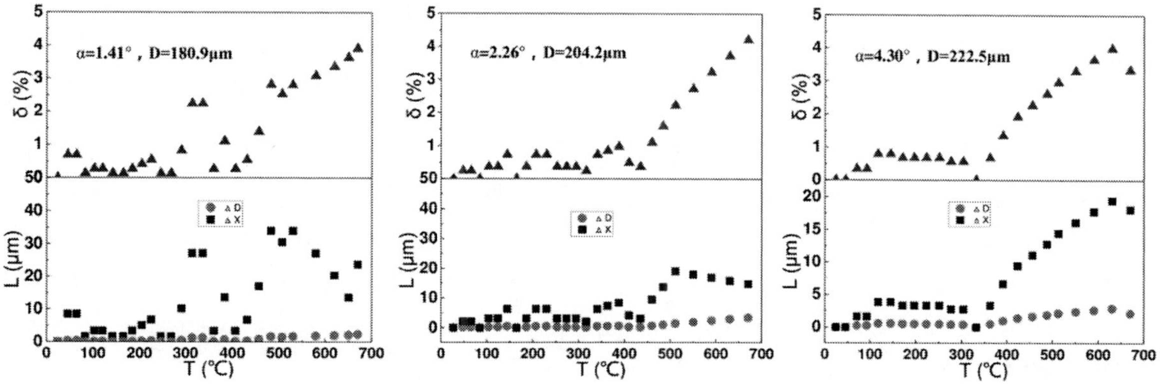


Figure 11. Results calculated based on experimental dynamic model

focus position and the inner side of the glass windows 1 and 2 ($x$ and $y$), the difference of the light spot equivalent radius on the inner side of the glass windows 1 and 2 ($H$ and $h$, $h = H - \frac{\Delta D}{2}$) and the equivalent radius of the OMV ($H_0$). Therefore, we have the following equations.

$$\frac{H - H_0}{h - H_0} = \frac{y}{x} \tag{8}$$

$$x + y = k \tag{9}$$

$$\beta = \arctan\left(\frac{h - H_0}{x}\right) = \arctan\left(\frac{H - H_0}{y}\right) \tag{10}$$

where $\beta$ is the light divergence angle under high-temperature condition; k is the distance of the inner sides of the glass windows 1 and 2.

The changes of $\Delta X$ and $\Delta D$ under high-temperature condition to those under atmospheric condition are shown as follows:

$$\Delta X = \frac{y - x}{2}, \text{ or } \Delta X = y - \frac{k}{2}, \text{ or } \Delta X = \frac{k}{2} - x \tag{11}$$

$$\Delta D = 2\Delta X \tan\beta = \Delta D' \tag{12}$$

Eq. 11 to Eq. 12 are the formulas of the experimental dynamic model.

Using the above established experimental dynamic model, the $\Delta X$, $\Delta D$ and $\delta$ of four cases in Table 2 were calculated, and the results are shown in Figure 11. It can be found that these three parameters increase obviously with the gas temperature increasing, but cannot meet the two function growth mode as the results of theoretical model. That may be caused by the reasons listed below: 1) the fluctuation of gas flow in the pipeline causes the unstable gas temperature field; 2) the beam quality analyzer is slightly influenced by other vibration sources; 3) the aerosol tube is affected by the gas flow fluctuation. But according to the experimental results, the maximum values of $\Delta X$ and $\Delta D$ are very small (less than 50 μm and 5 μm at the temperature below 700 °C, respectively) compared with the equivalent diameter of the OMV. Consequently, the ratio of the volume of the OMV is less than 5%. To sum up, the gas temperature has greatly weakened influence on the OMV, which can be ignored in actual technical development.

Table 2. Optical parameters of experimental model

| number | light divergence angle /° | spot equivalent diameter /μm |
|---|---|---|
| 1 | 1.41 | 180.9 |
| 3 | 1.94 | 285.1 |
| 4 | 2.26 | 204.2 |
| 5 | 4.30 | 222.5 |

### 3.2.3 Experiments

The performance of the optical sensor is investigated at different temperatures up to 650°C. The experimental results showed that the temperature of the optical sensor is around 50°C with the gas temperature in the range of 200-300 °C, and is going to increase about 10 °C at the gas temperature of 400-500 °C. The temperature is in the controllable range with a maximum value of 70 °C under the gas temperature of 650 °C. Therefore, the developed optical sensor can be used for the detection of particles in high-temperature conditions for a long time.

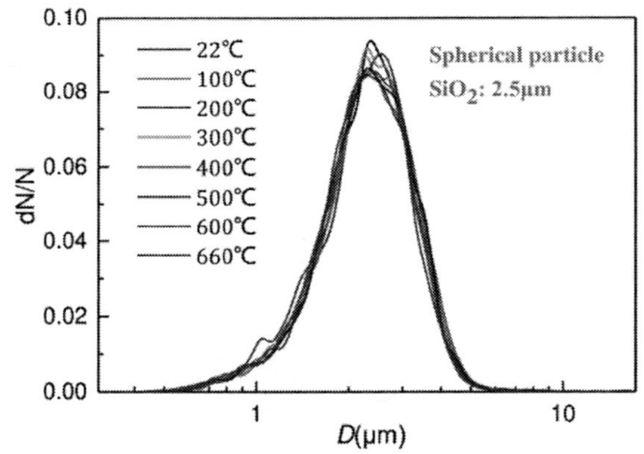


Figure 12. The results obtained under high-temperature condition

In the experiment of particle detection, the standard silica spherical particles with a strong temperature resistance ($D_{particle}$=2.5 μm) were tested at different gas temperatures. The testing time is 60 s. Figure 12 shows the test results. It can be found that all of the particle size distributions are in good agreement and have a good normal distribution, which meets the test results of single particle size, and the peak particle size is about 2.5 μm. Consequently, the optical sensor worked well for the particle detection under higher-temperature condition.

## 4 Conclusion

The emission of high-temperature flue gas causes serious air pollution and partial waste of available heat. In order to meet the relevant emission standards, additional purification units (high-temperature ceramic filters, etc.) are installed in the industrial process to collect the dusts in the flue gas. As the purification units are important for controlling the dust concentration, it is necessary to calculate the separation efficiency through a real-time measurement of the dust concentration in the flue gas in the upstream and downstream of the purification equipment.

In present work, a single particle measurement method and device suitable for high-temperature condition were developed. In order to complete the particles measurement at 650 °C, the beam images in optical sensing system under different temperature conditions were measured in real time by CMOS sensor. It was found that the beam images were deformed when the light divergence angle increased to more than 10°, and the deformation became more serious as the temperature increased. Depending on the principle of heat transfer, thermal expansion along the airflow direction caused by heat conduction between the duct and the gas is the main cause of this phenomenon. By changing the aerosol tube material t o one with an ultra-small linear expansion coefficient (Invar: 4J36), the optical system can be applied to the condition with temperature up to 700 °C, and in particular the light divergence angle extends to 30°.

According to the principle of gas state equation, the optical parameters (refractive index) of the gas at different temperatures change with the change in its state parameters. According to the results calculated by the formula of Ruegur's atmospheric refractive index, it was found that the refractive index of the gas decreases with the increase in temperature, which leads to the increase in the divergence angle of the beam in the aerosol tube, and shift of the position of the optical beam. Based on the geometrical optics theory and the geometric mathematical relationships, the theoretical dynamic model of the optical measuring volume with the change of the gas temperature was established. By calculating the amount of the beam focus offset and the volume of the optical measuring volume at different temperatures, it was found that the optical measuring volume increases with the temperature increasing. However, when the gas temperature was 650 °C, the increase rate of optical measuring volume was extremely small, so the influence of the temperature on the optical system can be negligible.

The performance of the whole detection system was tested by using 2.5 μm standard spherical silica particles. The satisfying measuring results were obtained and they could be applied to the actual on-site measurement at the temperature of 650 °C.

## Acknowledgement

The authors thank the supports of "The National Key Research and Development Program of China (NO.2016YFB0601100)"; and would also like to thank the laboratory and staff of the Beijing Key Laboratory of Process Fluid Filtration and Separation, China University of Petroleum.

## Reference

[1] J. Meyer, H. Umhauer, K. Reuter, A. Schiel, G. Kasper, M. Förster, Concentration and size measurements of fly ash particles from the clean gas side of a pressurised pulverised coal combustion test facility, Powder Technology, 180 (2008) 57-63.

[2] H. Leibold, F. Dirks, V. Rüdinger, Particulate emissions from a LLW incinerator and off-gas cleaning with a new type of ceramic candle filter, Waste Management, 9 (1989) 87-94.

[3] B.L. Perkins, Incineration facilities for treatment of radioactive wastes: a review, Incinerators, (1976).

[4] D. Fino, N. Russo, A. G. Saracco, V. Specchia, Multifunctional Filter for Treatment of the Flue Gases from Municipal Waste Incinerators, Industrial & Engineering Chemistry Research, 44 (2005) págs. 9542-9548.

[5] Scheibner, Wolters, Schumacher hot gas filter long-term operating experience in the nuon power buggenum IGCC power plant, (2002).

[6] J.M. Wang, Shell coal gasification technology and its application in China, Coal Quality Technology, (2006).

[7] Q. Xu, Z. Ji, L. Yang, Performance assessment of high temperature flue gas on-line particle size analyzer, Ciesc Journal, 63 (2012) 3506-3512.

[8] B.Y.H. LIU, V.A. MARPLE, K.T. WHITBY, N.J. BARSIC, Size distribution measurement of airborne coal dust by optical particle counters, Aihaj, 35 (1974) 443-451.

[9] R. Maus, F. Löffler, Size analysis of airborne bacteria by an optical particle counter and an impactor, Journal of Aerosol Science, 25 (1994) 141-142.

[10] B. Sachweh, H. Büttner, F. Ebert, Improvement of the resolution and the counting accuracy of an optical particle counter by fast digital signal recording, Journal of Aerosol Science, 20 (1989) 1541-1544.

[11] E.W. Max Born, Principles of Optics, Cambridge University Press, Cambridge, 1999.

[12] S. Hochrainer, Further developments of the particle counter sizer PCS-2000, Journal of Aerosol Science, 31 (2000) 771-772.

[13] B.S. Li, Z.L. Ji, L. Mei, An approach for correcting particle size distributions measured by the optical particle counter welas 2000, Aip Conference, 2010, pp. 686-689.

[14] J. Stier, M. Quinten, Simple refractive index correction for the optical particle counter PCS 2000 by Palas, Journal of Aerosol Science, 29 (1998) 223-225.

[15] H. Umhauer, Particle Size Distribution Analysis by Scattered Light Measurement Using an Optically Defined Measuring Volume, Journal of Aerosol Science, 14 (1983) 765-770.

[16] B. Sachweh, H. Umhauer, F. Ebert, H. Buttner, R. Friehmelt, In situ optical particle counter with improved coincidence error correction for number concentrations up to 107 particles cm-3 quantification and elimination of the effect of particle shape and structure, Journal of Aerosol Science, (1998) 1075-1086.

[17] D.I.J. Raasch, D.I.H. Umhauer, Errors in the Determination of Particle Size Distributions Caused by coincidences in optical particle counters, Particle & Particle Systems Characterization, 1 (1984) 53-58.

[18] H. Buettner, Measurement of the Size of Fine Nonspherical Particles with a Light-Scattering Particle Counter, Aerosol Science & Technology, 12 (1990) 413-421.

[19] D. Holve, S.A. Self, Optical particle sizing for in situ measurements Part 2, Applied Optics, 18 (1979) 1632-1645.

# 508245 An Experimental Investigation of Single Droplet Drying Above Boiling Point

**Abstract:**

An experimental method has been developed to enable dried particle structure, properties and drying rate to be predicted based on droplet drying history, particularly focusing on effects driven by boiling including bubble nucleation, growth and collapse. The method involves a single droplet drying rig which will be used to investigate drying kinetics above boiling point and across a broad range of drying conditions, drop size and composition space. The drying rig was improved as compared to previous apparatuses reported in literature, in that it is designed to measure droplet drying rates and morphological changes at high temperatures and under conditions that more closely resemble those found in industrial applications. The aim is to conduct drying experiment for droplets as small as 300μm diameter, in an air stream which temperature may be as high as 350°C.

Drying kinetics and morphological evolution will be determined, particularly focusing on the second drying stage as well as inflation/deflation periods. Initially, the constructed drying rig is checked and evaluated by conducting experiments on droplets of pure water. Afterwards, experiments will be conducted on solutions of different materials such as Hydroxypropyl Methylcellulose (HPMC), sodium sulphate and sucrose. An assessment of morphologies, transformations and drying behaviour under isothermal drying would be completed at a wide space of process condition, initial water content, initial drop size and composition space. The investigation will look to link material properties, particle size, initial moisture content and drying history to morphology. Results will be compared with single droplet drying models reported in literature.

An initial investigation has focused on mapping the morphological drying behaviour of HPMC, sodium sulphate and sucrose. It has shown that the three materials show different behaviour; sucrose showed deflation and inflation cycles, sodium sulphate droplets supersaturated and crystallized while HPMC showed inflation and rupture. Differences in drying behaviour is due to difference in material and transport properties. Therefore, the next stage of this work is focused on conducting some physical properties measurement at different water concentrations and temperature. In addition, different materials will be dried on the rig and the drying behaviour will be determined. The experimental data from this work will be used to develop, a universal morphology regime map which will link material properties, particle size, initial moisture content and drying history to morphology.

# 508482 Characterising Powder Flow in Dynamic Processes

**Abstract:**

Flow inconsistency of cohesive powders is a major issue for process industries. It is often difficult to predict and control flow of cohesive powders to enable efficient and reliable process operations particularly for dynamic processes such as mixing and filling. Many quantitative and qualitative flow measurement techniques exist that are used for measuring powder flowability under quasi-static conditions. However, there are few available techniques capable of assessing powder flowability under dynamic conditions. The FT4 Powder Rheometer (Freeman Technology), which drives a rotating blade along the vertical axis through a column of powder contained within a cylindrical vessel, is one of the few available devices that is capable of measuring powder flow in a dynamic state [1]. The flow resistance of the rotating blade as it penetrates the powder bed is characterised by flow energy, however, there is lack of understanding in how to relate this measurement to process design.

This work focuses on determining the relationship between the torque and shear stress in the FT4. The effect of particle size on the external flow profile around the blade has been investigated experimentally for glass beads using a high speed camera. The material investigated shows a decrease in downward specific flow energy with increasing particle size; indicating that the increased shear resistance in the bulk for smaller particle sizes exists under dynamic conditions. Particle Image Velocimetry (PIV) analysis indicates an increase in the volume of particles displaced by the blade with increasing particle size for both upward and downward blade movement. An increase in impeller tip speed leads to a reduction in flow energy for this material, and a reduction in the volume displaced by the blade. The prevailing stresses and strains as well as the volume of particles displaced by the blade are estimated by Discrete Element Method (DEM). DEM allows for the determination of a relationship between the flow energy and/or shear stress with particle properties and blade velocity and direction. Furthermore, Positron Emission Particle Tracking (PEPT) analysis of the FT4 has been carried out to determine the radial and axial displacement of particles with respect to the bed height, and to validate the DEM results.

# 508671 Impact ofÂ Non-Spherical Projectiles on Granular Media

**Abstract:**

Recently, several studies involving numerical and experimental methods have focused on the study of impact dynamics in both dry and wet granular media. An example of such instances can be an animalâ€™s foot impacting sand during its locomotion. Understanding the physics behind the interaction between granular media and animal feet can be helpful in optimizing locomotion parameters for bio-inspired robots. However, most of the studies involving impact dynamics considered spherical projectiles under different conditions, while characteristic models should involve more complex shapes. Impacting different geometries with conserved density, volume, and velocity on a granular bed may experience contrasting drag forces upon penetration. This is due to the difference in the surface areas coming in contact with the granular media. In this work, non-spherical geometries of following shapes were considered: cylinder, triangular prism, hexagonal prism and cuboid. These geometries were impacted on a loosely packed non-cohesive dry granular bed with the same impact velocities. The effect of surface area in contact with the granular media was analyzed for impact, with the help of penetration depth measurements. We also tested whether the power law for spherical projectiles is applicable for non-spherical projectile impacts.

# 508664 Computational and Experimental Shear Cell Study with Rigid Cylindrical Particles

**Abstract:**

Flow properties of particles are commonly assessed through shear cells. Both the Schulze ring shear tester (SRST) and Jenike shear tester (JST) measure the interparticle stresses which are important to the design of many granular flow processes. This work uses precision-cut rigid cylinders of various aspect ratios and sizes to compare the measured stresses between both testers. It is found that the shearing results for both testers are similar, which is important for future work involving flexible biomass materials.

Previous work has sought to use shear cell results to validate various aspects of discrete element method (DEM) simulations. The SRST is recreated in a DEM simulation with rigid glued-sphere cylindrical particles to assess the contact detection method by comparing the experimental and computation results. Also, a parameterization study is performed to examine the effects of coefficient of restitution, coefficient of friction, and other particle properties on the measured steady-state stress. Overall, this serves as an efficient validation of the contact detection method which will be used to simulate complex particle shapes.

# 508865 Defluidization Behaviour of Industrial Reactive Powders at High Temperature


**Abstract:**

Fluidized bed reactors are used in a wide range of industrial chemical processes, and they are particularly useful in high temperature systems thanks to their ability to provide high heat transfer rates and rapid solids mixing which lead to isothermal and more controllable conditions. Their applications spread the environmental, chemical, energy and process industries. For example, they are a key technology in the petroleum industry, catalytic chemicals synthesis processes, combustion and gasification of solid fuels (coal, wastes and biomass), metals productions and many more.

Use of fluidized beds in combustion-type operations may lead to particles sintering and bed agglomeration, which again may result in defluidization, unscheduled process downtime and additional costs. However, the details of the physical mechanisms behind such phenomena are still not entirely understood. Despite the countless researches that have been done so far, such processes still disclose unrevealed features and challenge the researchers worldwide.

The objective of this paper is to assess the effect of high temperature on the flow properties and fluidization behaviour of mixtures of carbon coke and two different titanium ores, namely synthetic and natural rutile. To this end, a simple approach for the early detection of agglomeration in the regimes of incipient fluidization -where the carbon combustion causes defluidization- will be followed. Such a monitoring approach is based on the simultaneous measurements of local temperatures and the pressure drop across the particles bed.

# 507739 Manufacturing and Characterization of Spherical Blend PBT-PC Particles for Additive Manufacturing

**Abstract:**

Polymer powder-based additive manufacturing technologies often yield excellent mechanical and functional part qualities. One of these technologies is selective laser sintering (SLS), a process employing micron-sized polymer particles, which are selectively fused by a laser [1]. While hardly any boundaries regarding part design exist, there are quite some restrictions concerning the variety of commercially available SLS materials. For the time being, the most widely used polymeric material for SLS is polyamide 12 (PA12) with a market share of roughly 95% [2]. Hardly any blend materials are available; mostly , the processing of multi-material systems via SLS is achieved by simply mixing different polymer powder materials [3].

In this contribution, a new approach to blend particles is introduced. Via co-grinding and subsequent rounding, manufacturing of spherical particles consisting of semi-crystalline polybutylene terephthalate (PBT) and amorphous polycarbonate (PC) thermoplastic is possible. The polymers are jointly comminuted and mixed in a planetary ball mill and the product particles are rounded in a heated downer reactor [4,5]. The thereby achieved intermixed, but not transesterified, PBT-PC composite particles are characterized regarding size and morphology via electron microscopy. Intermixing of the polymers is investigated via staining techniques and Raman spectroscopy. Thermal properties of the blend particles are analyzed via DSC, also in regard to transesterification in SLS process-like conditions. Furthermore, flowability and size distribution of the blend particles are examined. This way, we show the high potential of our process route [4] for the production of novel polymer blend particle systems for SLS.

References:

[1] I. Gibson, D. Rosen, B. Stucker, Additive manufacturing technologies, 2. ed., Springer, New York, NY, 2015.

[2] T.T. Wohlers, Wohlers Report 2016, Wohlers Associates, Fort Collins, Col., 2016.

[3] D. Drummer, K. Wudy, F. Kühnlein, M. Drexler, Polymer Blends for Selective Laser Sintering: Material and Process Requirements, Phys. Procedia. 39 (2012) 509–517. doi:10.1016/j.phpro.2012.10.067.

[4] J. Schmidt, M. Sachs, C. Blümel, B. Winzer, F. Toni, K.-E. Wirth, W. Peukert, A novel process route for the production of spherical LBM polymer powders with small size and good flowability, Powder Technol. 261 (2014) 78–86. doi:10.1016/j.powtec.2014.04.003.

[5] J. Schmidt, M. Sachs, S. Fanselow, M. Zhao, S. Romeis, D. Drummer, K.E. Wirth, W. Peukert, Optimized polybutylene terephthalate powders for selective laser beam melting, Chem. Eng. Sci. 156 (2016) 1–10. doi:10.1016/j.ces.2016.09.009.

# 506658 Creating Tuneable Agglomerates via 3D Printing

**Abstract:**

It is essential to offer a universally accepted way to systematically evaluate agglomerate breakage propensity and mechanisms. Discrete Element Method (DEM) simulation is a commonly used tool but is limited by the lack of identical, controlled agglomerates to test and validate simple models, let alone replicate the complex structure of real industrial agglomerates.

Our research presents a systematic approach to produce 3D printed agglomerates with tuneable properties. Agglomerates with different mechanical properties and structures were designed and printed by an Objet500 Connex 3D printer. The tensile and shear strength of inter-particle bonds was characterized using Instron tester. Breakage results of agglomerates with simple cubic structures and the design of agglomerates with irregular structures were presented. This 3D printing method will allow more rigorous testing of agglomerate breakage models. Discrete Element Method (DEM) has been used to simulate the agglomerate crushing using the Timoshenko Beam Bond Model (TBBM) with bond properties matching the 3D printed agglomerates. For both two agglomerate structures used in this research, the simulation and experiment showed similar breakage patterns. However, the simulation results significantly underestimated the compressive loads of spherical random structured agglomerates. In order to improve the simulations of compressive granule breakage, the influences of bond geometry definitions and non-linear behaviour of the bond contacts need to be explored.

# 508699 Effect of Particle Size Distribution and Wettability on Penetration Behavior, Granule Formation and Granule Properties in Single Drop Granulation

**Abstract:**

In wet granulation, single drop granulation (where one drop forms one granule) can simplify the complexity and reveal the fundamentals of the interaction between the liquid and the powder bed. Combining drop penetration time, granule formation mechanisms and the characterization of granule properties is a novel study to achieve a comprehensive map of the single drop granulation process, which is relevant in particulate applications like pharmaceuticals, detergents, agricultural chemicals and catalysts.

In this work, single drop impact of liquid on a static powder bed was studied to investigate the granule formation mechanism, droplet penetration time, and the characterization of granule properties (morphology, internal structure and content uniformity). Water was used as the liquid binder and two pharmaceutical powders, microcrystalline cellulose (MCC) (mean particle size ~54.8 Î¼m, contact angle with water ~36Â°) and acetaminophen (APAP) (mean particle size ~9.74 Î¼m, contact angle with water ~89°), were mixed to make heterogeneous powder beds with varying compositions. The complete drop impact and penetration was recorded with a high-speed camera. The granule morphology, its internal structure and content uniformity were characterized by the prism method with image analysis, micro-CT and UV-vis spectrometry, respectively. Besides the current MCC and APAP system, MCC and APAP with other particle sizes and relative contact angles will also be studied. It is believed that the mean particle size of the powder bed is predominant in influencing the granule formation mechanism, drop penetration time, and granule properties.

# 508017 Atomically Deposited Sintering Aids: Assessing the Effects of Alumina Particle ALD on the Sintering and Performance of SOFC Electrolytes and Dental Ceramics

**Abstract:**

The crystal structure of zirconium oxide ($ZrO_2$) can be tailored by the incorporation of aliovalent dopants, most commonly, yttrium oxide ($Y_2O_3$). The concentration of dopant dictates the thermodynamically stable crystal phase where 3 mol%
Y
conducting cubic phase. To achieve suitable mechanical properties and ionic conductivity, Y-TZP and YSZ ceramics must be near theoretical density, requiring high sintering temperatures. Previous researchers have successfully lowered the required Y-TZP and YSZ sintering temperatures through incorporation of aluminum oxide ($Al_2O_3$) particles. While this technique has proven to reduce the sintering temperature in both systems, the presence of particulate $Al_2O_3$ inclusions in densified parts reduces the homogeneity of the material which may have deleterious effects on densification and microstructure. In this work, we report on the use of atomic layer deposition (ALD) to conformally coat individual Y-TZP and YSZ particles with the desired amount of $Al_2O_3$. Constant-rate-of-heating (CRH) experiments were conducted using a horizontal push-rod dilatometer to extract the activation energy of sintering and also gain mechanistic insights into the active diffusion mechanisms as a function of ALD film thickness. Transmission electron microscopy (TEM) and scanning electron microscopy (SEM) were used to characterize the as-deposited thin $Al_2O_3$ films and the effects of the $Al_2O_3$ films on grain growth and microstructure during densification. Ionic conductivity and flexural strength tests were conducted to determine the effect of $Al_2O_3$ thin films on relevant material properties. $Al_2O_3$ incorporation by ALD is shown to have unique phenomenological effects on the densification of Y-TZP and YSZ as compared with conventional $Al_2O_3$ incorporation using particle-based approaches.

# 508644 Effect of Particle Size Distribution and Operating Parameters on Conduction and Convection Heat Transfer Mechanisms in Rotary Drums

**Abstract:**

Granular materials undergo process steps that include transportation, drying, heating and chemical or physical conversion. In several instances, these materials need to be heated or cooled during processing, and rotary drums (kilns) are the most commonly used process equipment for this purpose. It is not always possible to conduct experiments using different materials and consider all process conditions to study the thermal mechanism due to the inherent complexity of these processes. Therefore, computational modeling and simulations have been extensively used to understand the heat transfer mechanism in rotary drums. To maintain simplicity and to overcome the challenges of simulating polydispersed particles, a common practice is to assume that the polydispersed particles with a narrow size distribution behave like monodispersed particles thermally. However, no effort has been made to validate this assumption. This work focuses on understating the effect of polydispersed particles on the conduction and convection heat transfer mechanisms in indirectly heated rotary drums. It also aims to investigate the effect of rotation speed, fill level, particle size and their interaction effects on the conduction and convection heat transfer mechanisms in rotary drums. For this, the discrete element method (DEM) (using MFIX-DEM, an open-source computational fluid dynamics solver suite) simulation technique is used to analyze the thermal behavior of spherical silica beads, which are the most common catalyst support material.

# 507974 Selective Hydrogenation of Citral over Supported Pt Catalysts on Various Substrates

**Abstract:**

Highly dispersed platinum (Pt) nanoparticles (NPs) were deposited on various substrates by atomic layer deposition (ALD) for the reaction of citral hydrogenation. The substrates included multi-walled carbon nanotubes (MWCNTs), silica gel ($SiO_2$), commercial Î³-$Al_2O_3$, and ALD-prepared porous $Al_2O_3$ particles (ALD-$Al_2O_3$). Based on TEM and CO-chemisorption analysis, the Pt NPs on all substrates were around 1.3 nm, and Pt/MWCNTs, Pt/ALD-$Al_2O_3$ and Pt/Î³-$Al_2O_3$ showed similar Pt dispersion (75%-85%). But the Pt dispersion on $SiO_2$ was low (63%). In the reaction of citral hydrogenation, Pt/$SiO_2$ showed the highest activity due to the strong acidity of $SiO_2$ and the conversion of citral reached 82% after 12 hr with a selectivity of 58% of unsaturated alcohols (UA), including geraniol and nerol. Among all catalysts, Pt/MWCNTs showed the highest selectivity of UA (65%) due to its unique structure and electronic effect. In the cycling test, Pt/MWCNTs and Pt/ALD-$Al_2O_3$ catalysts were more stable than Pt/$SiO_2$ which could be attributed to the different interactions between the catalysts and the solvent.

# 508345 Numerical Simulation of Particle Classification in a Classifier Based on Coanda Effect

**Abstract:**

Efficient and precise classification of particles has been an important issue in many industrial applications as well as in academia. A relatively new particle classifier based on 'Coanda effect' was developed and reported to have many favorable features (Liang 2010). They include simultaneous multiple classification, simple design with no moving parts, and easy scale up to higher production rates. However, this classifier has attracted little attention so far, and our understanding of underlying mechanism associated with classification performance needs to be improved.

The purpose of this study is to investigate the detailed fluid flow and particle transport characteristics in a Coanda classifier using CFD (Computational Fluid Dynamics). The main focus is placed on the effects of working conditions (inlet flow velocity, particle diameter) and numerical method for turbulence dispersion of particles on the classification performance. As a numerical method, three different approaches are considered: k-ε model, k-ε model combined with DRW (Discrete Random Walk) model, LES (Large Eddy Simulation). In the first approach, standard k-ε model is adopted to obtain the mean flow and then particle trajectory is calculated based on the mean flow without consideration of turbulence dispersion. However, in the second approach using DRW model, the instantaneous flow velocity is constructed under the assumptions of isotropic turbulence in order to include turbulence effect. Finally, in LES, the instantaneous flow field is dynamically and accurately obtained at each time step. Numerical simulations are carried out for two-dimensional mean flow in case of k-ε model, but for three-dimensional unsteady flow in LES. Particles are tracked using a Lagrangian approach based on one-way coupling. We consider talc particles with density of 2,860 $kg/m^3$ and diameter of 1 to 20 μm.

The Coanda classifier is mainly composed of three inlet channels, three outlet channels, and Coanda block. Inlet channels are divided into one main inlet and two control inlets. The main air jet containing particles is supplied with the velocity of 35 m/s through the main inlet. Two control air jets with velocities of 32 m/s (lower) and 40 m/s (upper) are also supplied through the control inlets for enhanced control.

The trajectories of moving particles obtained from the mean flow (k-ε model) are analyzed for different diameters. The main jet containing particles deflects toward the curved Coanda block due to Coanda effect as expected. Fine particles ($d \leq 2\mu m$) follow the flow streamlines and enter the lower outlet channel. However, coarse particles ($d \geq 8\mu m$) are detached from the streamlines and go into the upper outlet channel. This observation agrees well with the previous result (Liang, 2010).

Even though the mean flow and corresponding particle trajectories provide useful information, realistic particle passage do not take a form of single curve but they change in time due to flow turbulence. Therefore, particle trajectories are simulated again, turbulence dispersion being considered with DRW model. Due to turbulence dispersion, particles of the same diameter are scattered around the particle pathline obtained from the mean flow. As a result, particles of the same diameter can not be collected into a single outlet. For example, some particles (about 5%) of 1 μm enter the middle outlet even though most of them (about 95%) enter the lower outlet designed for collecting fine particles. This means that flow turbulence deteriorates the accuracy of classification.

Finally, LES simulations are performed to understand dynamic behaviors of fluid flow and particle transport. Instantaneous flow fields such as vorticity contours clearly show unsteady fluid motions inside the classifier. It is also shown that the generation of unsteady vortices leads to the scattering of particle trajectory.

A more detailed analysis including the classification efficiency will be given in the final presentation.

**Acknowledgements**

This research was supported by the National Research Foundation of Korea (NRF) under grant number 2017M2A8A4018482 and 2015R1D1A1A0105 9675.

**References**

Liang, S. (2010). *Chem. Eng. Comm.*, **197**, 1016-1032.

# 503491 Numerical Simulation of the Secondary Air Distribution Layer in the Gas Solid Fluidized Bed Based on Flunt Software

**Abstract:**

Due to the continuity beneficiation in the practical production, the gangue would accumulate on the distributor forming the gangue layer, which was called Secondary Air Distribution Layer and has vital effect on the gas distribution. In this paper, the ventilation characteristics of the secondary air distribution layer was analyzed. In addition, FLUENT software was used to study the effect of the modeled layer for the fluidity characteristics (e.g., pressure drop, bed height, volatility, etc.) in the gas-solid fluidized bed with the complex layer. The results showed that the complex layer could promote the secondary distribution of the airflow in the bed, which effectively inhibited the production of large bubbles and reduced the bed pressure drop fluctuation. Thus, itâ€™s of significance to provide synergy effect on the fluidization stability in the fluidized bed. Based on the above analysis, the introduced complex structure would be convenient to achieve the coal separation effectively.

# 507949 CFD-DEM Simulation of the Fluidization of Non-Spherical Particles in Fluidized Bed

**Abstract:**

In chemical and energy industries, fluidization concerning non-spherical particles (such as rod-like particle, disk-like particles, ellipsoidal particles, etc.) is often encountered, but research in this area is minimal. Yet understanding fluidization of non-spherical particles is crucial in many relevant industrial process. To investigate the fluidization of non-spherical particles, a CFD-DEM coupling algorithm is employed in this research to simulate a bubbling fluidized bed. The non-spherical particles are modeled by super-ellipsoids, multi-sphere models, and other particle models. The simulations results show that, the fluidization of non-spherical particles has significant differences from that of the spherical particles. The local void fraction and particle's projected face area perpendicular to fluid flow direction are directly related to the orientations of the non-spherical particles, which has a significant influence on the particle drag force and fluidization. Due to the significance of particle orientation, we further investigate the factors that influence the distribution of particle orientations. The results show that particle orientation depends on the fluidization time, fluidization velocity, and structure of the fluidized bed.

# 508320 CFD and DEM Simulation of the Cold Spray Process for Surface Coating with Fine Particles

**Abstract:**

The coating and functionalization of technical surfaces has been become increasingly important in the last two decades for various applications, since of its manifold possibilities for process optimization, as friction reduction, corrosion resistance or improved wetting behavior. Cold spray is a thermal spraying method for surface coating. It was developed in the late 1980s in Novosibirsk [1] and has gone through continuous optimization on certain applications since then. For example showed our previous study a reduction of the friction properties on steel pins, microstructured with fine Titania particles by an adapted spraying method [2]. In the cold spray process particles are dispersed in a gas stream, which is heated up to several hundred degrees and accelerated to supersonic velocities in a Laval nozzle. The particles impact on a substrate with about 800 m/s and form a strong bonding due to the high kinetic energy which is transferred to deformation and heat in the contact zone. In contrast to other thermal spraying methods, the temperatures occurring during the process are far below the melting temperatures of the used materials, which makes the process suitable also for thermal sensitive powders. The quality of the produced coating layers depends highly on the properties of the powder and substrate as well as on the nozzle design and the process parameters as initial temperature and particle velocities. Due to the high gradients, the high occurring velocities and the small size of the particles, typically in the range of 1 to 50 μm, the contact zone of impacting particles can only be measured indirectly and with destructive methods. Therefore, still not all phenomena which lead to a strong bonding in the contact area and a high quality coating are completely understood. For the study of these mechanisms two different simulation approaches have been used:

a) Simulation of particle and gas flow with Computational Fluid Dynamics (CFD)

For a better understanding of the spray process, the nozzle, the adjacent flow areas and the substrate were computed with a CFD method. The trajectories of single particles were calculated by an Euler-Lagrange approach. For the flow the k-Ï‰ SST model was used. Local temperatures in the substrate were calculated by the law of heat conduction. The impact conditions dependent on the nozzle geometry and the particle properties. With the local temperature distribution in the substrate and in the particle, precise impact conditions could be estimated. A study has been conducted to vary the parameters of the spray process, temperature and pressure as well as the particle properties as shape factor, density and size. Moreover the nozzle geometry has been varied in the simulation to optimize the spray process. The simulation results are compared to images of coated surfaces obtained by Scanning Electron Microscopy (SEM) for validation.

b) Simulation of particle-surface impact behavior with Discrete Element Method (DEM)

The DEM is widely used for investigations of particular processes. The advantage of this method is the detailed calculation of particle-particle or particle-wall contacts to obtain the important parameters as forces and deformations. The contact force in the particle wall contact is also dependent on the surface morphology. Therefore the surface topography of two real technical surfaces, a polished and an unpolished Titanium substrate were measured by Scanning Probe Microscopy (SPM) imaging with a nanoindenter (TI Premier, Hysitron). The surface topography was discretized and implemented in a DEM model. A parameter study has been conducted to determine the distribution of forces and contact angles occurring at the particle impact. Different particle sizes and impact velocities were studied.

The cold pray process dependents on various parameters which affect each other. Since of the high gradients and the small scales occurring during the process experimental measurements are often difficult. With this numerical study different parameters could be considered uncoupled of the other parameters. The results can be used for a better process understanding and an optimization of commercial cold spray systems.

[1] Alkhimov, A. P., Papyrin, A. N., Kosarev, V. F., Nesterovich, N. I., Shuspanov, M. M.: Gas dynamic spraying method for applying a coating; US Patent, 5302414; 1994

[2] Buhl, S., Schmidt K., Sappok D., Merz, R., Godard, C., Sauer, B., Kerscher, E., Kopnarski, M., Antonyuk, S., Ripperger, S.: Surface structuring of case hardened chain pins by cold-sprayed micro particles to modify friction and wear properties; Particuology 21, 2015

# 508568 Two-Fluid Validation of Constituive Models for the Simulation of Cylindrical Particles

**Abstract:**

Two-fluid simulations are a popular tool in the modeling of large processes. The constitutive models that close the governing equations are based in granular kinetic theory and typically assume that the particles are spherical. In reality, very few processes utilize spherical particles, hence there is a need for constitutive models that describe more complex shapes. Previous work used discrete element method (DEM) simulations to create a solid stress[1,2] and collisional dissipation rate[3,4] model for cylindrical particles that can replace the current spherical models. These new models have been added to MFiX and this work focuses on their validation. Cylindrical particles of various aspect ratios and solid densities were used in cratering and hopper discharge experiments. Simulations are designed to recreate the experiments in order to compare the results for crater depth and hopper discharge rate. The model effectiveness is assessed with regards to assumptions made concerning particle rotation and preferential alignment.

1. Guo Y, Wassgren C, Ketterhagen W, Hancock B, James B, Curtis J. A numerical study of granular shear flows of rod-like particles using the discrete element method. *J Fluid Mech*. 2012:1-26.

2. Berzi D, Thai-Quang N, Guo Y, Curtis J. Stresses and orientational order in shearing flows of granular liquid crystals. *Phys Rev E*. 2016;93(4):40901.

3. Buettner KE, Guo Y, Curtis JS. Using the Discrete Element Method to develop Collisional Dissipation Rate Models that Incorporate Particle Shape. *AIChE J*.

4. Berzi D, Thai-Quang N, Guo Y, Curtis J. Collisional dissipation rate in shearing flows of granular liquid crystals. *Phys Rev E*. 2017;95(5):50901.

# 509002 The Comparsion of Coarse Grained CFD-DEM for Simulating the Dense Bubbling Fluidized Bed" Requires Immediate Action

**Abstract:**

CFD-DEM is an effective tool to study the hydrodynamics in dense bubbling fluidized bed. However, the expansive cost of calculating the particles motion is not easy for its application in industrial application or dense particles system. Thus several coarse grained models (CGM) were put forward to simplifying the calculation of particles and decreasing the particles motion, such as PSM, SPA, MSPA,CGEE, EMMS-DPM where a group of particles are set as a parcel or coarse particle to reduce the computation. In this study the experiment and simulation test conducted to test five coarse grained methods. The result indicated that the models increasing the gravity force, pressure gradient force and drag force linearly with the coarse grained ratio (CGR) can be effective to ensure similar macro behavior between original system and coarse graining system. The simulation data of the solid fraction distribution and bed expansion ratio reflected good agreement as experiment. But the bubbles structure became less clear as increasing the CGR. Furthermore, the power spectrum density (PSD) and Wavelet analysis of the pressure signal were used and showed that SPA and EMMS-DPM gave better agreement result in term of dominant frequency and energy ratio at mesoscale feature. Finally the computational efficiency for different models was compared and the models of EMMS-DPM, PSM and CGEE have higher computation efficiency than SPA and MSPA relatively if the collision model of spring-damping were adopted.

# 509153 Stability Analysis of Uniform Gas Solids Flow

**Abstract:**

Uniform distribution of gas solids flow should be ensured to gain sufficient performance of fluidized beds. The highly chaotic entropy dissipation across solids bed will trigger non-uniform fluidization, although distributor plays an important role in obtaining homogenous gas stream. Hence, total solids pressure loading (Î¦$_T$), the pressure drop ratio between solids bed and distributor, is introduced here to present their interaction. The boundary between uniform and non-uniform fluidization can be detected by stability analysis based on Prigogineâ€™s minimum of entropy production principle, and this criterion for *N* parallel paths can be simplified into calculating the second derivative of pressure drop through one path with respect of Î¦$_T$. Furthermore, a phase diagram, illustrating the effects from operational parameter (Î¦$_T$), geometrical characteristics of distributor ($C_d$) and property of solids (*Ga*) on instability of uniformity, is attained to draw some useful conclusions of robust uniform fluidization.

# 509309 Numerical Simulation of Particle Sedimentation Related to Nuclear Safety By CFD-DEM Algorithm

**Abstract:**

**Numerical Simulation of Particle Sedimentation related to Nuclear Safety by CFD-DEM Algorithm**

**Byoungcheol Hwang**[1], **Kiyofumi Moriyama**[1], **and Hyun Sun Park**[1*]

[1]*Division of Advanced Nuclear Engineering (DANE), POSTECH, Pohang, Korea*

**corresponding author: hejsunny@postech.ac.kr*

## Abstract

The Discrete Element Method (DEM) is a technique to treat particle-particle interactions, and is used in many industries such as civil engineering, powder processing, pharmacy, and so on. Coupling of this method with Computational Fluid Dynamics (CFD) is needed in problems where the particles interacts with fluid significantly, such as bubble column reactor, mineral processing, and so on.

The simulation of important phenomena in nuclear power plant accidents that involves significant core damage and melting, i.e. so called severe accidents, is also one of the topic where the CFD-DEM is useful. After the Fukushima accident in 2011, the severe accident analysis and preparation of counter measures to mitigate the accident consequences became regulatory requirements in many countries. One of the measures for the cooling and stabilization of the molten core is the flooding of the cavity under the reactor vessel where the dropping core melt is caught in a water pool. It is expected that the molten core breaks up into particles of several millimeter sizes and accumulates on the floor, then makes a particle bed, which is hopefully coolable. In this situation, the assessment for the coolability of the particle bed is very important, and a number of experimental and analytical approached suggest that the internal/external structure of the particle bed affects strongly the coolability. The CFD-DEM is a promising candidate for an analysis method that can clarify such details of the particle bed formation.

We tested the applicability of the multiphase CFD-DEM code, an open-source developed by CFDEM®coupling in 2011, for the simulation of particle sedimentation and particle bed formation under water. Our focus in the present work is on the prediction of the particle falling characteristics and particle bed shape. Although the real condition during a severe accident includes very high temperature particles (>1000K) and boiling water (two-phase flow) in a large scale, we started with a simulation of much simple experiment cases without heat transfer or gas phase, with spherical particles only as the first step for the validation of the method. The experimental data with solid particles of various diameters (4 mm, 6 mm) and densities (3600, 6000, 7800 kg/m$^3$), particle release nozzle diameters (20, 30, 40 mm) were referred to. The results of numerical simulation were compared with the experimental data in terms of particle falling time, maximum particle dispersion angle, particle average falling velocity, particle bed distribution, and so on. The particle falling time was well predicted by the simulation with the maximum deviation 5%. The maximum particle dispersion angle by the simulation significantly deviated from the experimental data probably due to the influences of initial disturbances. The particle average falling velocity was overestimated. The final particle bed profile by the simulation showed agreed with the experimental observation accurately in the peripheral region, but the height of the center part was underestimated. The fluid dynamic drag model and the particle-particle contact model would need improvement for better accuracy. The present work confirmed the validity of CFD-DEM algorithm on the application for the particle sedimentation in a simple liquid pool. The particle-pool interaction including the gas phase will be the next step.

**Keywords:** CFD-DEM; particle-fluid interaction; particle sedimentation; nuclear safety; severe accident

# 509440 Application of Extedned Discrete Element Method to the Melting Process of a Packed of Particles

**Abstract:**

### Application of eXtedned Discrete Element Method to the melting process of a packed of particles

Mehdi Baniasadi, Maryam Baniasadi, Bernhard Peters, Xavier Besseron

Université du Luxembourg, Faculté des Sciences, de la Technologie et de la Communication, Luxembourg

e-mail: mehdi.baniasadi@uni.lu, maryam.baniasadi@uni.lu, bernhard.peters@uni.lu, xavier.besseron@uni.lu, web page: http:/www.xdem.de

## ABSTRACT

The phenomena of melting takes place in many industrial processes, such as metal processing, environmental engineering and thermal energy storage systems. The problem of modeling of solid-liquid-gas flow including physical and chemical conversions of particles such as melting is the subject of intense research in industrial processes. These systems are difficult to model with numerical analysis due to the complex interaction between the different fluids and granular phases condering heat, mass and momentum excahnge. Basically models used to predict multiphase flows with particles are Eulerian-Eulerian and Lagrangian-Eulerian. The difference between the two models is based on different treatment of the movement of the particles. The particles are tracked individually in Lagrangian-Eulerian approach while they are considered as a continuum medium in Eulerian-Eulerian. Lagrangian-Eulerian is closer to the real physical processes rather than the Eulerian-Eulerian model for granular flows with particleâ€"particle interaction[1].

In this paper, our main focus is on the modelling of the melting process for a packed bed of particles subject to a fluid flow using a numerical technique based on Lagrangian-Eulerian so-called eXtended Discrete Elemnet Method (XDEM)[2]. In this approach, the particles are resolved as discrete phase coupled via heat, mass and momentum transfer to the surrounding fluid phase. The Discrete Element Method (DEM) coupled by heat transfer is used in order to evaluate temperature distribution, melting rate as well as velocity, oriantion and position for each particle individually. In addition the approach allows to be linked with continues numerical approaches such as Computational Fluid Dynamics (CFD) in order to solve fluid fields. The model has been validated by comparing predicted results with existing experimental data for melting of a single ice particle as well as a packed bed of ice particles in the horizontally flowing water. Fluid-particle and particle-particle intractions has also been considered. In addition, the model has the capability to be extended to the packed bed of particles with different size and properties to produce different liquid phases.

## REFERENCES

[1] Dierich, F., Nikrityuk, P. A. & Ananiev, S. 2D modeling of moving particles with phase-change effect. *Chem. Eng. Sci.* **66,** 5459â€"5473 (2011).

[2] B. Peters, Thermal Conversion of Solid Fuels (WIT Press, Southampton, 2003).

# 514248 Real-Time Interpretation of Chaotic Characteristics and Instability of Dense Gas Solid Fluidization

**Abstract:**

The establishment of instability monitoring system targets at analyzing the chaotic characteristics and evaluating the instability of the fluidization process for dense gas solid fluidization. Experimental pressure fluctuation and pressure drop measurements were interpreted from dense gas solid fluidization system based on chaotic time-series analysis. The pressure fluctuation signals were filtered in terms of attenuating the noise signals. The filtered pressure signals were analyzed with respect to attractor reconstruction. Based on the reconstructed attractor, the maximum Lyapunov exponent and entropy were investigated and estimated under different bed heights and air velocities. Results show that the maximum Lyapunov exponents udner different bed heights and air velocities are positive, which indicates that dense gas solid fluidization exhibits chaotic characteristics. With the increase of bed height, estimated entropy of dense gas solid fluidization increases with the increase of bed height under identical air velocity, while it varies slightly with the increase of air velocity when the air velocity alters from minimum fluidization velocity to 1.36 times of the minimum fluidization velocity.

# 508510 Glucan Particles As Potential Carriers of Natural Flavonoids for Treatment of Idiopatic Inflammation Diseases

**Abstract:**

One of the novel methods of pharmaceutical formulation is the loading of the drug into biocompatible microparticles, preferably extracted from natural sources. This approach offers several potential advantages, *e.g.* increased bioavailability (the ratio of the administered drug, which actually enters the bloodstream) compared to conventional drug delivery methods, enabling the option of oral administration even for parenteral drugs. Multiple principles of the loading process are possible – drug encapsulation, adsorption onto the particle surface or precipitation inside the pores a hollow particle. The presented work focuses on precipitation of a drug from an organic solvent inside glucan particles. Glucan microparticles are ellipsoid porous shells, which are obtained from cell walls of *Saccharomyces cerevisiae* (baker´s yeast). The particle size is 2-5 micrometers. Cell walls of glucan particles are mainly composed of β-1,3-glucans. In order to prevent inflammatory reaction in organism, the yeast organelles are removed by a series of alkaline and acidic extraction steps. Subsequently, present lipopolysacharides are extracted using different organic solvents.
Lymphatic cells are mainly present in inflammation area in organism where they are responsible for progress of inflammation. Glucan particles (GPs) interact with lymphatic cell´s receptors on their surface and after that β-1,3-glucans are phagocytosed by the lymphatic cells. Therefore, GPs represent an effective oral delivery system for transporting of biological active substances. Curcumin is an anti-inflammatory flavonoid, which occurs in plants of family *Zingiberacaea.* In our project, curcumin was delivered to an inflammation in organism using GPs. Loading of curcumin into GPs was enabled via precipitation of curcumin from organic solvents. Curcumin dissolved in organic solvents was added to the GPs and through pores penetrated into cavity of glucan shell. Next, the solvent was removed and the curcumin was precipitated in glucan microparticles. GPs with the flavonoid inside were observed using confocal microscopy and SEM (scanning electron microscopy). The achieved content of curcumin was successfully analyzed by other methods, which can not be mentioned directly due to article in preparation. Furthermore, the curcumin-loaded GPs were tested *in vitro* on macrophage like cells and *in vivo* on laboratory mice in order to evaluate their effectivity. The obtained results show that these composites are promising materials for biomedical applications such as efficient treatment of inflammatory diseases (e.g. idiopatic inflammatory bowel disease, rheumatoic arthritis).

# 508601 Use of Advanced Imaging Techniques in Tablet Disintegration Study

**Abstract:**

Certified testing of tablet disintegration and dissolution provides us useful information. However, there are many more that are hidden due to techniques' restrictions. Such information is e.g. mechanism of disintegration, change of surface morphology or uniformity of the substances' distribution in the tablet. Therefore, there are many techniques used to obtain the remaining information such as Magnetic Resonance Imaging (MRI), Fourier Transform Infra-Red (FTIR) or Raman spectroscopy and microscopy, Terahertz spectroscopy, Scanning Electron Microscopy (SEM) and other.

Presented work uses tablet made by direct compression of binary mixture of ibuprofen and lactose (based on previous results there was 0.2 - 0.5 w/w of ibuprofen content used) pressed at different compaction pressures (low, medium and high). The tablets were subjected to several solutions and were studied by techniques that have insight into the tablet (MRI) and ones that can see the surface (FTIR and SEM).

MRI experiments provided us with several interesting and several expectable results. As the content of ibuprofen was increasing so was increasing the dissolution time and the wetted layer was decreasing. Similar results were observed when the compaction pressure was increased. When the tablets were submerged to dissolution limiting solutions (saturated lactose solution and solution saturated with both lactose and ibuprofen) the disintegration was severely limited and some tablets began to increase their volume. However, in case of 0.2 w/w ibuprofen tablets it didn't matter whether there was dissolution discriminating solution or not, it disintegrated anyway.

FTIR microscopy showed, that not only the amount, but also the shape and size of the particles from which the tablets were compacted can be determined. These findings were further supplemented by the dry state information obtained by Scanning Electron Microscopy after different dissolution times. Further the FTIR-ATR imaging technique helped to understand the basic mechanisms of the tablet disintegration and to study the dissolution and water penetration into the model ibuprofen-lactose tablets.

# 508632 Correlation Study between Liquid Penetration and Mechanical Properties of Pharmaceutical Tablets


**Abstract:**

Main phenomenon influencing treatment of patient in pharmacy is amount of active pharmaceutical ingredient dissolved in medium such as water or blood. Even though this isa key step to achieve desirable medical effect it is not the first one occurring.The first step is governed by surface properties such as wettability and roughness and the first contact point is followed by liquid penetration which is highly dependent not onlyon type of used liquid, ionstrength, liquid density but also on mechanical and structural properties of tablets such as porosity, structure strength or particle bonds.Through many analytical methods we can describe each of thesesteps individually. Aim of this study is to find proper correlation amongseveralmethods such as sessile drop method, texture analysis (TA) and magnetic resonance imaging (MRI) of tablet during dissolution.

For experiments binary mixture of tabletwas used. Tablets consist of lactose (hydrophilic) and ibuprofen (hydrophobic)with variable content (from 0.2 to 0.5 w/w ratio) and were also variable in compaction pressure used on them (from low â€“ 5 kN through 10kN as middle to 18 kN as high).Three different solutions were used according to their saturation with primary substances to prevent the tablet from dissolving. Surface properties were evaluated on contact angle and soaking rate of liquid, both measured by sessile drop method. Texture analysis was performed by recording the material resistance during dry and wet state of tablet. Then the data was processed and trends were obtained.Tablets during the MRI experiments were submerged into dissolution medium and the proceeding wetting front was measured. Later the images were processed via image analysis and trends were obtained.

Results obtained from surface experiments show that higher the ibuprofen content is the more hydrophobic surface appears to be and therefore the contact angle is increasing. Contact angle is independent on manufacturing pressure but soaking rate on the other hand depends strongly.MRI images show that this technique is not suitable for rapidly disintegration tablets, where the calculated error is within experiment conditions. On the other hand, tablet with prolonged dissolution are well measured. Results from MRI experiments also state that local saturation with substances supress dissolution of tablet. Texture analysis of tablets show that they undergo immediately structural weakening after being poured with solution independently on type of solution. Rate of this structural change mainly depends on solution's acidity. Thenthe tablet decreases its resistance in time which can be interpreted as relaxation rate. With increment of ibuprofen or saturation with primary substances this rate decrease, which means that dissolution is suppressed. Correlation between MRI and relaxation rate was done. Results shows that the correlation is mainly dependent on texture analysis measurement. On the other hand, correlation between MRI and soaking rate appears to be more tentative.

Experiment characterising surface, structural and liquid penetration properties were investigated by three different techniques. Obtained results were evaluated and put to mutual correlation. From liquid penetration point of interest there is correlation between MRI and contact angle method. Also, a correlation between MRI and texture analyser was found.

# 508642 In Situ Drug Amorphisation By Microwave Irradiation Stabilized By Mesoporous Silica

**Abstract:**

One of the common issues in the pharmaceutical industry is the poor solubility of the newly developed active pharmaceutical ingredients (APIs). The solubility might be significantly enhanced by changing the form of the API from crystalline to amorphous form. The amorphous state is characterized by lower thermodynamic stability and thus better dissolution properties leading to better bioavailability. In addition, the amorphous API in a dosage form (e.g. tablets) must also remain stable and not recrystallize for at least the shelf life of the drug product (usually 2 - 3 years). Various methods are being utilized to prepare and stabilize amorphous formulations in the industry, however they usually involve difficult formulation steps or suffer from poor efficiency, low stability or can be patent obstructed.

In this work a stable amorphous formulation is prepared simply using the standard wet and dry granulation process or direct tableting. Crystalline form of API is used during the initial process. The amorphous form is then prepared *in situ* and â€œon demandâ€ either during manufacturing, or by the end user by applying microwave radiation (MW).

To stabilize the API in an amorphous form after the MW, multi-structured mesoporous silica particles were added to the formulation. The particles were prepared using an emulsion synthesis of octylamine in tetraethyl orthosilicate. Using ethanol as an inhibitor instead of the commonly used nitric acid as a catalyst reduces the silica primary nano-particles nucleation rate. The formed nuclei grow larger and then self-assemble around the octylamine droplets and form hollow micro-particles. The prepared microparticles were characterized using the Scanning electron microscopy (SEM) and the Transition electron microscopy (TEM). To characterize the porous structure, nitrogen sorption (BET) analysis was conducted. The results clearly show the particle assemblies contain macropores (cavities), mesopores and micropores and possess very high specific surface area and also very large pore volume. Together with the thermal and chemical stability of silica these properties make the particles well suitable for preserving the melted API in an amorphous form after the MW.

As a proof of concept, initially powder mixtures and later simple tablets composed of Ibuprofen (model API), silica micro-particles and a filler excipient were experimentally investigated. The *in situ* amorphization of ibuprofen was conducted using a commercial microwave oven. The heating properties of the oven, i.e. the spatial distribution of microwave heating was characterized. Various microwave power input and varying composition of the mixtures (excipient/Ibuprofen ratio) were tested with respect to the melting point of the Ibuprofen. The change from the crystalline to amorphous structure due to MW heating was analyzed by x-ray diffraction (XRD), differential scanning calorimetry (DSC) and FTIR-ATR techniques. The dissolution profiles of ibuprofen from both MW untreated and treated tablets were measured using the High Performance Liquid Chromatography (HPLC).

Based on these initial results, fully formulated tablets were prepared with various API dosage strength using a tablet press. Same compression force and same dwell time was used for all of the tablets. The final tablet composition includes excipient capable of absorption of the MW, which ensures a uniform heating in the whole volume of the tablet. Next it contains the silica microparticles, which by capillary forces absorb the API once melted and stabilize it after cooling down.

Lastly, the tablets contain standard filler, lubricant and disintegrant excipients. After the MW application the tablets change their structure due to the melted API intake into the silica micro-particles, forming visible pores in the tablet structure. This has a notable impact on the dissolution and hardness of the resulting tablets. For the final tablet characterization SEM, Texture Analysis, FTIR and Magnetic Resonance Imaging were used.

# 508646 Manufacture of Personalised Medicines By API Printing on Porous Tablets

**Abstract:**

Personalised medicine offers an opportunity to produce tailor-made medicaments for patients. In order to realize this vision, it is necessary to find an alternative approach to the conventional tablet design and production. One that will replace the existing mass production of drugs offering only a few dose options of the active pharmaceutical ingredient (API), while at the same time it must meet the strict reproducibility and a defined relationship between composition and release rate of active substances requirements.

One of such approaches is to print the API on demand on a solid substrate directly when it becomes needed. Appropriate dosing of API and the possibility to even combine several APIs into a single medicament improves the variability of the dosages available for prescription and reduces the overall number of pills the patient has to take. The goal of this work is to find a simple, automated and cost-effective method of preparing personalised medicaments, which could take place directly in the pharmacy shops or at home.

The method in this work consists of a precise dosing of a dissolved API into an already prepared substrate â€“ porous placebo tablets with high sorption capacity. The tablets are prepared using direct compression in a tablet press from a simple physical mixture of a binder, filler, disintegrant and lubricant. As a filler, mesoporous silica particles were used. The high pore volume provides the free space for API sorption into the tablets and the small size of the mesopores ensure the API is deposited in an amorphous form.

The prepared particles used in the formulation were characterized using Transmission electron microscopy (TEM), Dynamic light scattering (DLS) and nitrogen sorption analysis (BET). The prepared tablets hardness and uniformity was characterized both prior and after loading. The tablets after loading were characterized using the X-ray powder diffraction and the Fourier transform infrared spectroscopy to determine the API form and deposition uniformity, followed by dissolution tests to confirm the applied dosage strength and to measure the release profile. So far, several BCS Class II APIs were successfully infused into the substrates using precise on-demand dosing. Impregnation of various doses of API were accurate. Later characterization revealed no crystalline API were present and the dissolution profiles were very satisfactory.

# 509441 Understanding Phase Transition of Acetaminophen in the Bulk and Surface of Acetaminophen

**Abstract:**

Pharmaceutical solids processing often lead to the generation of new phases in the bulk and on the surface of the processed drugs. Phase transformation of the API can potentially change the physiochemical properties and influence the drug performances including solubility, dissolution and stability. To detect these changes, analytical tools such as XRD are usually employed. However, these techniques remain suitable for bulk analysis. Whereas, phase transition on the surface remains hard to detect due to the lack of literature, researches in this area and the absence of surface analytical techniques. The aim of this work is to investigate different phases (pure crystalline, partially crystalline partially amorphous and amorphous) of Acetaminophen drug by combining bulk and surface analytical techniques. For bulk analysis, FTIR and XRD have been employed. The former shows no significant differences between Acetaminophen various phases. Whereas the latter confirmed the presence of different phases indicated by a remarkable variation in the absorbance intensity the peak width. Where, high and sharp peaks obtained for the pure crystalline Acetaminophen, fewer peaks observed for semi crystalline Acetaminophen and a single broad peak observed for the amorphous Acetaminophen. For the surface analysis, Nitrogen K near-edge X-ray absorption fine structure (NEXAFS) spectra of a series of NEXAFS indicates that the precise determination of IPs and Ï*f** shape resonance energies is sensitive enough to distinguish between the different phases of Acetaminophen. DFT calculations provides information on the molecular orbitals involved during electronic excitation occurred as a result of NEXAFS experiment.

# 509599 Treatment of Cystinosis through Vitamin E Modified Silicone Hydrogel

**Abstract:**

Cystinosis is an orphan disease caused by a genetic mutation that leads to deposition of cystine crystals in many organs including cornea. Ophthalmic manifestation of the disease can be treated with hourly instillation of cysteamine eye drops. The hourly eye drop instillation is tedious to the patients leading to poor compliance and additionally, significant degradation of the drug occurs within one week of opening the bottle, which further complicates this delivery approach. Our work focuses on designing a contact lens to treat the disease with improved efficacy compared to eye drops. Our goal is to design a lens that is safe and that can deliver a daily therapeutic dose of cysteamine to the cornea while retaining drug stability. We show that cysteamine diffuses out rapidly from all lenses due to its small size. The rapid release of cysteamine and other ophthalmic drugs can be prolonged significantly by integration of vitamin E aggregates inside of the polymer matrix of silicone hydrogels. The increase in release duration increases as square of the volume fraction of the loaded vitamin E fraction which leads to high efficacy and suggests that the mechanism of attenuation in drug release is formation of nanosized barriers. The increase in duration depends on the type of the contact lens. The barriers are not effective in single-phase hydrogel lenses while the effect in the bi-phasic silicone hydrogels depends on the type of the contact lens. Vitamin E incorporation increases the release duration of both ACUVUE®OASYS® and ACUVUE® TruEye™ but the effect is more pronounced in TruEye™ likely due to the low solubility of vitamin E in the lens matrix and higher aspect ratio of the barriers. This result along with the high aspect ratio seen through SEM imaging in silicone hydrogels suggests that barriers could be forming at the interface of the silicone and hydrogel phases. The presence of vitamin E has an additional beneficial effect of reduction in the oxidation rates, likely due to a transport barrier between the oxygen diffusing through the silicone channels and drug located in the hydrogel phase. Based on this study, both Acuvue®OASYS® and ACUVUE® TruEye™ can be loaded with vitamin E to design a cysteamine eluting contact lenses for effective therapy of cystinosis. The lenses must be worn for about 4-5 hr. each day, which is less than the typical duration of daily-wear. Vitamin E loaded contact lenses could be very useful for treating many other ophthalmic diseases as well.

# 517683 Systems Integration for Dry Granulation Based Continuous Pharmaceutical Tableting

**Abstract:**

Over the past decade, the pharmaceutical industry has seen significant progress in the mechanistic understanding of relevant unit operations, availability of multiple sensor technologies for inline monitoring of critical quality attributes and advances in data analytics to support manufacturing decision-making. These have enabled a systematic realization of continuous manufacturing for pharmaceutical tablets using the direct compaction route. However, the limitations and difficulty in handling of some particulate materials provide the impetus for investigating granulation routes to the continuous manufacturing of tablets.

In this work, the integration of dry granulation into a continuous tableting line is investigated. Specifically, the Process Analytical Technology (PAT) tools required for inline and at-line monitoring of the tableting line, and the semi-mechanistic models for the roller compactor and the tablet press are discussed. While the Kawakita model is used for the tablet press, the roller compaction process model is consisting of an integration of material balance and actuator dynamics from Hsu et. al (2010) with the modified Johansonâ€™s model presented by Liu & Wassgren (2016). The PAT tools, mechanistic equations, and automation system are integrated for measurement validation and data accuracy using data reconciliation in a structured ISA-95 network using DeltaV. The experimental data is presented from the operation of the continuous dry granulation line-based pilot plant at Purdue University.

# 509248 Modeling Granular Material Mixing and Segregation Using a Multi-Scale Model

**Abstract:**

A multi-scale model is presented for predicting the magnitude and rate of powder mixing and segregation in blenders. The model combines diffusion and percolation correlations from small-scale discrete element method (DEM) simulations with advective flow field information from large-scale finite element method simulations. The multi-scale model predictions closely match results from both large-scale DEM-only simulations and experiments, but they are much more computationally efficient than DEM-only simulations. Therefore, in contrast to DEM, the multi-scale model is well suited for predicting blending in industrial-scale systems.

# Modelling Granular Media using Dynamical Density Functional Theory

T. Hurst, B. Goddard & R. Ocone

## Motivation

Granular media is ubiquitous in industrial and physical problems. One of the characteristic qualities of granular media at the microscopic level is inelastic hard body collisions, which drive macroscopic behaviour. Modelling granular media is a complex and interesting problem. Solid-like phenomena make it difficult to approximate a granular system with a continuum model, while chaotic behaviour and inelastic collapse in hard particle systems introduce modelling challenges on the microscopic scale. We investigate uses of dynamical density functional theory (DDFT), augmented with data gained from microscopic models, for granular media simulation.

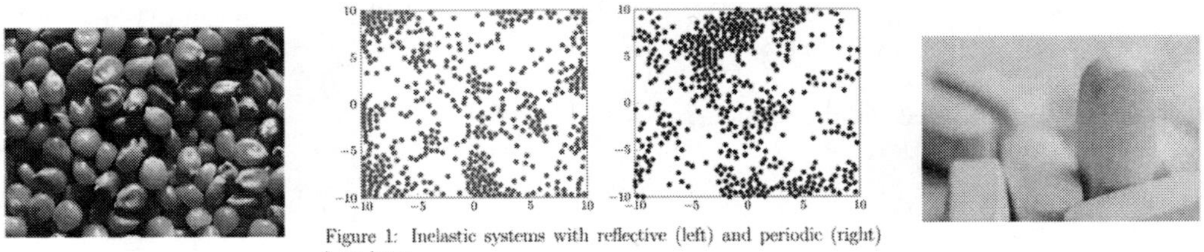

Figure 1: Inelastic systems with reflective (left) and periodic (right) boundaries

## Dynamical Density Functional Theory

To derive a DDFT, we start with a set of Langevin equations for $N$ hard particles of diameter $\sigma$ in $d$ dimensions with positions $\boldsymbol{r}^N$ and momenta $\boldsymbol{p}^N$:

$$\frac{d\boldsymbol{r}^N}{dt} = \frac{\boldsymbol{p}^N}{m}, \quad \frac{d\boldsymbol{p}^N}{dt} = -\nabla_{\boldsymbol{r}^N} V(\boldsymbol{r}^N, t) - \gamma \boldsymbol{p}^N + (mk_bT\gamma)^{1/2}\boldsymbol{G}(t), \tag{1}$$

on the hard sphere domain, $\mathcal{D}_N = \{(\boldsymbol{r}^N, \boldsymbol{p}^N) \in \mathbb{R}^{2dN} : \forall i,j, \|\boldsymbol{r}_i - \boldsymbol{r}_j\| \geq \sigma\}$, where $V(\boldsymbol{r}^N, t)$ includes external and interparticle potentials, and $\boldsymbol{G}(t)$ is an optional stochastic force term. The Kramer's equation is then a partial differential equation for its distribution function $f^{(N)}(\boldsymbol{r}^N, \boldsymbol{p}^N, t)$:

$$\frac{\partial f^{(N)}}{\partial t} + \frac{1}{m}\boldsymbol{p}^N \cdot \nabla_{\boldsymbol{r}^N} f^{(N)} + \nabla_{\boldsymbol{r}^N} V \cdot \nabla_{\boldsymbol{p}^N} f^{(N)} = \gamma \nabla_{\boldsymbol{p}^N} \cdot \boldsymbol{p}^N f^{(N)} + \gamma k_b T \nabla^2_{\boldsymbol{p}^N} f^{(N)}. \tag{2}$$

We integrate out all but one position and momenta in the Kramers equation, to find:

$$\frac{\partial}{\partial t} f^{(1)} = \mathcal{L}_1 f^{(1)} + \mathcal{L}^{2,\dots,N}_{\text{part}}(f^{(k)}) + \mathcal{L}^{(1)}_{\text{diss}} f^{(1)} + \mathcal{L}_{\text{coll}}(f^{(2)}), \tag{3}$$

where $\mathcal{L}_1$ introduces effects from external potentials, $\mathcal{L}^{(1)}_{\text{diss}}$ effects due to dissipation and friction, $\mathcal{L}^{2,\dots,N}_{\text{part}}$ effects from interparticle potentials, and

$$\mathcal{L}_{\text{coll}}(f^{(2)}) = -\sigma^{d-1} \int_{\mathbb{R}^d} d\boldsymbol{p}_1 \int_{\mathbb{R}^d} d\boldsymbol{p}_2 \int_{S_1^{d-1}} (\boldsymbol{p}_1 - \boldsymbol{p}_2) \cdot \omega f^{(2)}(\boldsymbol{r}_1, \boldsymbol{r}_1 + \sigma\omega, \boldsymbol{p}_1, \boldsymbol{p}_2, t) d\omega, \tag{4}$$

is a general collision term that arises due to the hard-sphere domain, when assuming only binary collisions. By taking the first two $\boldsymbol{p}$-moments and upon additional assumptions we arrive at a closed set of equations to model the dynamics of the system, a DDFT [1, 2]:

$$\frac{\partial}{\partial t}\rho = -\nabla \cdot (\rho \boldsymbol{v}), \tag{5}$$

$$\left(\frac{\partial \boldsymbol{v}}{\partial t} + \boldsymbol{v} \cdot \nabla \boldsymbol{v}\right) + \gamma \boldsymbol{v} = -\frac{1}{m}\nabla \frac{\delta \mathcal{F}[\rho]}{\delta \rho} - \sigma^{d-1} g_\sigma \int_{S_1^{d-1}} \rho(\boldsymbol{r} + \sigma\boldsymbol{\omega}/2)\rho(\boldsymbol{r} - \sigma\boldsymbol{\omega}/2)$$
$$\times \boldsymbol{\omega} \cdot \left\{\frac{k_B T}{m}\mathbf{1} + \boldsymbol{v}(\boldsymbol{r} + \sigma\boldsymbol{\omega}/2) \otimes \boldsymbol{v}(\boldsymbol{r} + \sigma\boldsymbol{\omega}) - \boldsymbol{v}(\boldsymbol{r} + \sigma\boldsymbol{\omega}/2) \otimes \boldsymbol{v}(\boldsymbol{r} - \sigma\boldsymbol{\omega}/2)\right\} \mathrm{d}\boldsymbol{\omega}, \tag{6}$$

where $\mathcal{F}$ is the well-studied Helmholtz free energy operator, which can include particle-particle interactions and external potentials, $\rho$ is the density, $\boldsymbol{v}$ is the local velocity and $g_\sigma$ is the value of the correlation function at $\sigma$, which must be determined. We choose to model $g$ using statistical data from microscopic simulation.

## DDFT Simulations

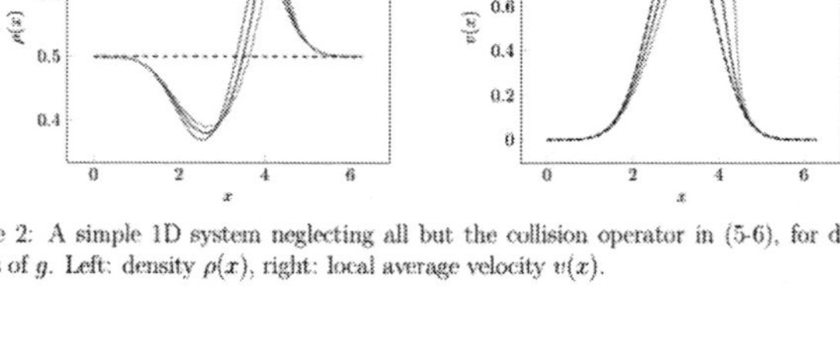


Figure 2: A simple 1D system neglecting all but the collision operator in (5-6), for different values of $g$. Left: density $\rho(x)$, right: local average velocity $v(x)$.

## Data-Driven Modelling of $g$

A system of $N$ hard spheres of diameter $\sigma$ in $d$ dimensions with positions $\boldsymbol{x}^N$ and velocities $\boldsymbol{v}^N$ obey

$$\frac{d\boldsymbol{r}^N}{dt} = \boldsymbol{p}^N, \quad \frac{d\boldsymbol{p}^N}{dt} = V(\boldsymbol{r}^N, t), \tag{7}$$

with a pairwise collision condition: if particles $i$ and $j$ are in contact, i.e. $\|\boldsymbol{r}_i - \boldsymbol{r}_j\| = \sigma$ and $(\boldsymbol{r}_i - \boldsymbol{r}_j) \cdot (\boldsymbol{p}_i - \boldsymbol{p}_j) < 0$, we instantaneously change the velocities via the rule:

$$\boldsymbol{p}_i^{out} = \boldsymbol{p}_i^{in} - \frac{1+\alpha}{2}\boldsymbol{\nu}_{ij} \cdot (\boldsymbol{p}_i^{in} - \boldsymbol{p}_j^{in})\boldsymbol{\nu}_{ij}, \quad \boldsymbol{p}_j^{out} = \boldsymbol{p}_j^{in} + \frac{1+\alpha}{2}\boldsymbol{\eta}_{ij} \cdot (\boldsymbol{p}_i^{in} - \boldsymbol{p}_j^{in})\boldsymbol{\eta}_{ij}, \tag{8}$$

where $\boldsymbol{\eta}_{ij} = \frac{\boldsymbol{r}_i - \boldsymbol{r}_j}{\|\boldsymbol{r}_i - \boldsymbol{r}_j\|}$ and $\alpha \in [0, 1]$ is the coefficient of restitution. For simple $V$, the Newton equations can be modelled using Event Driven Particle Dynamics (EDPD). The cell method and bounded increasing priority queues [3] reduce the computational effort to $\mathcal{O}(1)$ per collision. Using statistical methods for synthetic data (e.g. emulation [4]) we gain an understanding of the correlation function $g$ near equilibrium, which we can include in (6-7).

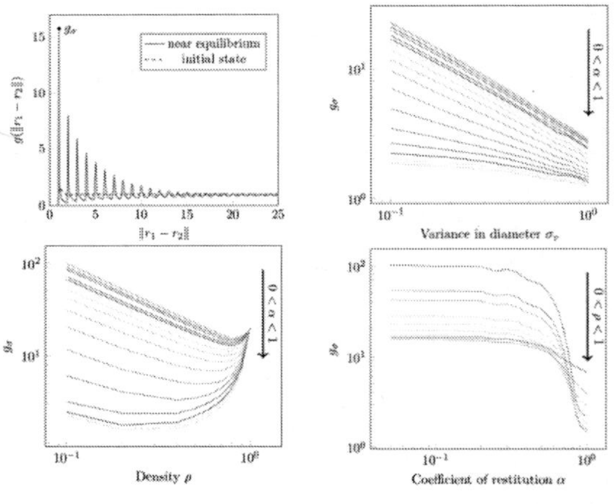


Figure 3: Top left: $g(\|r_1 - r_2\|)$ near equilibrium for a 1D system with $\rho = 0.5$ and $\alpha = 0.5$, using 10000 samples of EDPD simulations. The other three graphs show how $g_\sigma$ changes for different parameters.

## Future Work

We wish to include additional moments in the DDFT derivation, simulate DDFT in higher dimensions including interparticle potentials, and use EDPD in higher dimensions to calculate $g$.

[1] Archer, A., *J.Chem.Phys.*, **130**, 014509, (2009).
[2] Goddard, B. et. al., *Phys.Rev.Lett*, **109**, 1-5, (2009).
[3] Paul, G., *J.Comp.Phys.*, **221**, 615-625, (2011).
[4] O'Hagan, A., *Reliab.Eng.Syst.Saf.*, **91**, 1290-1300, (2006).
The support from EPSRC (grants EP/N034066/1, EP/L025159/1 and EP/L016508/01) is kindly acknowledged.

# 509555 A Study on Partially Wetted Particle Collisions with a Wet Wall


**Abstract:**

For the production of poly-ethylene, a gas-phase polymerization reaction is executed in a fluidized bed reactor. The polymerization reaction produces an excess of heat and the reactor is unable to efficiently remove the excess heat. The introduction of an Inert Cooling Agent (ICA) solves the temperature problem but also has a downside. Introducing the ICA in the reactor influences the particle collisions, leading to substantially different flow behavior. Besides the reduction of the restitution coefficient, the introduction of the liquid can result in adhesion instead of collision. This adhesion results in the formation agglomerates in the reactor and reduces the efficiency of the process. Due to the complex physics of this problem, Computational Fluid Dynamics (CFD) is a suitable method to gain insight in this phenomenon.

In order to study this complex system of partially wetted particle collisions a Direct Numerical Simulation (DNS) is used for incompressible flow. To model the coupling between the solid phases and the fluid phase, a second order accurate Immersed Boundary Method (IBM) is used (Deen et al.,2012). The interface between the fluids is tracked by a Volume Of Fluid (VOF) method (Baltussen et al., 2017). The contact-line dynamics are solved by imposing a contact angle as a boundary condition (Patel et al., 2017).

The simulated results will be compared with dry particle collisions on a wet wall (Tang et al., 2017) and with wet particle collisions on a wet wall.


**References:**


1. Deen, N.G., Kriebitzsch, S.H., van der Hoef, M.A. and Kuipers, J.A.M., Direct numerical simulation of flow and heat transfer in dense fluidâ€"particle systems. Chemical Engineering Science, 81 (2012): 329-344.

2. Baltussen, M. W., Segers, Q. I. E., Kuipers, J. A. M., and Deen, N. G., Cutting bubbles with a single wire. Chemical Engineering Science 157 (2017): 138-146.

3. Patel, H.V., Das, S., Kuipers, J.A.M., Padding, J.T. and Peters, E.A.J.F., A coupled Volume of Fluid and Immersed Boundary Method for simulating 3D multiphase flows with contact line dynamics in complex geometries. Chemical Engineering Science, 166 (2017): 28-41.

4. Tang, Y., Buck, B., Heinrich, S., Deen, N.G., Kuipers, J.A.M., Interface-resolved simulations of normal collisions of spheres on a wet surface. American Institute of Chemical Engineers, (2017).

# 507223 Experimental and Simulation Investigation on Arching Behaviour of Two Biomass Materials from a Wedge Shape Hopper

**Abstract:**

The interest in solid biomass has been increasing over the last decades for their potential as renewable energy sources, to be used in thermochemical processes to produce energy or gaseous fuels or biochemical processes to produce liquid fuels like ethanol. Particle elasticity and particle fibrous shapes are the main characteristics of biomass and both may be a possible cause of deviation of the biomass behavior from the Coulomb model.

One problem that happens during silo discharge is the formation of stable arches. This problem could be addressed through the knowledge of flow properties of bulk solids. However, common design methods for storage units to ensure flow, are based on Jenike analysis and have not yet been fully proven to work for biomass bulk solids. In particular Barletta et al. [1,2] measured the arching behavior of biomass materials in a wedge shape silo and compared the results with the method proposed by Jenike. In this research the critical outlet diameter that leads to arch free silo were experimentally measured for sawdusts and wood powders. These results were compared with the DEM simulation results and the model proposed by Jenike.

Realistic mechanical properties of biomass plays a very important role in DEM simulation. In order to find the mechanical properties of the two biomasses, the approach used by [3] were followed to measure the modulus of elasticity. Furthermore, to measure the coefficients of static and rolling frictions as well as the coefficients of restitution between biomass particle materials and also biomass particles with silo wall, DEM simulation of the Schulze shear cell were conducted. These coefficients were changed to have the best fitting shear stress profile between simulations and shear tester results.

After defining the material mechanical properties by the Schulze and the uniaxial tester, the discharging behavior of these two biomasses from a wedge shape silo were conducted by using EDEM software. Good agreement between experimental and simulation results of the critical opening diameter of arch free silo were observed. However, the critical opening diameter obtained by using the Jenike approach was higher than both simulation and experimental results.

References

[1] D. Barletta, M. Poletto, An assessment on silo design procedures for granular woody biomass, Chemical Eng. Transaction, 2013. doi:10.3303/CET1332369.

[2] D. Barletta, R.J. Berry, S.H. Larsson, T.A. Lestander, M. Poletto, Alvaro Ramirez-Gomez, Assessment on bulk solids best practice techniques for flow characterization and storage/handling equipment design for biomass materials of different classes, Fuel Process. Technol, 2015. doi:10.1016/j.fuproc.2015.06.034.

[3] M. Stasiak, M. Molenda, M. BaÅ„da, E. Gondek, Mechanical properties of sawdust and woodchips, Fuel, 2015. doi:10.1016/j.fuel.2015.07.044.

# 508500 Experimental and Numerical Study of Discharge Rates for Cohesive Particles in Hopper Geometries

**Abstract:**

## Introduction:

In many industrial applications silos are used for the storage of particulate materials and products. The properties of particles, the wall material, as well as the silo geometry determine the flow characteristics inside the shaft and hopper. The flow profile can be crucial, especially for perishable goods where dead zones can lead to rotting and health hazards. Jenike (1964) established a method to determine the critical hopper angle to prevent funnel flow as well as the minimal discharge orifice size to ensure no blocking or bridging of particles. Besides the critical silo parameters to create a reliable material flow in the first place, the discharge rate of the particles is another important step to determine the corresponding process.

## Overview and research questions:

During the last century, many empirical correlations of the mass flow rate for hoppers emerged. One widely used and expanded model is the Beverloo model for simple flat orifices with coarse and mostly cohesionless particles (Beverloo et al., 1961), which has been adapted for angled hoppers (Rose & Tanaka, 1956). Additionally, modifications were included for fine powders, which can be affected by air counterflow, induced by the vacuum due to the discharging particles (Crewdson et al., 1977).

Contrary to empirical correlations, Tomas (1991a, 1991b) derived an analytical approach of the outflow velocity for fine particles, which also includes the mentioned air flow due to the induced vacuum. By addressing the forces on an incremental particle bridge inside a hopper, differential equations for the stationary and instationary outflow velocities could be established. The underlying concept was also successfully expanded for vibrated geometries (Kache, 2010). One advantage of the model is that flow characteristics of the particles are used in the formulas. The cohesion and friction of particles are considered by the minimal discharge orifice diameter, which can be obtained with shear cell measurements.

Schulze (2006) proposes that the induced vacuum and therefore the counterflow during discharge is difficult to estimate; therefore leading to unreliable calculated mass flows, especially for very fine and cohesive particles. Kache (2010) used a modified Tomas model to consider those air drag forces by proposing an lower estimated pore diameter for the counterflow inside the bulk solid. While this reduces the calculated influence of the vacuum, the model is able to predict the discharge mass flow of very fine calcite MX10 for an industrial scale silo decently, which is indicative of an overestimation of the counterflow in the Tomas model.

The mentioned studies about particle flow therefore lead to the following questions for fine particle systems: Firstly, how do drag forces due to induced air flow inhibit the particle discharge. More information about this induced air counterflow could lead to models usable for Discrete Element Method (DEM) simulations in cases where a packed particle flow takes place. This could result in more realistic results without explicitly solving the fluid flow. Another question is how cohesion affects the mass flow rate and the transition between instationary and stationary flow. The effects of cohesion on potential fluctuations of the stationary flow rate, the change in porosity and flow profiles are additional issues. Also the impact of the filling height is relevant, since this can affect the particle flow rate as well (Heinrici, 2006).

In this contribution the experimental and numerical study of discharge rates for cohesive particles in hopper geometries was performed. The tested particulate systems consist of nearly monomodal spheres. The size of particles is varied to obtain different cohesive to gravitational force ratios. The flow properties of the different particle systems are measured by ring shear cells, which allows the calculation of the corresponding critical silo dimensions with the Jenike method. The test hoppers are constructed accordingly. During discharge experiments, the mass flow rate is measured by simple scales.

Besides experiments, the empirical and analytical equations for the mass flow rate are further analyzed for small particles and powders inside hoppers. Especially the model of Tomas and the appropriate equations for the counterflow are addressed thoroughly. The calculated critical discharge orifice size is used for the model to calculate the outflow velocities. Possible fluctuations during the instationary and stationary discharge are considered as well.

DEM simulations of particle discharge are also carried out to further analyze the drag forces by comparing the experiments and equations with the simulation results. The dimensions of the simulated hoppers are in line with the critical dimensions according to Jenike. The appropriate modeling of the the vacuum induced air counterflow is discussed.

**References:**


**Beverloo, W. A., Leniger, H. A. und Van de Velde, J. (1961),** The flow of granular solids through orifices. Chemical Engineering Science, (15), pp. 260-269.

**Crewdson, B. J., Ormond, A. L. und Nedderman, R. M. (1977),** Air-impeded discharge of fine particles from a hopper. Powder Technology, (16), pp. 197-207.

**Heinrici, H. (2006),** Discharge Massflows. Seminar, Haus der Technik, Essen.

**Jenike, A. W. (1964),** Storage and Flow of Particulate Solids. Utah Engineering Experiment Station: Bulletin of the University of Utah, Bul. 123.

**Kache, G. (2010)**, Verbesserung des Schwerkraftflusses kohäsiver Pulver durch Schwingungseintrag, Dissertation, Otto-von-Guericke-Universität Magdeburg.

**Rose, H. und Tanaka, T. (1956),** Rate of Discharge of Granular Materials. The Engineer, (208), pp. 465-469.

**Schulze, D. (2006),** Pulver und Schüttgüter, Springer-Verlag Berlin Heidelberg, 2014.

**Tomas, J. (1991a)**, Habilitation Thesis, TU Bergakademie Freiberg.

**Tomas, J. (1991b)**, Modelling of instationary discharge behaviour of cohesive particulate solids out of bunkers, Chem. Technik, (43), pp. 307-309.

# 508635 Validating the Conductive Heating Time Scales of Particles in a Rotary Drum

**Abstract:**

Thermal treatment of granular materials in rotating drums is encountered in several industrial processes [1–6]. Many applications involve heat transfer where the temperature of the particles is raised to a given temperature such that the desired processes take place. In such instances, heat is exchanged not only between individual particles but also between the particles and external surfaces during the duration of the particle–particle or particle–surface contact. For a good product quality and efficient process, it is required to raise the temperature of the particles uniformly with a minimum processing time. Therefore, a good understanding of the thermal behavior of the solid bed is required, and it is essential to relate the properties of individual particles to the bulk properties of the granular material.

Most recently, Emady et al. [7] identified three relevant dimensionless time parameters, which depends on the particle properties and operating conditions, for conductive heat transfer from the drum walls to the particle bed. The three-time scales are: 1) the thermal time constant, $\tau$, which is the characteristic heating time of the particle bed; 2) the particle thermal time constant, $\tau_p$; and 3) the contact time between the particle at the wall and the wall, $\tau_c$. A regime map was developed showing the relationship between the ratio of thermal time constant to contact time, ($\tau/\tau_c$), and to $\varphi$ ($\varphi$ is the ratio of particle thermal time constant to contact time, $\tau_p/\tau_c$). These time scales help in predicting the total time taken by a bed of particles to reach the target temperature. By calculating $\tau_p$ and $\tau_c$ from the material and operating parameters, the characteristic heating time, $\tau$, can be predicted a priori.

In the current work, the effect of rotation speed on heat transfer in silica beads is investigated to validate the thermal time scales, identified by Emady et at. [7]. Experiments are performed using polydispersed 4 mm diameter silica particles to investigate the heat transfer mechanism inside a 3-inch radius and 3-inch long stainless-steel rotary drum. It is observed that the ratio of thermal time constant, $\tau$, to contact time, $\tau_c$, increases proportionally to $\varphi$. Also, DEM simulations were conducted using MFIX-DEM, an open source multi-solver suite to verify the applicability of the developed regime map to monodisperse particles.

**Reference:**

[1] T. Xu, P.D. Torres, L.W. Beck, J.F. Haw, Catalyst prepartion by impregnation and activity distribution, Chem. Eng. Sci. 39 (1984) 859–864.

[2] kurt E. Peray, J.J. Waddell, the Rotary Cement Kiln, Chem. Publ. Co. (1986) 1–5. doi:10.1007/s13398-014-0173-7.2.

[3] R.T. Bui, G. Simard, A. Charette, Y. Kocaefe, J. Perron, Mathematical modeling of the rotary coke calcining kiln, Can. J. Chem. Eng. 73 (1995) 534–545. doi:10.1002/cjce.5450730414.

[4] M. Rovaglio, D. Manca, G. Biardi, Dynamic modeling of waste incineration plants with rotary kilns, Chem. Eng. Sci. 53 (1998) 2727–2742. doi:10.1016/S0009-2509(98)00081-5.

[5] O.A. Ortiz, G.I. Suárez, A. Nelson, Dynamic simulation of a pilot rotary kiln for charcoal activation, Comput. Chem. Eng. 29 (2005) 1837–1848. doi:10.1016/j.compchemeng.2005.03.005.

[6] M.A. Martins, L.S. Oliveira, A.S. Franca, Modeling and simulation of petroleum coke calcination in rotary kilns, Fuel. 80 (2001) 1611–1622. doi:10.1016/S0016-2361(01)00032-1.

[7] H.N. Emady, K. V. Anderson, W.G. Borghard, F.J. Muzzio, B.J. Glasser, A. Cuitino, Prediction of conductive heating time scales of particles in a rotary drum, Chem. Eng. Sci. 152 (2016) 45–54. doi:10.1016/j.ces.2016.05.022.

# 508465 DEM Simulation and Analysis of the Effects of Compression Pressure on Appropriate Adhesive Force of Admixed Particle for Improving Packing Fraction

**Abstract:**

One technique for improving particle flowability is admixing smaller particles than original material particles (main particles). However, the mechanisms by which this technique improves flowability are not yet fully understood. In this study we examined compressed packing in a particle bed, which is affected by particle flowability. In order to estimate the mechanism of improvement, we investigated the effects of the compression pressure on appropriate adhesive force of admixed particle by DEM simulation.

The diameters of main and admixed particles were 400 and 20 nm, respectively. The physical properties of the particles were set based on silica and glass materials. The Hamaker constant of the main particles was kept constant while that of the admixed particle was changed from $2\times10^{-27}$ to $1\times10^{-20}$ J. The compression pressure was changed from 0.44 Pa to 176 Pa. From the bed height after a compression, we calculated packing fractions of main particle. As a result, the adhesive force of admixed particle at which shows the most improvement packing fraction decreased with decresing compression pressure. This result suggests that the appropriate adhesive force of admixed particles was determined by the the applied compressive pressure. In addition, this result was explained by magnitude relation between two forces. One is an adhesive force between admixed particles. The other is a shear force which acts on main particles.

# 507575 Elutriation, Particle Separation, Fines, Cold Flow, Energy, Fluidized Bed, Bubbling

**Abstract:**

Chemical looping combustion (CLC) technology is being pursued as a promising clean and high efficiency method of power generation. The basis of CLC technology is to supply the fuel reaction process with oxygen via â€œoxygen carriersâ€ that are oxidized in a separate air reactor in a recirculating cycle. One challenge with applying CLC to solid fuel is that typical combustion is estimated to produce 1% of unburnt carbon which has the potential to get cycled into the air reactor resulting in efficiency and emission penalties. Therefore, the US DOE has a set a goal to demonstrate that over 90% of the unburnt carbon can be separated from the oxygen carrier being transported back into the air reactor.

To meet this objective cold flow experiments are being performed on a 0.1 m meter diameter bubbling bed with a 0.9 m height to characterize elutriation rates for a binary mixture in a continuous process. A binary mixture of glass beads with a sauter mean diameter of 707 Î¼m and DayGlo plastic particles with a sauter mean diameter of 39 Î¼m were used to simulate the respective oxygen carrier and char ash fines. Steady state conditions were achieved by continuously feeding the mixture into the system with over flow ports at predefined bed heights to maintain constant heights. In addition to standard differential pressure measurements along the riser the time dependent weight of the mixture from the overflow port and the elutriated fines exiting the top of the riser were recorded via scales. In addition the separation efficiencies were determined by post-test sieving of the overflow material. Therefore the fines elutriation rates and efficiency could be analyzed, mapped, and optimized based on the following parameters: riser gas velocity, bed height, binary mixture feed rate, and concentration of fines in the binary mixture.

# 509582 Segregation during Die Filling with Air-Sensitive Powders


**Abstract:**

Die filling is a typical process step in the manufacturing of a wide range of particulate products (e.g. pharmaceutical tablets, catalyst pellets and powder metallurgical components) in various industries. Segregation of powder mixtures during die filling can have a detrimental effect on the quality of the end product. In this study, segregation tendency of powder mixtures with air-sensitive components during die filling was explored, for which binary mixtures of two powders with either different particle sizes or true densities were considered. A model die filling system with a segmented shoe and die was developed and used to quantitatively examine segregation during die filling. The results indicated that, during die filling, depending on the shoe speed, different segregation mechanisms effected, ranging from the percolation segregation mechanism often occurring in hoppers to the fluidization (air-current) segregation mechanism. The latter prevails when fine or light particles are involved, especially when the die filling process is primarily dominated by bulk flow.

# 507562 Reducing Segregation By Wet Granulation

**Abstract:**

In various industrial applications, powder additives are regularly added to bulk solid products to achieve desired functionality in subsequent processes or final products. However, these powder additives often have different particle sizes or size distributions from the main components (bulk solid products), which can result in segregation during solids handling, transportation or packaging. Segregation between the additives and the bulk solid products can decrease the process efficiency or cause product quality problems. One approach to minimize or even eliminate segregation is wet granulation. The wet granulation process aims to produce granulated additives having similar particle size distributions to the bulk solids so that size segregation can be prevented.

In this research, we present an example of an industrial application where the strong segregation tendency between the powder additive and the bulk solid products can be minimized by a wet granulation process. First, segregation of the dry mixture of the powder additive and bulk solid product was quantified by both experimental and recently developed theoretical approaches [1]. Then, a granulation process was developed and optimized to produce additive granulates with similar particle size distribution to the bulk solid product. Finally, theoretical calculations on the segregation tendency for the mixture of the granulated additives and bulk solid product were performed, showing that segregation is significantly reduced and the homogeneity of the mixture is preserved.

[1] Y. Fan, C. Schlick, P. Umbanhowar, J. Ottino and R. Lueptow, "Modelling size segregation of granular materials: the roles of segregation, advection and diffusion," *Journal of Fluid Mechanics,* vol. 741, pp. 252-279, 2014.

# 509639 Experimental Study of 3D Printed Surface on Granular Mixing in a Rotating Drum

**Abstract:**

Improving mixing processes in rotating granular flows is still a present issue with typically low productivity and slow rates of mixing achieved in current operation, these problems are further exasperated by the parallel phenomenon of segregation for systems cont particles with different properties. Use of a travelling sinusoidal wave boundary on a circular quasi-2D drum has been proposed as a method of alleviating these issues.Experiments consisting of monodisperse and bi-disperse mixtures of different sizes were tested in both a circular and wavy cross sections at different rotational speeds in order to investigate how an alternative boundary configuration influences flow behavior and mixing dynamics.

# 508697 The Importance of the Residence Time Distribution for Designing Solids Handling Process Plants

**Abstract:**

Most of the intermediate and final products in the chemical or polymer industry are handled as bulk solids. Unit operations for these materials need be designed based on their material properties. A typical example is the use of the flow properties such as the internal friction and wall friction of bulk solids for mass flow of silos.

Continuous production processes become more and more important and call for the accurate performance. Key information for designing and dimensioning of the continuous process in industrial sized plants is the residence time distribution of certain vessel and apparatus but also of the interaction of the entire system. For bulks solids, the residence time distribution of such an apparatus is strongly influenced by the product itself.

Therefore a model has been applied to predict or define the residence time distribution of process silos (drying, degassing, blending) taking into account the measured product properties. Based on the results of the single equipment the effect of combining subsequent apparatus can be simulated by using the convolution integral method.

The model and also the results will be presented for the downstream pellet degassing or drying process in polymer production plants where continuous operation can be efficiently applied. The results have been used for an improved selection of equipment type, size and number.

# 507992 Engineering of Assemblies of Inorganic Particles

**Abstract:**

Nano-particles, of particle size less than 1 μm, tend to be difficult for users to handle. Therefore, the particles are often assembled into larger granules for ease of handling and to allow controlled delivery in their desired functions. Engineers who manufacture granules want to be able to control and predict the structures formed and ultimately predict the functional performance of the granular materials.

This project studies the use of spray drying to engineer titanium dioxide ($TiO_2$) pigmentary granules of controlled morphology and mechanical properties. The first stage considers the characterisation of $TiO_2$ slurries ($TiO_2$ pigments dispersed in water) in order to develop an overall understanding of the feed materials for the subsequent spray drying experiments. Following, these materials were spray dried and the product morphologies assessed. Techniques for controlling product morphologies were explored via the manipulation of the particle-particle interaction strength in the $TiO_2$ slurry feed, as well as via the drying process conditions (e.g. drying temperature, slurry feeding rate etc.). Empirical Models have been used to better understand the drying transformation (structural evolution of the droplets) throughout the entire spray drying process. Finally, the resulting granule morphologies were correlated with the mechanical properties such as the granule compressibility. The relationship(s) between the slurry characteristics, processing aspects of the spray drying and the morphologies of the produced granules is being developed.

The processing of non-coated and alumina-coated $TiO_2$ slurries has been investigated as a function of the slurry pH. SEM images of the non-coated particles (in slurry) showed that the primary particles were elliptic with diameter of 300 ± 100nm, while the coated particles were found to be highly aggregated with diameter of 5 μm. Zeta potential measurements confirmed the iso-electric point (IEP) of the non-coated particles to be pH = 4.5 and pH = 8.5 for the alumina-coated particles. Sizing, sedimentation and rheology studies demonstrate the formation of weak aggregated structure at the IEP for non-coated slurry. However, the behaviour of the alumina-coated slurry was very similar with no dependence on the slurry pH.

Spray drying of the non-coated slurries produced spherical granules in the size range 20-60 μm. However, these spherical granules were found to be weak in strength. The slurry pH and processing conditions had only a minor effect on either morphology or size of the granules. The alumina-coated slurries formed spherical granules at non-IEP pHs, however buckled elliptical morphology was observed when drying the slurry at the IEP. The mechanical strength of the granules made from the alumina coated-slurries were found to be stronger than non-coated slurries.

# 508122 Granulation of Soft Porous Crystal Particles

**Abstract:**

Metal-Organic frameworks (MOFs) are new porous crystals consisting of metal ions and organic ligands formed thorough the self-assembly process with coordination bond in between them [1]. Remarkable features of MOFs are a regular pore structure, a larger surface area, and tunable pore size by choosing the combinations of metal ions and organic ligands. In addition to these features, specific types of MOFs show a flexibility in their structure that leads to a structural transition in response to external gas pressure, and they are also referred to as soft porous crystals (SPCs) [2]. A typical adsorption behavior of SPCs shows a step-wise increase in adsorbed amount due to the structural transition from closed non-porous structure to an open porous one at a certain pressure, which is called a gate adsorption phenomenon. This peculiar adsorption behavior is promising for a variety of applications such as gas storages, gas separations, and molecular sensors. For these applications, it is of crucial importance to control gate adsorption phenomenon. As an interesting method, controlling the particle size is a simple approach for tuning the gate adsorption phenomenon [3]. Here, because the SPC particles are usually obtained with the size of a submicron order, a low handling performance is problematic, and accordingly establishing the granulation process of SPC particles is needed for the industrial applications.

In the present study, we focused on Zeolitic Imidazolate Framework-8 (ZIF-8) which is classified as an SPC. ZIF-8 consists of zinc ions and 2-methylimidazolate linkers, and exhibits intra-framework flexibility which stems from a rotation of linkers. We employed a high shear mixing granulator to establish the granulation process of ZIF-8 particles (650 nm) with water and hydroxypropyl cellulose (HPC) as a binder. As a result, we successfully obtained the granules of ZIF-8 with the size of ca. 200 μm under the condition of 5 wt% of HPC and 42 wt% of water. BET surface areas calculated from $N_2$ adsorption isotherms at 77 K were 1696 and 1540 $m^2/g$ for ZIF-8 particles and the granules, respectively. Also, the reposed angle and compressibility decreased from 50.2° to 37.8° and from 42% to 7 %. These results suggest that we obtained the granules with a less loss of the adsorption capacity and improvement of the fluidity. We also investigated the effect of the concentration of HPC and the size of ZIF-8 particles on the granulation process, and the control of the granule size and the granule size dependence of the gate adsorption will be discussed.

Reference

1. S. Kitagawa et al., *Angew. Chem. Int. Ed. Engl.,* **43,** 2334 (2004).

2. S. Horike et al., *Nat. Chem.,* **1,** 695 (2009).

3. Y. Sakata et al., *Science,* **339,** 193 (2013).

# 509413 Continuum Modeling of Granular Segregation in Hopper Discharge Flows

**Abstract:**

Predicting segregation during hopper discharge is important for bulk solids handling as segregation of particles with different sizes can result in severe consequences in product quality. However, due to the complexity of the flow kinematics in the hopper and the nature of granular segregation, little progress has been made in modeling segregation in hopper discharge. In this work, GPU (Graphics Processing Unit) based DEM (Discrete Element Methods) simulations are performed to understand the segregation mechanism and to develop a description of hopper flow kinematics. Then, a continuum model based on a modified transport equation is developed to capture the material transport and segregation within the hopper. Using a hopper filling model to provide the initial particle size distributions after filling, this model is capable of predicting the particle size distribution within the hopper and in the discharged materials during the discharge process. The model is solved using a Finite Element Method algorithm, and the predictions agree well with the DEM results. The model is potentially capable of predicting hopper discharge segregation at the industrial scale and its implications can help with hopper design for minimizing segregation.

# 508256 Head Formation in Bidisperse Shallow Granular Flows

**Abstract:**

Size-segregation plays an important role in many industrial and natural systems. This is particularly relevant for shallow granular flows, where segregation has a significant influence on the flow behavior. Applications range from industrial chutes and hoppers to the prediction of natural hazards.

In these situations, particles segregate vertically by size, with larger particles on top of smaller particles. As the surface velocity of such flows is larger than the mean velocity, the larger material is transported to the flow front. This causes size-segregation in downstream direction, resulting in a flow front composed of purely large particles. As large particles are generally more frictional, the mobility of the flow front is reduced, resulting in a so-called bulbous head.

One of the main challenges of simulating granular flows is the enormous number of particles they contain, which makes discrete particle simulations too computationally expensive to be practically useful. Continuum methods are able to simulate the bulk flow and segregation behaviour of such flows, but have to make averaging approximations to reduce the huge number of degrees of freedom to the relevant continuum fields. We use a depth-averaged model to predict the flow profile for such flows, based on flow height, depth-averaged velocity and particle-size distribution [1]. Small-scale periodic discrete particle simulations are used to determine the material parameters of the continuum model.

In this talk, we show that the bulbous head structure emerges naturally from this depth-averaged continuum framework, and that the long-time behaviour of this solution of the depth-averaged continuum model converges to a novel travelling wave solution [2]. Furthermore, we validate a simple 1D continuum level simulation against a computationally expensive 3D particle simulation, where we see surprisingly good agreement between both approaches, considering the approximations made in the continuum model. We conclude that the travelling distance and height of a bidisperse granular avalanche can be well predicted by our continuum model.

REFERENCES

[1] M. J. Woodhouse, A. R. Thornton, C. G. Johnson, B. P. Kokelaar and J. M. N. T. Gray, J. Fluid Mech., 709, 543-580 (2012)
[2] I.F.C. Denissen, T. Weinhart, A. Te Voortwis, S. Luding, J.M.N.T. Gray and A.R. Thornton, under review with J. Fluid Mech. (2017)

# 507893 Local Segregation Forces and Velocities in Granular Flows

**Abstract:**

Size, density, and shape segregation for flowing bidisperse granular mixtures depends on the species concentration, relative particle properties, and flow conditions. However, the physics of segregation is not well understood at the particle level or the local flow level. We have used 3D Discrete Element Method (DEM) simulations to measure 1) the segregation force on a single large spherical test particle tethered to a â€œvirtualâ€ spring in a shearing bed of small particles; and 2) the local segregation velocity in the macroscale flow of size-bidisperse and density-bidisperse spherical particles as well as for length-bidisperse rod-like particles. At the particle level in the dilute limit, changing the density of the large test particle such that the buoyancy force matches the test particle weight for different size ratios suggests an upward size segregation force that is a quadratic function of the particle size ratio. Based on this, the net vertical force on a single large particle with zero vertical velocity in a sea of shearing small particles is the sum of this upward size ratio dependent force, the upward buoyancy force, and the downward weight of the particle. Focusing on macroscale flows away from the dilute limit, the local segregation velocity for one species is approximately proportional to both the local shear rate and the local concentration of the other particle species for size-bidisperse mm-size spherical particles. The coefficient of proportionality in this relationship, which can be thought of as a segregation length scale, depends logarithmically on the ratio of the particle sizes up to a size ratio of three. Surprisingly, even though the linear dependence of the segregation velocity on the shear rate and concentration derives from a kinetic sieving model for size-bidisperse particles, it also works just as well for density-bidisperse spherical particles and length-bidisperse rod-like particles. In these cases, instead of a logarithmic dependence on the size ratio, the segregation length scale is a logarithmic function of the particle density ratio and rod length ratio, respectively. While these results provide significant insight into the local nature of segregation for particles of various types, the connection between the forces on an individual particle and the local segregation velocity needs further study.

# 507950 Spatiotemporal Distribution of Granular Kinetic Energy in Rotating Tumblers with Axial Segregation

**Abstract:**

Axial bands of particles having different sizes often occur in long rotating tumblers. However, the mechanisms of the formation and evolution of the axial bands is not yet clear. Here, axial segregation of granular mixtures in a long tumbler is investigated through DEM (Discrete Element Method) simulations. We find that the spatiotemporal distribution of the granular kinetic energy in the tumbler is nonuniform as the axial segregation appears. Small particle bands have lower kinetic energy than large particle bands. Furthermore, the time-averaged kinetic energy of all particles increases as the number of bands increases. This suggests that more mixed particles at the interfaces between bands result in a lower overall kinetic energy of particles. Thus, it appears that the kinetic energy related to the purity of the bands and interfaces plays a role in the formation and evolution (coarsening) of the axial bands in long tumblers.

# 510390 The Divide between Academia and Industry in Modeling Gas-Solids Reacting Flows

**Abstract:**

Developing new and efficient commercial scale multiphase reactors is extremely difficult as one needs to consider the complex interactions over a wide range of both temporal and spatial scales encountered in these systems (from molecules to macroscale). Therefore, it is important to codify and automate the knowledge collectively acquired so far, reserving human resources for the creative solutions that build upon codifying past learnings. Computational science (algorithms, theory and modeling, computer science, etc.) combined with exponential growth in computing hardware has great potential to provide us valuable tools for improving the effectiveness of existing plants but also design new reactors. Industry is in desperate need of fast simulation tools that are comparable to the accuracy of the single-phase reacting flows to advance the adoption of multiphase flow reactors and primarily to reduce the risk of scale-up. On the other hand, academia and national laboratories are focused on improving numerical methods, submodels or certain physics to push the state-of-the-art and predictability of the models. However, there is tremendous need for rapid integration of various capabilities that exploit latest computer hardware improvements, robust model reduction techniques, etc. that would improve the computational tools available to the industry to deliver the next generation of efficient and cost-effective reactors. These developments typically fall into the valley of death where industry and software vendors feel these efforts are far-fetched while the academic world considers it incremental. One way to address this divide is to develop mechanistic upscaling framework to coarse-grain information from more detailed modeling approaches to fast models accessible to the industry with relatively minimal information loss.

I will present such an approach using a gas-phase fluidized bed polymerization reactor as an example. First, I will provide an overview of the various models currently used at the different scales in gas-solids reacting flows. The coupling across the scales will be introduced through few different approaches:

1. Discrete-Continuum Coupling using Discrete Element Method for particles
2. Upscaling data from Discrete Element Modeling results to continuum based Computational Fluid Dynamics (CFD)
3. Upscaling information from CFD to simplified reactor network models

In conclusion, the presentation will elucidate the approaches and opportunities in bridging the gap between academic advances and industry needs to model and advance our understanding of gas-solids reacting flow reactors.

# 507029 A CFD-DEM Analysis of Particle Attrition in a Cyclone

**Abstract:**

Cyclones are used to separate entrained particles from the flue gas streams. Particles entering a cyclone are subjected to a radial centrifugal force field, driving them to the cyclone walls, where they experience collisional and rapid shearing stresses. For frangible particles this could lead to attrition and dust generation, whilst for hard particles wear of cyclone wall could occur.

In this work we address the issue of attrition of manganese di-oxide particles used in the CLC process. A CFD-DEM analysis of particle motion and interactions is used to calculate the frequency of collisions, impact velocities and shear stresses. These are then coupled with a model of particle breakage due to impact and of surface wear, in order to predict attrition in a cyclone. The outcome of the work is presented, providing a correlation of the attrition extent as a function of particle properties and operating conditions.

## DEM Simulation of Cylinders and Capsules in a Fluidized Bed

*Oleh Baran**
*Thomas Eppinger*
*Kuanjin Han*

Material presented at World Congress on Particle Technology in Orlando on April 23, 2018

### Abstract

Fluidization of non-spherical particles of cylindrical or close to cylindrical shape is part of many industrial processes, especially in relation to biomass fuel production, fluidized bed gasifiers and pharmaceutical tablet coating processes. In these applications accurately accounting for the details of particle shape is important for safe and efficient design of industrial equipment. In this work the numerical model based on the Discrete Element Method (DEM) fully coupled with CFD is used to predict the effect of particle aspect ratio and presence of sharp edges in the particle's shape on the fluidized bed hydrodynamics. The model is implemented within commercial software Simcenter STAR-CCM+ version 12.06 developed by Siemens PLM. The model assumes particles are cylinders or spherocylinders (capsules). It accurately resolves the collisions between cylindrical particles or between capsules without employing conventional method of "spheres glued together" for non-spherical particles.

Cylindrical Particles

Capsules

The model is then used to simulate cylindrical particles or capsules of the same volume, mass, aspect ratio, and frictional properties in a cylindrical container with air inlet at the bottom and pressure outlet boundary at the top with the range of inlet velocities both below and above minimum fluidization velocity. The results for pressure drop, particle packing and particle preferred orientation are analyzed for both cylinders and capsules and compared to the correlations reported in the literature. The results for the percentage of particles aligned with fluid flow direction in bubbling flow regime indicate the presence of surprisingly longer timescale for establishing steady state preferred orientation of particles. The effect of the absence of the sharp edges in capsules compared to cylinders on the hydrodynamics of the fluidized bed is discussed in details.

## Introduction

The computational models for the flow of elongated particles in fluidized bed can bring valuable insight in understanding the effect of particle shape on fluidization state inside industrial reactors and ultimately help designing cost effective fluidization processes and equipment. Notably Discrete Element Method (DEM) is recognized as promising computational approach for modeling the flow of non-spherical particles. For example, (Borzsonyi and Stannarius 2013), (Guo, et al. 2013) (Abreu, Tavares and Castier 2003) (Williams and Philipse 2003) use DEM to study interesting aspects on packing and shear resistance of rods without coupling to fluid flow and (Mema, et al. 2017), (Zhong, et al. 2009), (Mahajan, et al. 2017), (Cai, et al. 2016) use DEM coupled with CFD to investigate the flow of elongated particles in fluidized bed. Experimental data for validating DEM-CFD simulations of above systems are also becoming available recently, see, for example, (Buist, et al. 2017), (Liao 2013). However, DEM by itself and DEM-CFD coupling for non-spherical particles are known to be computationally too expensive to be applied to even smaller scale industrial scale problems.

In this work new DEM-CFD model is introduced for simulating cylinders or capsules in a fluidized bed which offers the computational speed advantages due to particular choice of modeling the contacts between non-spherical particles and applicability to complex geometries advantage due to special cell clustering approach in 2-way coupling between solid and fluid phases.

## Contact model for arbitrarily shaped discrete elements

Resolving collisions between particles in the distinct feature of DEM, see Fig 1 below.

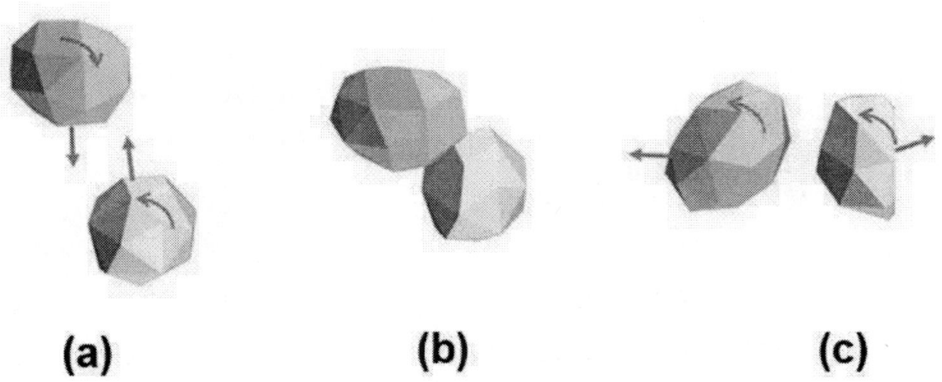

**Figure 1. Collision between non-spherical particles. (a) Before collision. (b) During collision. (c) After collision. Blue arrows show particle's linear velocities and green arrows show particle's rotational velocities**

Generalizing well validated and tested soft sphere approach for DEM on non-spherical particles is described next. Within this approach particle collision (see Fig 1 (b)) takes finite time or many DEM timesteps and can be described by the physics of spring-dashpot system as shown on schematics in Fig 2.

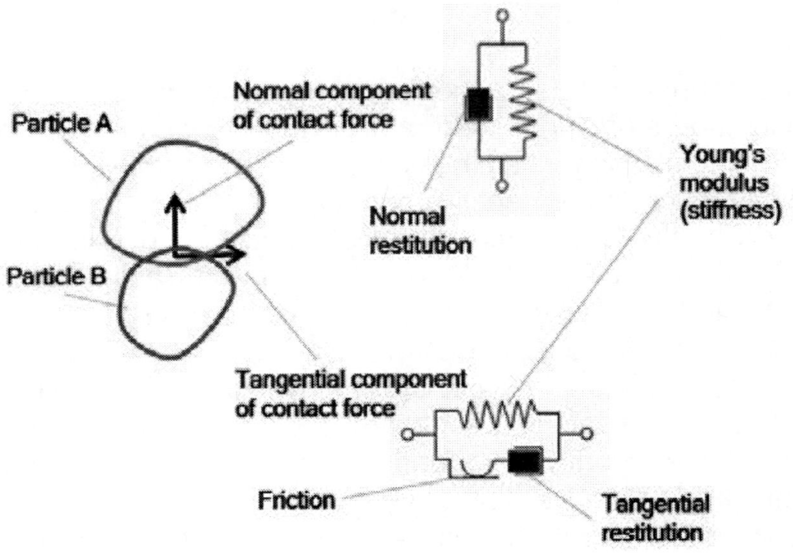


**Figure 2. Schematics for soft collision approach in DEM**

Within this new soft collision approach colliding Particle A and Particle B are rigid bodies that do not deform. During collision the overlap between these two arbitrary shape objects is monitored and contact force which depends on this overlap is calculated. Total contact force is a sum of normal and tangential components. According to (Feng, et al., 2012) the direction of the normal component can be uniquely determined and it provides the maximum rate of slowing down two impacting particles.

As shown in (Feng, et al., 2012) the elastic component of the normal contact force must conserve mechanical energy and, in order to do so, it must satisfy several requirements described next.

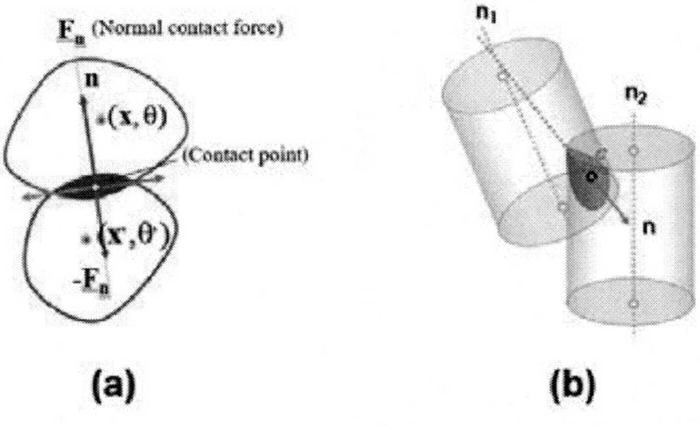


**Figure 3. (a) Two colliding bodies with reference points. (b) Collision between axi-symmetric objects**

In a fixed global coordinate system two points on the two colliding bodies have coordinates $\boldsymbol{x}$ and $\boldsymbol{x}'$ and the orientations of these two bodies are described in terms of variables $\Theta$ and $\Theta'$, see Fig. 3 (a). The normal elastic component of the contact force is the function of position and orientation of two bodies, $\boldsymbol{F}_n = \boldsymbol{F}_n(\boldsymbol{x}, \boldsymbol{x}', \Theta, \Theta', \ldots)$ and it must obey the following "Energy Conservation Theorem". Contact model will be energy conserved if there exists a scalar potential function $\phi(\boldsymbol{x}, \Theta)$ such that

$$F_n = -\nabla_x \phi(x, \Theta) = -\frac{\partial \phi(x, \Theta)}{\partial x}$$

For more specific cases of axi-symmetric objects such as cylinders or capsules, see Fig. 3 (b), the following "Normal Contact Theorem" applies. For a pair of axi-symmetric bodies in contact with contact point $c$, the normal contact line $n$ , which passes the two central lines, can be uniquely determined. In particular,

$$c = x + \frac{n \times M}{|F_n|} + \alpha n$$

Here $M = -\nabla_\Theta \phi(x, \Theta) = -\frac{\partial \phi(x,\Theta)}{\partial \Theta}$ is a contact moment.

With above described approach the DEM simulations of various shape particles were conducted and validated first under conditions that do not require coupling to fluid. One such validation example involves piling up particles of different shape. The final state of the simulated process is shown in Fig 4 below.

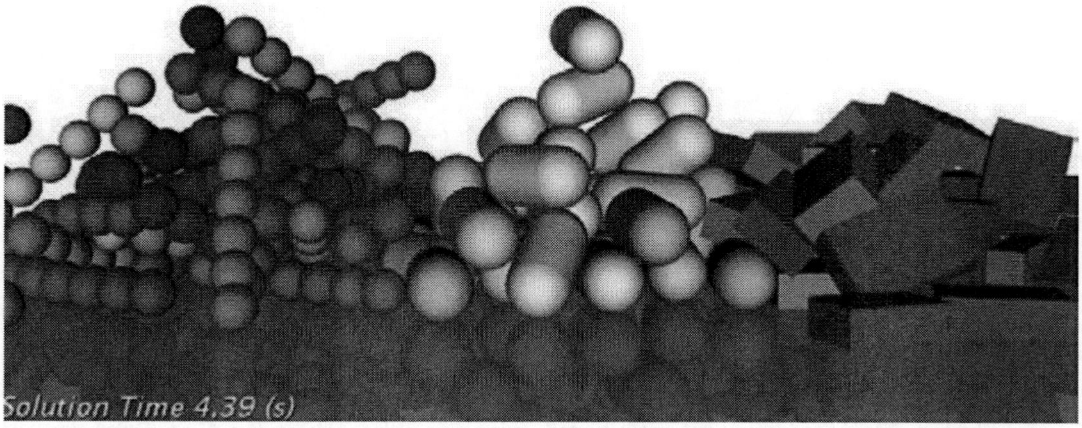


**Figure 4. Interaction between different particle types. Left: blue clumped particles (flexible rods). Middle: Yellow capsules (spherocylinders). Right: Green polyhedral particles (boxes)**

## DEM-CFD two-way coupling for non-spherical particles

The novel source smoothing using cell clustering method for 2-way coupling between DEM and CFD is used in all simulations discussed next. This source smoothing algorithm considerably reduces the mesh size dependence in DEM-CFD coupling. This algorithm was first introduced in (Baran, Greening and Kodl, Numerical Simulations of Fluidized Beds in Fine-Grid Geometries Using Source Smoothing with Cell Clustering Method for DEM-CFD Coupling 2016) and (Baran, Eppinger, et al. 2016).

The drag force on capsules or cylinders is set to be a combination of Wen-Yu model and Ergun based model

$$\overrightarrow{F_{drag}} = \frac{1}{2}\rho_f C_d A_d |\vec{u} - \vec{v_p}|(\vec{u} - \vec{v_p}); \qquad C_d \text{ is a function of } Re_p = \frac{\rho_f |\vec{u} - \vec{v_p}| d}{\mu}$$

$$C_d = \begin{cases} \frac{4}{3}\left(\frac{150(1-\epsilon)^2}{\epsilon Re_p} + 1.75\right); & \text{if } \epsilon < \epsilon_{\min};\ Ergun \\ \frac{24 + 3.6 Re_p^{0.687}}{Re_p}\epsilon^{-3.65}; & \text{if } \epsilon \geq \epsilon_{min};\ Wen\,Yu \end{cases}$$

$\epsilon_{min}$ is the cutoff void fraction (=0.8)

$A_d$ is a projected area of particle. For cylinders and capsules projected area is recalculated every DEM timestep as the orientation of particle relative to the fluid flow is continuously changing, see Fig. 5.

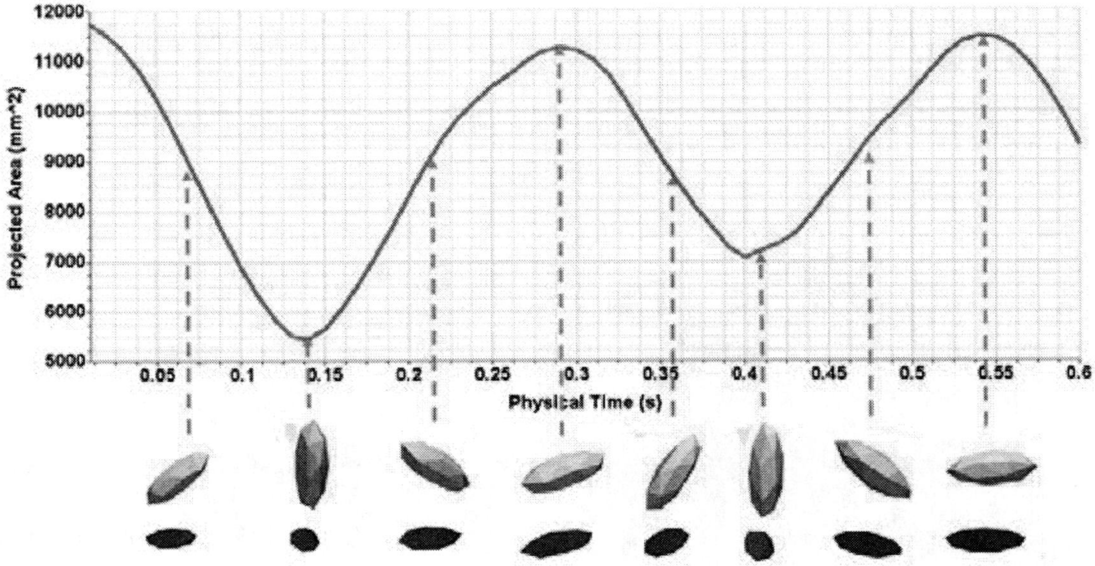


**Figure 5. Projected area of elongated particle rotating in constant fluid flow coming from bottom to top. Orientation of particle for selected points in time are shown below the bottom axis along with the shape of projected area colored in black.**

## Fluidized bed setup

The simulations of capsules and cylinders of various aspect ratio in fluidized bed are discussed next. Geometry includes simple cylindrical container of diameter 0.8 m and height 2 m with velocity inlet at the bottom and pressure outlet at the top with gravity pointing downwards, see Fig. 6. Total simulated time is 20 seconds. The inlet velocity is zero for the first second during settling phase, then it is linearly ramped over the next four seconds to the final maximum inlet velocity $v_{inlet} = 4\ m/s$ and stays at this value until the end of simulation. In all simulated cases the number of particles is always 28320. The density of material particles are made from and their equivalent sphere diameter are fixed to have values $2200\ kg/m^3$ and $20\ mm$ respectively. This means regardless the shape of particle the total amount of solids is always the same. The shape and dimension of particles used in this study are given in Fig. 6. For the base case of capsules with aspect ratio 2 (short capsules) the maximum inlet velocity is 1.22 times minimum fluidization velocity as this study focuses mostly on the effect of particle shape on the hydrodynamics in a fluidized state regime.

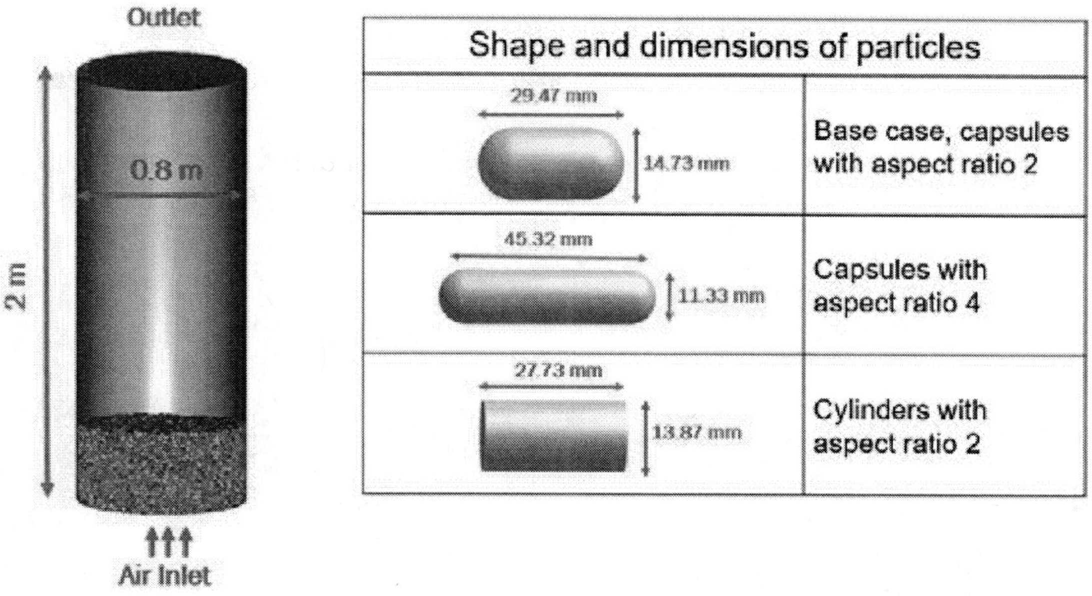


**Figure 6. Fluidized bed setup and types of particles used in simulation.**

All simulations are set to monitor the following quantities.

- Pressure Drop between inlet and outlet is calculated and compared with Ergun prediction for inlet velocities below fluidization
- Bed height is calculated as the vertical distance between inlet and center of mass of all particles times 2
- Void fraction is obtained by averaging in volume cells below fill level. It is used for evaluating the Ergun prediction for pressure drop for the inlet velocities below minimum fluidization velocity
- Alignment of cylinder's or capsule's orientation with the direction of fluid flow, see Fig. 7

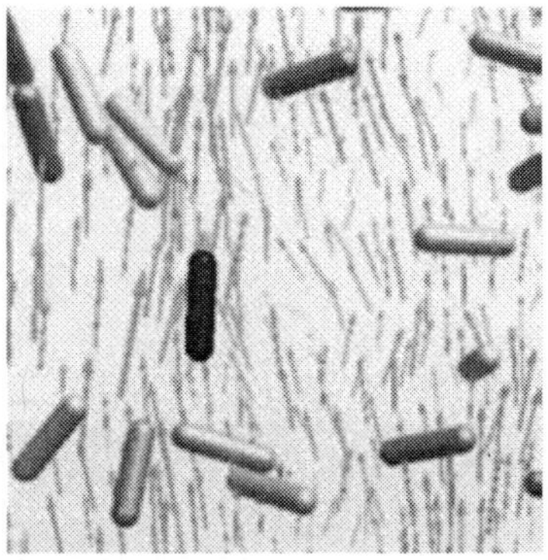

**Figure 7. The fluid direction is shown by vectors. Particle aligned with fluid direction is shown in dark green**

In particular, the angle between particle's main symmetry axis and local direction of fluid, angle of attack, is evaluated for every particle at the end of each fluid timestep and the status of "aligned" is set for each particle with angle of attack smaller than 25 degrees. Number of aligned particles or fraction of particles aligned with flow as a function of time is then available for analysis.

## Simulation results

The snapshots for particle-fluid states in fluidized state for three shapes of particles are shown in Fig. 8

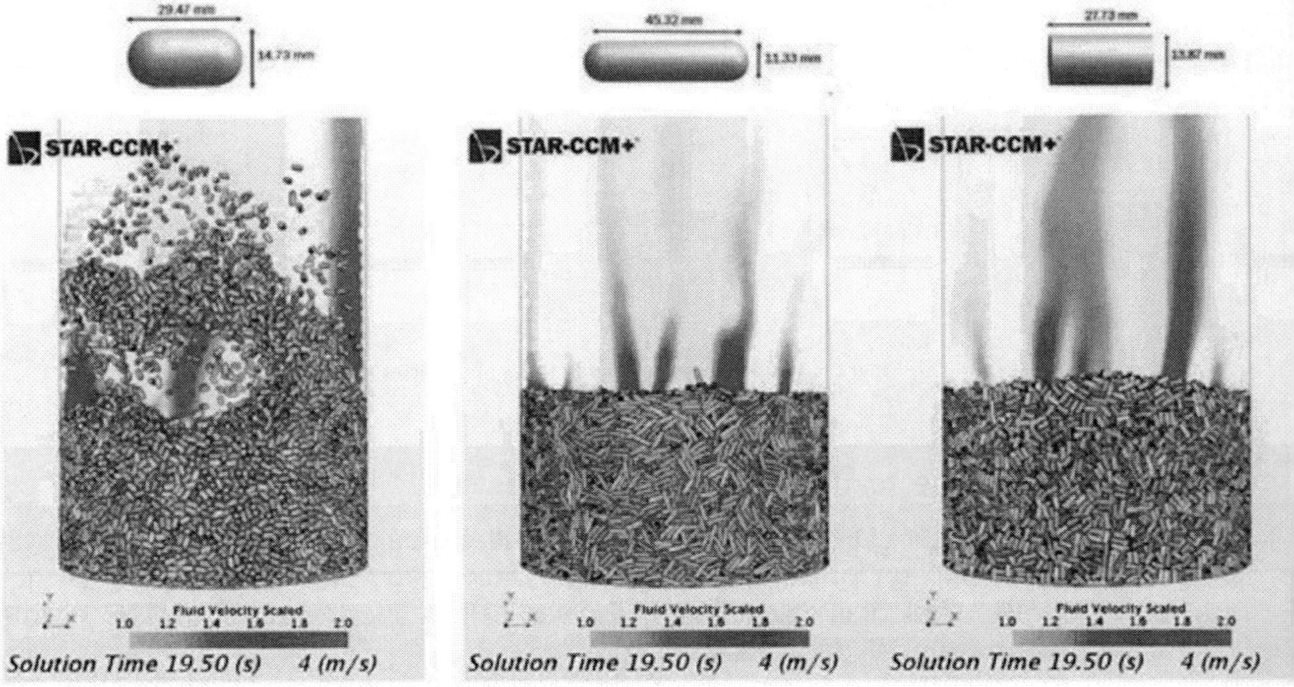


**Figure 8. Rendering of particles in 3D (yellow) and fluid velocity magnitude (blue) on cross-section plane for cases of capsules with aspect ratio 2 (left), capsules of aspect ratio 4 (middle), and cylinders of aspect ratio 2 (right).**

3D visualization of the magnitude of fluid velocity using resampled volume technology is shown in Fig. 9.

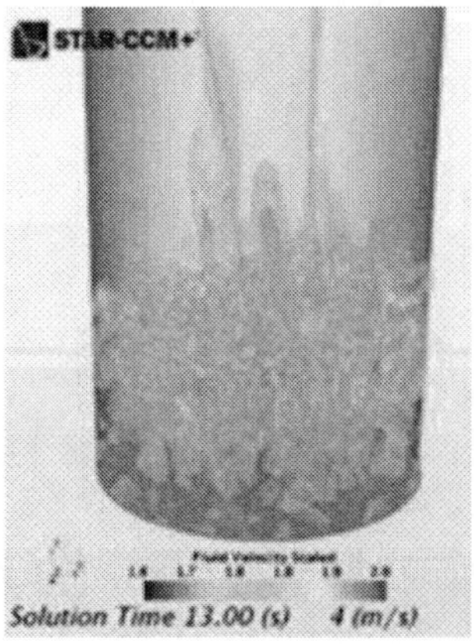


**Figure 9. 3D rendering of fluid velocity for the case with capsules of aspect ratio 2**

The pressure drop as a function of time for each considered case is shown in Fig. 10.

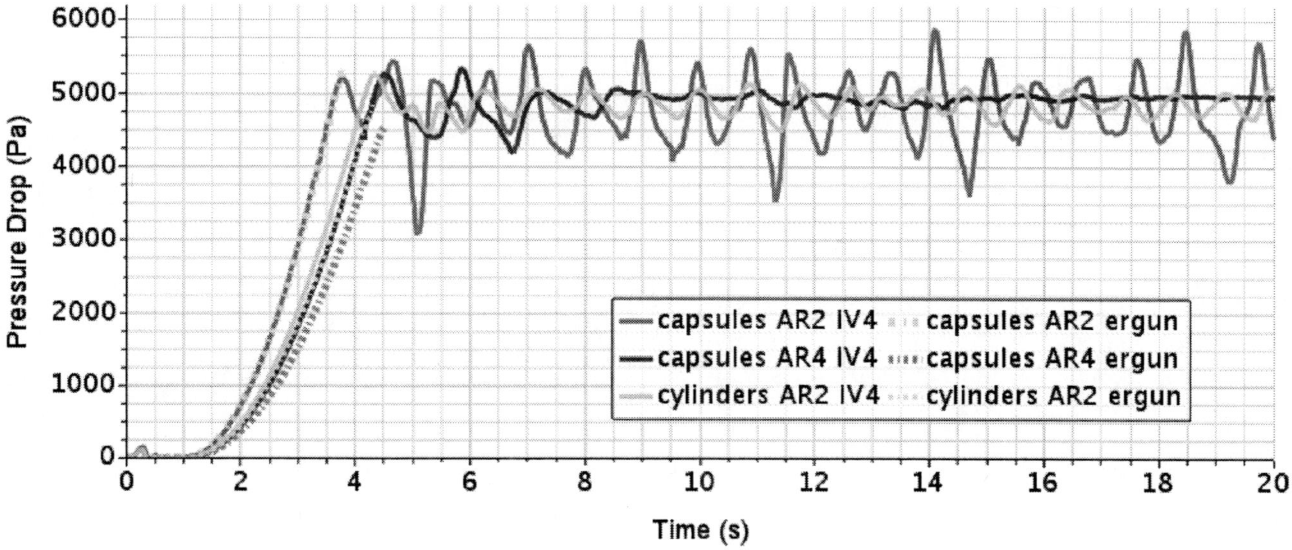


**Figure 10. Pressure drop as a function of time**

The results show good agreement with Ergun prediction for the inlet velocities below minimum fluidization velocities (which correspond to time interval between 1 s and 3.8 seconds). They also show much higher fluctuations of pressure for the case capsules with aspect ratio 2 compared to both other cases. This is consistent with visual analysis of the animations for each case where bubbling flow regime was prevailing for short capsules and channeling flow regime was prevailing for both long capsules and short cylinders.

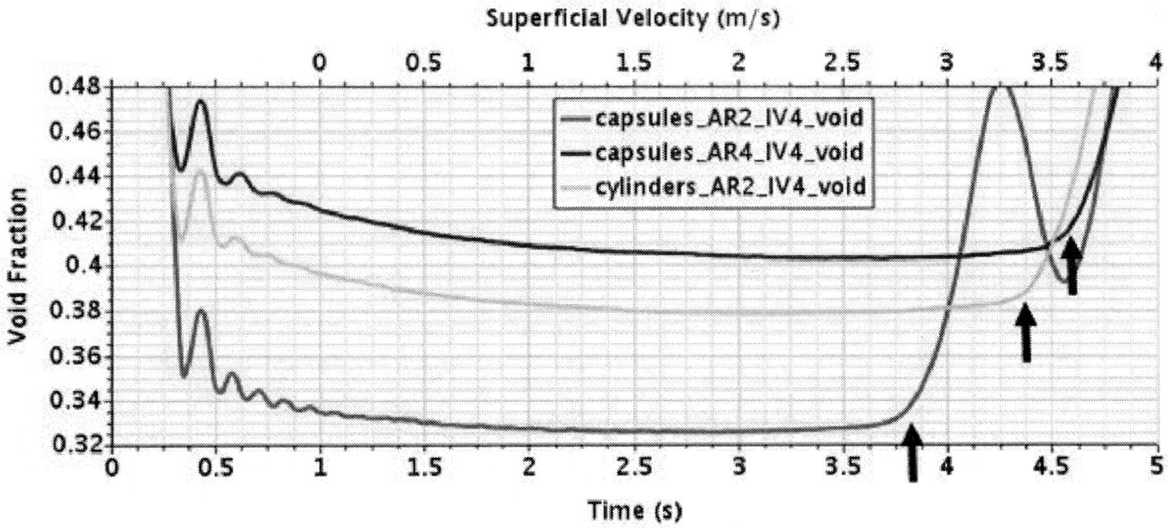


**Figure 11. Void fraction below fill level as a function of time (bottom axis) and superficial velocity (top axis). Black arrows indicate the state of the onset of fluidization.**

Fig. 11 shows the results for void fraction of packed particles. These results are used to estimate the minimum fluidization velocity as the threshold inlet velocity above which void fraction starts rapidly increasing. In particular, the minimum fluidization velocity is 2.7 m/s for short capsules, 3.5 m/s for long capsules and 3.2 m/s for short cylinders. The values of void fraction were also used in calculating the Ergun prediction for pressure drop plotted in Fig. 10.

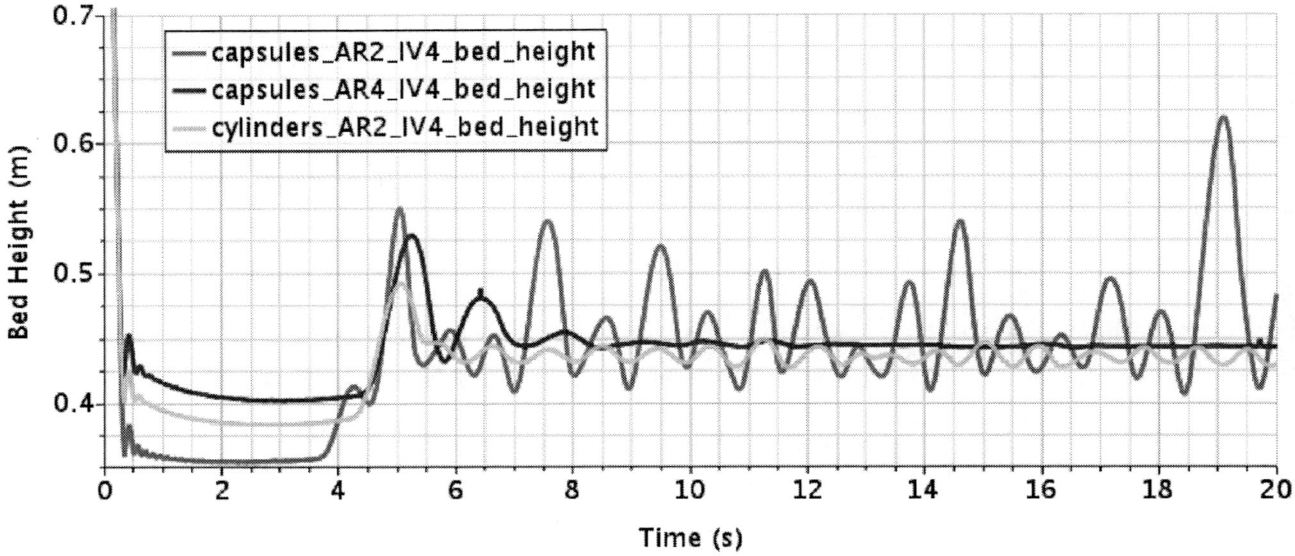


**Figure 12. Pressure drop as a function of time**

Bed height as a function of time is shown in Fig 12. It is consistent with previous results indicating more dense packing of short capsules compared to long capsules and short cylinders. It also shows much higher fluctuation in bed height in fluidized regime for short capsules compared to other two cases.

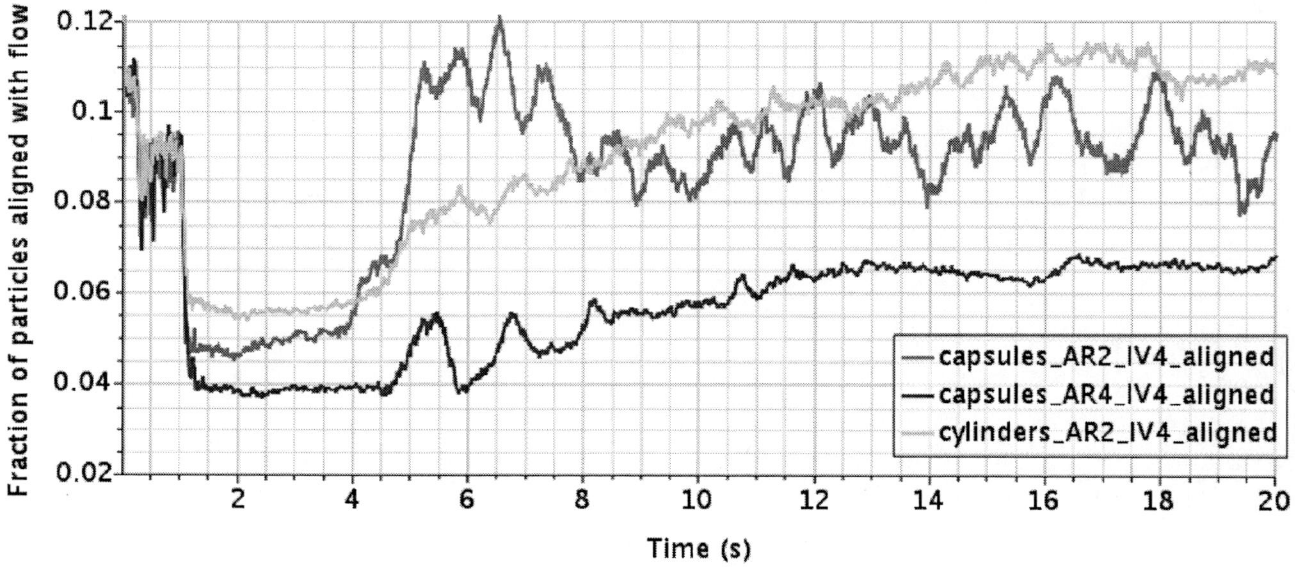


**Figure 13. Ratio of aligned with the flow particles to all particles as a function of time**

Finally, Fig. 13 shows the results for the number of aligned with the flow particles as a function of time. Unlike previous results for pressure, bed height, and void fraction, the steady state orientation of particle has much higher dependence on the aspect ratio of particles. In particular both short cylinders and short capsules end up with more particles aligned with flow direction in steady state compared to long capsules. These results will be discussed in more details next.

## Discussion

A fully coupled DEM-CFD method for simulating capsules or cylinders in fluidized bed is introduced and used to study the effect of particle shape on fluidization in fluidized bed. This algorithm is implemented within commercial software Simcenter STAR-CCM+ which allows running simulations massively in parallel. One of the values of this algorithm is computational cost being bigger but still comparable with the cost of running the equivalent simulation with spherical particles. In particular, it takes less than 2.5 hours to simulate 20 seconds of physical time on 36 cores for any of the considered in this study shape of particles.

The results confirmed that with the increase of particle's aspect ratio

- Packing density decreases
- Initial bed height increases
- Minimum fluidization velocity increases
- Channeling flow regime is replacing bubbling flow regime for the same inlet velocity
- Takes longer to get steady state in particle's orientation

Similar changes but of lesser degree are observed when changing shape from capsule to cylinder while keeping the same aspect ratio. All recorded observation are consistent with the similar simulation or experimental results reported in the literature (Williams and Philipse 2003), (Buist, et al. 2017), (Cai, et al. 2016), (Mahajan, et al. 2017).

## References

Abreu, Charlles R.A., Frederico W. Tavares, and Marcelo Castier. 2003. "Influence of particle shape on the packing and on the segregation of spherocylinders via Monte Carlo simulations." *Powder Technology* 134: 167– 180.

Baran, Oleh, Doran Greening, and Petr Kodl. 2016. "Numerical Simulations of Fluidized Beds in Fine-Grid Geometries Using Source Smoothing with Cell Clustering Method for DEM-CFD Coupling." *AIChE.* San-Francisco.

Baran, Oleh, Thomas Eppinger, Ravindra Aglave, and Simon Lo. 2016. "A fully coupled CFD-DEM algorithm for S-L mixing in stirred vessels." *AIChE.* San-Francisco.

Borzsonyi, Tamas, and Ralf Stannarius. 2013. "Granular materials composed of shape-anisotropic grains." *Soft Matter* 9: 7401.

Buist, Kay A, Pavithra Jayaprakash, J.A.M. Kuipers, Niels G. Deen, and Johan T. Padding. 2017. "Magnetic Particle Tracking for Nonspherical Particles in a Cylindrical Fluidized Bed." (AIChE Journal) 63: 5335.

Cai, Jie, Zhengbiao Peng, Charley Wu, Xiaobao Zhao, Zhulin Yuan, Behdad Moghtaderi, and Elham Doroodchi. 2016. "Numerical Study of the Orientation of Cylindrical Particles in a Circulating Fluidized Bed." *Ind. Eng. Chem. Res.* 55: 12806-12817.

Feng, Y.T., K. Han, and D.R.J. Owen. 2012. "Energy-conserving contact interaction models for arbitrarily shaped discrete elements." *Comput. Methods Appl. Mech. Engrg.* 205: 169-177.

Guo, Y., C. Wassgren, B. Hancock, and J. Curtis. 2013. "Granular shear flows of flat disks and elongated rods without and with friction." *Physics of Fluids* 25: 063304.

Liao, Lingzhi. 2013. "Influence of particle shape and bed height on fluidization." thesis, University of Florida.

Mahajan, Vinay, Tim M.J. Nijssen, Barry W. Fitzgerald, Jeroen Hofman, Hans Kuipers, and Johan T. Padding. 2017. "Fluidization of spherocylindrical particles." *Powders & Grains* 140: 06019.

Mema, Ivan, Vinay V Mahajan, Barry W Fitzgerald, Hans Kuipers, and Johan T Padding. 2017. "Effect of lift force on dense gas-fluidized beds of non-spherical particles." 12th International Conference on CFD in Oil & Gas, Metallurgical and Process Industries.

Williams, S. R., and A. P. Philipse. 2003. "Random packings of spheres and spherocylinders simulated by mechanical contraction." *Phys. Rev. E* 67: 051301.

Zhong, Wenqi, Yong Zhang, Baosheng Jin, and Mingyao Zhang. 2009. "Discrete Element Method Simulation of Cylinder-Shaped Particle Flow in a Gas-Solid Fluidized Bed." *Chem. Eng. Technol.* 32: 386-391.

# APPLICATION OF CPFD METHOD IN THE SIMULATION OF VERTICAL DENSE PHASE PNEUMATIC CONVEYING OF PULVERIZED COAL

*Yong Jin*[†], *East China University of Science and Technology, Shanghai, PRChina*
*Haifeng Lu, East China University of Science and Technology, Shanghai, PRChina*
*Xiaolei Guo, East China University of Science and Technology, Shanghai, PRChina*
*Xin Gong*[‡], *East China University of Science and Technology, Shanghai, PRChina*

## Abstract

Flow stability is a key issue to be addressed for the development of dense phase pneumatic conveying. Flow patterns are closely related with flow stability. From the view of experimental materials, angle of repose, packing density and tapped density were determined by PT-X powder tester. And flowability was determined by FT4 Powder Rheometer. The experimental results shew that the tested pulverized coal was a kind of Geldart's group C powder with strong cohesion. This study contributes with a computational particle fluid dynamic (CPFD) simulation which is based on the numerical solution of continuity and momentum balance equations in a three-dimensional (3-D) horizontal pipeline. A novel correction model was developed, describing interaction among particles and between fluid and particles decided by the experiment mentioned before. CPFD approach predicted the particle velocity and pressure drop were generally below 5% for the stable cases. What's more, the twelve different flow patterns were predicted by the CPFD approach embedded in the correction model, which was confirmed by the electrical capacitance tomography (ECT) in the experiment facility set up by our own. With the help of our correction model, it could reveal the mechanism of the change of flow patterns.

Keywords: CPFD; flow patterns; flow stability; ECT; correction model;dense

## Highlights

Multiphase particle-in-cell model is used to predict dense phase pneumatic conveying.
Pressure drop and velocity profiles show good agreement with experiments.
A improved model was proposed to give the right simulation results.

## 1. Introduction

The practical application of the pneumatic conveying technology of particulate materials are extremely extensive, especially for the field of pulverized coal pressurization and gasification process. High pressure dense phase pneumatic conveying of pulverized coal is one of the key technologies of entrained flow gasification technology. The stability of transportation directly affects the gasification temperature, gasification products and the safety of the gasifier. Depending on the materials to be conveyed, operational conditions and facility geometries, transportation can

† Present author: Yong Jin, PhD student, E-mail: mhzxjinyong@126.com
‡ Corresponding author: Prof. Xin Gong, E-mail: gongxin@ecust.edu.cn

occur in different flow regimes, which usually can be categorized as dilute or dense. Despite wide applications, efficient pneumatic system design remains challenging and it is complicated by interactions between the particle phase and the turbulent carrier flow as well as particle-particle and particle-wall interactions. Because of these complications, a fundamental understanding of the complex physics, involved in gas-solids flows is crucial for the proper design, control and optimization of any pneumatic conveying systems.

For the dense phase pneumatic conveying system, transportation spped is low, the solid concentration is high and the flow patterns are complex, where the transportation requirements and conditions of the system are quite different from the general conveying system. Nowadays, many researches focused on the pressure drop analysis [1], flow patterns [2], signal analysis [3], where the research methods are experiment. However, the complex interactions among the particle phase and fluid phase are difficult for people to explore by experiments, like particle-particle and particle-fluid micro interactions, resulting the macro particle flow. Numerical simulation emerges at a historic moment. The computational particle fluid dynamics (CPFD) numerical scheme incorporated with the multiphase-particle-in-cell (MP-PIC) method is a recently developed model for gas-solids flows. To date, this method has been used to predict certain processes; however, to the authors' best knowledge, published information is unavailable on the application of this model in dense pneumatic conveying. Previous applications of the MP-PIC method include bubbling and circulating fluidized beds (CFBs), spouting beds, fluidized bed gasifiers (FBG), fluidized beds (FB) for carbon capture, gas/liquid/solid fluidized beds, Rayleigh-Taylor mixing layers and etc.

As in all other Eulerian-Lagrangian approaches, the MP-PIC method treats the gas as a continuous phase and the particles in a discrete manner. Solving a transport equation called a Liouville equation for the particle distribution function predicts the particle dynamics. The particle distribution function contains particle properties such as particle spatial location, particle velocity, particle mass, and time. The MP-PIC method uses the concept of parcels (numerical particles) in order to reduce the number of particles involved in the computations; this provides a significant acceleration of simulation speed. A parcel is a collection of several real particles with the same properties (species, size, density, temperature, etc.). Unlike DEM models that calculate particle-to-particle force with a spring–damper model and direct particle contact, the MP-PIC methodology models particle collision force for each particle as a spatial gradient. A particle normal stress model is developed from this concept to describe the particle collisions.

Therefore, this work uses the MP-PIC method to perform three-dimensional simulations of the dilute-phase pneumatic conveying of plastic pellets in a horizontal circular pipe.. In addition to investigating the validity of the method for pneumatic conveying applications, we also develop a model, describing interaction among particles and between fluid and particles decided by the experiment for the accurate prediction of pressure drops and flow patterns in dense phase pneumatic conveying. Moreover, some of the major model parameters deemed important in predicting pneumatic conveying characteristics, namely particle-particle and particle-wall collision parameters. The CPFD numerical methodology incorporated in the commercially available Barracuda® 15 code is the platform used for the modelling and simulations in this study.

## 2. Model description

### *2.1. Governing equations*

The governing equations of the models for describing gas and solid phase are presented below. The gas phase equations consist of mass, momentum and energy conservation equation.

The gas phase continuity equation without interphase transfer is given by.

$$\frac{\partial \theta_f}{\partial t} + \nabla \cdot (\theta_f \mathbf{u}_f) = 0 \quad (1)$$

Where $\mathbf{u}_f$ is gas velocity and $\theta_f$ is gas volume fraction.

The gas phase velocity components in the *x*, *y* and *z* directions are given by time-averaged Navier-Stokes equations.

$$\frac{\partial(\theta_f \mathbf{u}_f)}{\partial t} + \nabla \cdot (\theta_f \mathbf{u}_f \mathbf{u}_f) = -\frac{1}{\rho_f}\nabla \mathbf{P} - \frac{1}{\rho_f}\mathbf{F} + \theta_f \mathbf{g} \quad (2)$$

Where $\rho_f$ is fluid density, **P** is fluid pressure, **g** is gravity, **F** is the rate of momentum exchange per volume between the fluid and particles phases.

The constitutive equation for the non-hydrostatic part of the gas stress tensor is given by

$$\tau_g = \mu_{eff}(\nabla \mathbf{u}_g + \nabla \mathbf{u}_g^T) - \frac{2}{3}\mu_{eff}\nabla \cdot \mathbf{u}_g \mathbf{I} \quad (3)$$

where $\mu_{eff}$ is the shear viscosity, which is the sum of the laminar shear viscosity and the turbulent viscosity ($\mu_{t,g}$). The large eddy simulation (LES) model is used for calculating the gas phase turbulence. In LES, large eddies are calculated from the flow equations and the subgrid turbulence is captured with a Smagorinsky subgrid scale model.

$$\mu_{t,g} = C_S \rho_g \Delta^2 \left| \nabla \mathbf{u}_g + \nabla \mathbf{u}_g^T \right| \quad (4)$$

where Δ, $\Delta = (\delta x \delta y \delta z)^{1/3}$ is the subgrid length scale. Smagorinsky constant = 0.01 is used for the current simulations.

The interphase momentum transfer is given through an interphase viscous drag force ***F***.

Gidaspow [4,5] proposed a drag function by blending Wen-Yu and Ergun functions; this model is used as the drag function in the present study. A switching function is used in order to prevent possible numerical problems due to sharp transitions or discontinuities in the blended model. In the present study, $\theta_{cp} = 0.6$ is used.

$$D = \begin{cases} D_1 & \theta_p < 0.75\theta_{cp} \\ (D_2 - D_1)\left(\dfrac{\theta_p - 0.75\theta_{cp}}{0.85\theta_{cp} - 0.75\theta_{cp}}\right) + D_1 & 0.85\theta_{cp} \geq \theta_p \geq 0.75\theta_{cp} \\ D_2 & 0.85\theta_{cp} < \theta_p \end{cases} \quad (5)$$

Where, $D_1 = \frac{3}{8} C_d \frac{\rho_f |\mathbf{u}_f - \mathbf{u}_p|}{\rho_p r_p}$ , $D_2 = 0.5\left(\frac{C_1 \theta_p}{\theta_f Re} + C_2\right)\frac{\rho_f |\mathbf{u}_f - \mathbf{u}_p|}{r_p \rho_p}$

The dynamics of the particle phase is predicted by solving a transport equation for the particle distribution function (PDF), $\varphi(x, \mathbf{u}_p, m_p, \Omega_p, t)$ (Eq. (12)). Since the particles are non-vapourizing and chemically inert, the dependency of $\varphi$ on particle temperature and material composition is ignored in this study.

The PDF $\varphi(\boldsymbol{x}, \mathbf{u}_p, m_p, \Omega_p, t)$ is calculated from the Liouville equation

$$\frac{\partial \varphi}{\partial t}+\nabla(\varphi \mathbf{u}_p)+\nabla(\varphi A)=0 \tag{6}$$

Where $\boldsymbol{x}$ is the particle position, $\mathbf{u}_p$ is the particle velocity, $m_p$ is the particle mass, $\Omega_p$ is the particle volume and $t$ is the time

The trajectory of a given particle is solved by Newton's second law. The particle acceleration in the original MP-PIC method is

$$A=D_p(\mathbf{u}_f-\mathbf{u}_p)-\frac{1}{\rho_p}\nabla\mathbf{P}+\mathbf{g}-\frac{1}{\theta_p\rho_p}\nabla\tau \tag{7}$$

The particle volume fraction in the above equations is given by

$$\theta_p=\int\int\int \varphi\Omega_p d\Omega_p d\rho_p d\mathbf{u}_p \tag{8}$$

Moreover, $\theta_p$ and $\theta_f$ are related by

$$\theta_p+\theta_f=1 \tag{9}$$

$\tau$ in Eq. ( 7 ) is the particle stress function (also called particle normal stress or isotropic solid stress), which is used to model particle-particle interactions. Due to the difficulties in calculating the particle stress gradient for each particle in a dense flow, the particle-particle interaction forces are calculated as a spatial gradient (as a gradient on the grid) that is mapped back to the particles. The particle stress model used here is an extension of the model from Harris and Crighton [6].

$$\tau=\frac{P_s\theta_p^{\beta}}{\max[(\theta_{cp}-\theta_p),\varepsilon(1-\theta_p)]} \tag{10}$$

where $P_s$ is constant with units of pressure, $\beta$ is supposed to lie between 2 and 5 [7], and $\varepsilon$ is a small number on the order of $10^{-7}$ to remove singularity. The values used in the current simulations are 1, 3 and $10^{-8}$, respectively. However, the particle stress model has a negligible effect on the particle behaviour in the dense phase. The stress model will begin to affect particles as the particle volume fraction of a region begins to approach the close-packed level.

***2.2. Model modification***

For the particle force term, the external force term was modified by the the polydisperse system contact model of Liu et al [8], which was based on the Rumpf [9] equation, a model with elastic-plastic deformation was considered for the granular system:

$$\sigma_{t,iso}=\frac{1-\varepsilon}{\pi}\sum_{j=1}^{n} f_j d_j \sum_{i=1}^{n} k_j F_{ji} \Big/ \sum_{j=1}^{n} f_j d_j^3 \tag{11}$$

Here, the interaction force between particle $i$ and particle $j$ is $F_{ji}$ :

$$F_{ji}=F_{H0}(j,i)+h(R_{ji}^*)[F_{H0}(j,i)+F_N] \tag{12}$$

## 3 Computational details

The geometry, particle properties and operating conditions of the present simulations are based on self-established experimental facilities. work done by Tsuji and Morikawa [64]. These details including solution procedure are explained in the Subsections; 3.1–3.6. The fluid flow is isothermal, compressible and no reactions are involved. The effect of gravity is included.

### *3.1. Geometry and grid generation*

The closed-loop experimental system designed to investigate the dense phase flow of pulverized coal is illustrated in Figure 1. The mass flow rates of three gaseous streams feeding the conveying system were individually controlled to achieve the desired operating conditions. The ECT sensor was installed on the 50 mm I.D. vertical pipe to monitor flow patterns. Two pressure transducers were mounted at the top of the horizontal pipe to obtain the wall static pressure signals at a sampling frequency of 1 Hz for a total of more than 200 sample points. The ECT and pressure signals were both recorded in a computer data acquisition system. The test section with the ECT sensor was located near the pressure transducers. Dry compressed air was used as the conveying gas.

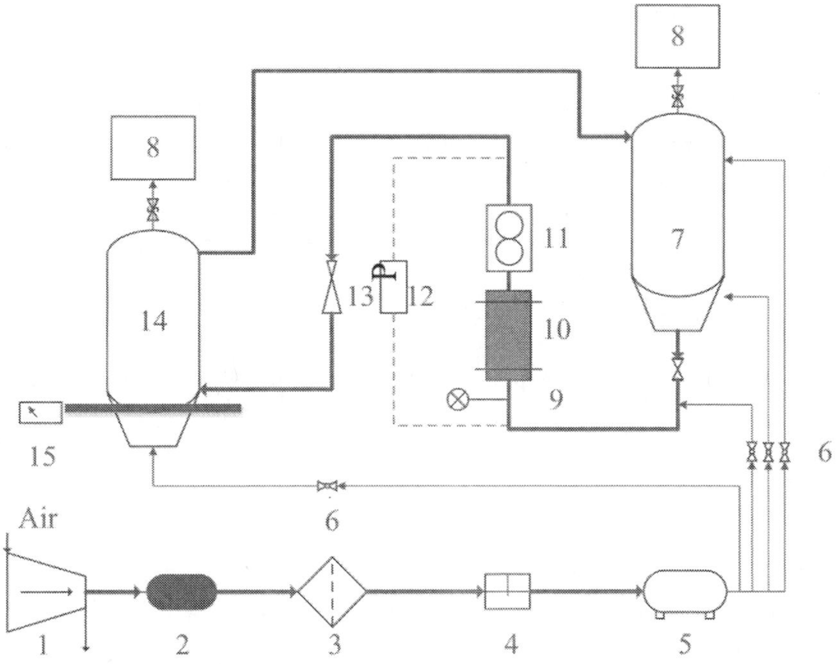


**Figure 1.** Schematic diagram of the vertical experimental facility.

1-air compressor; 2-buffer tank; 3-refrigeration compressed air drier; 4-filter; 5-gas distributor; 6-gas flowmeter; 7-feeding tank; 8-fabric filter; 9-pressure sensor; 10-solid mass flowmeter; 11-ECT sensor; 12-differential pressure sensor; 13-venturi; 14-receving tank; 15-weighing module.

A three-dimensional 4.2 m vertical circular pipe with an internal diameter of 50 mm was created (Figure 2 (left)). The geometry was meshed by a built-in module in Barracuda software. The mesh of the pipe cross section is shown in Figure 2 (right). The mesh had a total number of 89000 cells (5 × 5 × 5) and these were cube. The computational time was approximately 1–3 days for a high performance computer (Intel® Xeon® Processor X5650) characterized as 12 M Cache, 2.66 GHz, RAM 24 GB and 2 TB.

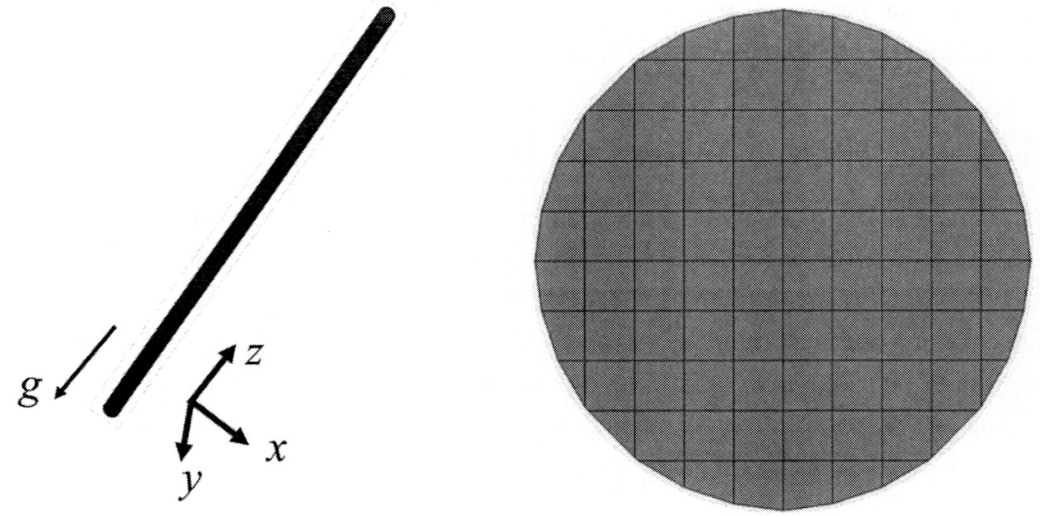


**Figure 2.** The computational domain (left) and grid of pipe cross section (right).

### *3.2. Material properties*

The particles were lignite pulverized coal with a particle density of 1530 kg·m$^{-3}$. Particles were considered to be spherical. The used particle size distribution and characteristic particle size are shown in Fig. 2. The air viscosity was 1.84469 × 10$^{-5}$ kg·m$^{-3}$.

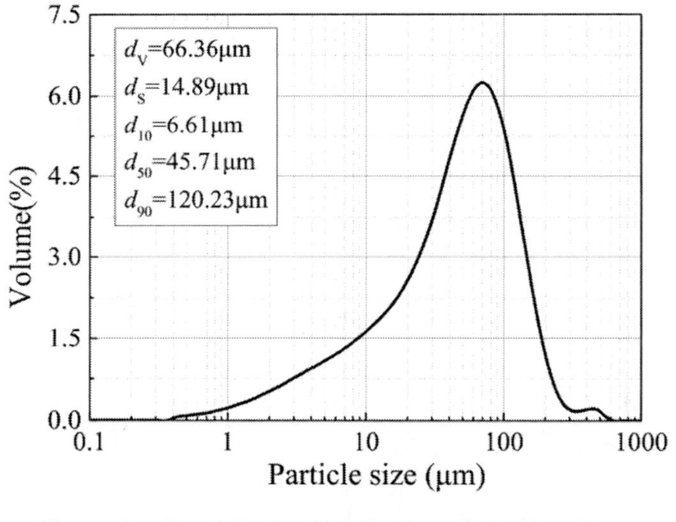


**Figure 3. Particle size distribution of plastic pellets.**

**Table 1** Material properties of pulverized coal

| ω(%) | $\theta_r$(°) | $\theta_f$(°) | $\theta_d$(°) | $\theta_s$(°) | $\rho_b$(g·cm$^{-3}$) | $\rho_t$(g·cm$^{-3}$) | $\rho_p$(g·cm$^{-3}$) |
|---|---|---|---|---|---|---|---|
| 3.35 | 50.09 | 28.09 | 22.00 | 60.70 | 0.4475 | 0.8593 | 1.490 |

According to the Carr's flowability index chart[10], the flowability index of our tested pulverized coal is 37, between 20 and 39, which are classified as the poor flowability powder, like Geldart's Group C powder. [11] Usually, this kind of powder shows great cohesive and there is a necessity for people to use detailed force equations to describe the motion of particles. However, in an MP-PIC scheme, each cell contains a number of numerical particles (parcels). Particle properties are mapped to and from the Eulerian grid to obtain grid-based properties for the particles; fluid properties are, in turn, mapped to discrete particle locations.

### *3.3. Initial and boundary conditions*

The pipe is initially filled with 1 atm air, no particles, and an air velocity of zero. The mass flow rates are calculated according to the superficial air velocities and solids loadings used in the experiments. As for the mass flow rate calculation, the air density at the inlet is 1.2 $kg \cdot m^{-3}$, which corresponds to 1 atm pressure at a temperature of 298 K, as specified at the inlet in the Barracuda setup. The solids have the same velocity to the air at the inlet. In Barracuda setup, it is possible to specify "the numerical particle feed per average volume" which controls the feed rate of parcels (numerical particles) into the domain. In the present study, it is specified as 1,000,000 which gives around 2–4 real particles/parcel for all the cases. At the outlet, a pressure boundary condition is adopted with a specified pressure of 1 atm. No slippage of the gas phase and a partial slip of the particles are assumed as the wall boundary conditions. A bounce-back condition is used to model the particle-wall interactions. It is assumed that 95% of the normal and tangential momentum of the particles are retained after the collision with the pipe wall, i.e., a normal-to-wall momentum retention = 0.95 and a tangent-to-wall momentum retention = 0.95 in the Barracuda setup.

### *3.4. Solver numeric and convergence criterion*

The equations are solved in a transient state. An initial time step of 0.0001 s is applied to the CPFD simulation. However, there is a built-in-model in Barracuda that uses a Courant-Friedrichs-Lewy (CFL) algorithm to automatically determine the proper time step based on the simulated meshes and parcels in order to maintain stability, accuracy and calculation speed [13]. The lower and upper limits for CFL are 0.8 and 1.5, respectively. Each simulation was run over 6 s (in real time) to achieve a quasi-steady state. No major changes were observed in the variables after 1 s. As for the level of accuracy, the limited residual values for convergence were $10^{-7}$, $10^{-8}$ and $10^{-7}$ for volume, pressure and velocity, respectively.

## 4. Results and discussion

The simulations are carried out with a certain set of model parameters, which will be published after the acceptance of our paper.

### *4.2. Prediction of pressure drop*

Having determined the importance of the correction models, the results from the simulations carried out with the new collision terms are compared with experimental data.

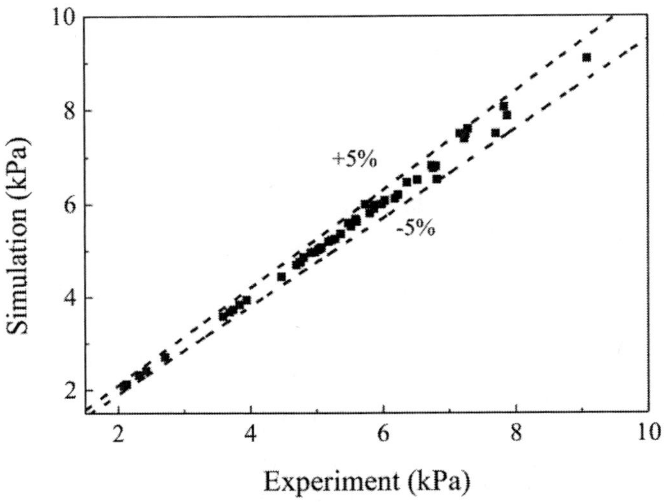


**Figure 4.** Comparison of 52 groups pf experimental data and simulation results

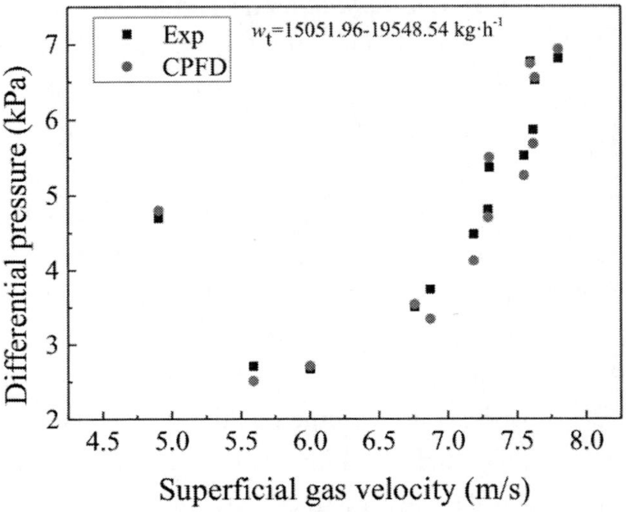


**Figure 5.** Phase diagram simulation (14 groups)

As shown in Figure 4, the predicted pressure drop results that the model provide better agreement with the experimental results. The predictions deviate from the experimental data is approximately 20%. Figure 5 gives the simulation of phase diagram.

### *4.2. Prediction of particle velocity*

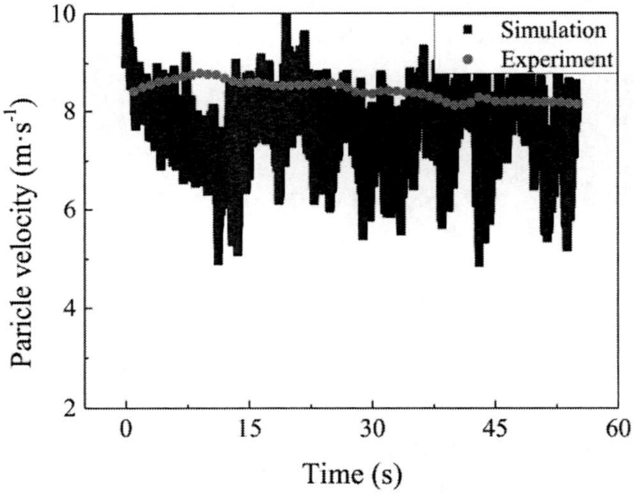


**Figure 6.** Comparison of simulated particle velocity and experiment velocity

Figure 6 shows that the deviation is about 10%.

### *4.3. Prediction of flow patterns*

Table 2 give the simulation results and experiment results which shows great agreement.

**Table 2** The relationship between superficial gas velocity and different flow patterns.

| | a | b | c | d | e | f | g | h | i | j | k | l |
|---|---|---|---|---|---|---|---|---|---|---|---|---|

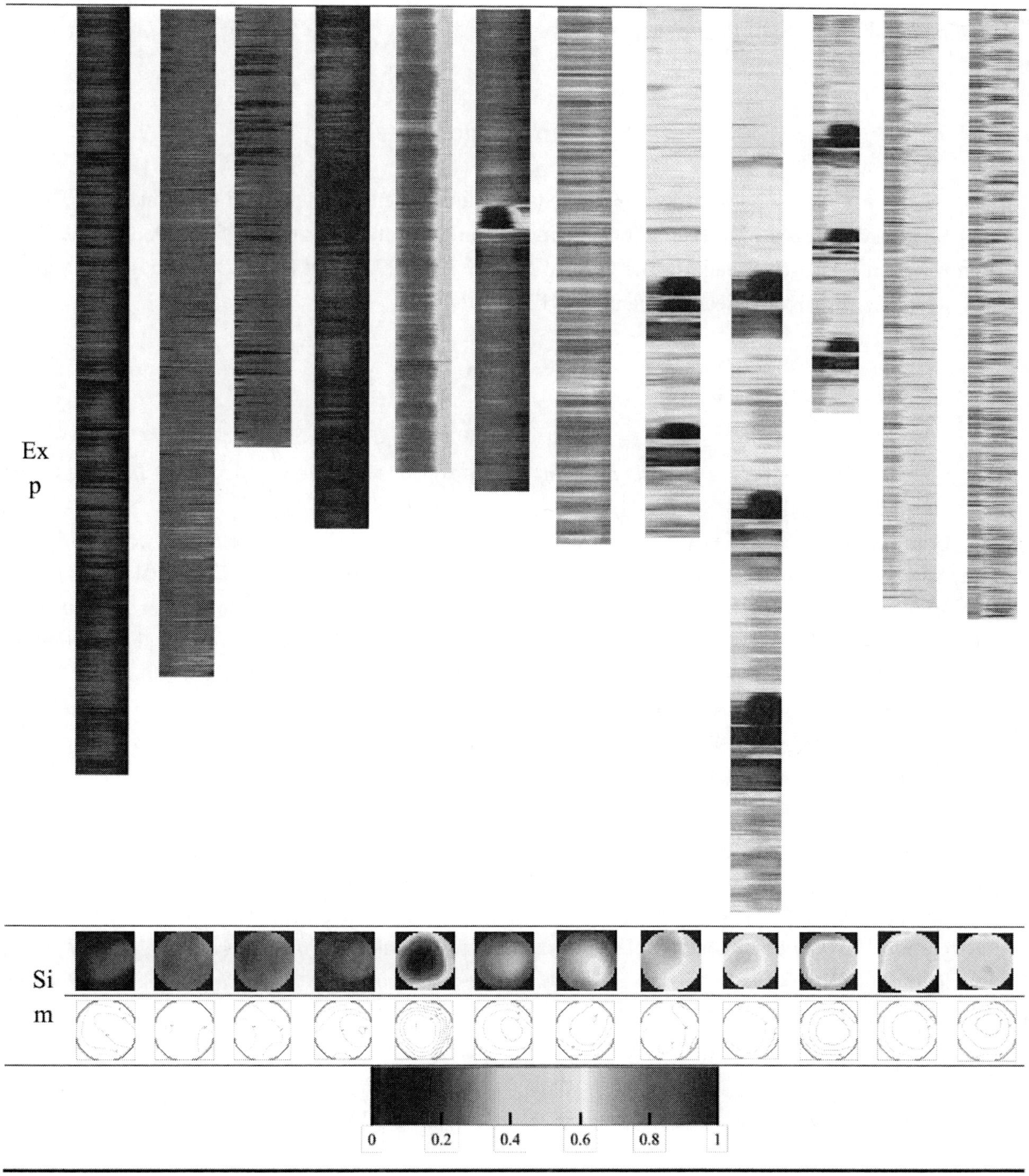


## 5. Conclusions

In this paper, a novel correction model was proposed to describe interaction among particles and between fluid and particles decided by the experiment mentioned before. CPFD approach predicted the particle velocity and pressure drop were generally below 5% for the stable cases. What's more, the twelve different flow patterns were predicted by the CPFD approach embedded in the correction model, which was confirmed by the electrical capacitance tomography (ECT) in the experiment facility set up by our own. With the help of our correction model, it may help people understand the change of flow patterns.

## Acknowledgements

This work was supported by the National Natural Science Foundation of China (21206041), and Shanghai Science and Technology Commission Project (14DZ1200100, 15DZ1200802), the Fundamental Research Funds for the Central Universities (222201717004, 222201514338) and the key scientific research project in Shanxi (MH2014-01).

## References

**1.** CW A, JH H, Hp S. Pressure Drop in Flow of Dense Coal-Air Mixtures. *Ind.Eng.Chem.* 1951;43(8):1837-1840.

**2.** Cong X, Guo X, Lu H, et al. Flow patterns of pulverized coal pneumatic conveying and time-series analysis of pressure fluctuations. *Chemical Engineering Science.* 2013;101:303-314.

**3.** Xingliang Cong. Study on the relationship between the flow type and the pressure signal in the pipeline. Shanghai, East China University of Science and Technology, 2012. (doctoral dissertation)

**4.** Wen CY, Yu YH. Mechanics of fluidization. *Chem.Eng.Progr.Symp.* 1966;62:100-110.

**5.** D.Gidaspow. *Multiphase Flow and Fluidization, Continuum and Kinetic Theory Description.* Boston: Academic Press; 1994.

**6.** Harris SE, Crighton DG. Solitons, solitary waves, and voidage disturbances in gas-fluidized beds. *J. Fluid Mech.* 1994;266:243 - 276.

**7.** Auzerals FM, Jackson R, Russel WB. The resolution of shocks and the effects of compressible sediments in transient settling. *Journal Fluid Mesh.* 1988;195:437-462.

**8.** Liu Y, Lu H, Poletto M, Guo X, Gong X. Bulk flow properties of pulverised coal systems and the relationship between inter-particle forces and particle contacts. *Powder Technology.* 2017.

**9.** Rumpf DIH. Grundlagen und Methoden des Granulierens. *Chemie Ingenieur Technik.* 1958;30(3):144-158.

**10.** Carr.R.L, Div.B. Evaluating flow properties of solids. *Chemical Engineering.* 1965;18(1):163-168.

**11.** D.Deldart. Types of gas fluidization. *Powder Technology.* 1973;7:285-292.

# DIFREX Reactor and Technology Solutions for Many Particle Types and Sizes in Catalytic and Non-Catalytic Reaction Systems

Subhash Dutta, Stephen Arnold, Aashish Gaurav, Jim Brenner
DIFREX LLC., 8900 Astronaut Blvd. Cape Canaveral, FL 32920 USA

Full understanding of particle flow and hydrodynamic behavior, characterization of the properties of these materials, and of their effects on the operation and performance of industrial reactors, is essential for successful industrial reactor design and scale-up. This discussion will focus on gas-solid reactor systems and will use the term "particle" or "solid" to represent both particulates and particles. Where the terms are distinguished, "particulates" are defined as fine particles usually suspended in gas and in dynamic and/or flow mode with a bulk gas flow stream, and are the solid types used in fluidized bed reactors of various kinds and in pneumatic transport lines. The term "particles", on the other hand, is used for relatively larger particles that are usually utilized in fixed or packed bed reactors and remain static or in flow mode in moving bed reactors. Schematics of the four basic industrial solid-gas reactor types are shown in **Figure 1**: packed bed, tubular fixed bed, bubbling/turbulent fluid bed, and circulating fluid bed.

## Reactor Types and Particle Technology

Common industrial solid-gas reactors are of two types: a packed/fixed bed reactor (PKB/FXB) or a fluidized bed reactor, with variations of each. A fluidized bed reactor uses solid particulates. Two prominent versions of the fluidized bed are a bubbling/turbulent fluidized bed (BFB) and a circulating fluidized bed (CFB). Impacts of solid-gas interactions, and effects of particulate flow and hydrodynamic behaviors of fluidized beds, on heat and mass transfer characteristics of reaction systems are discussed with examples, including designs and performances of the reactors.

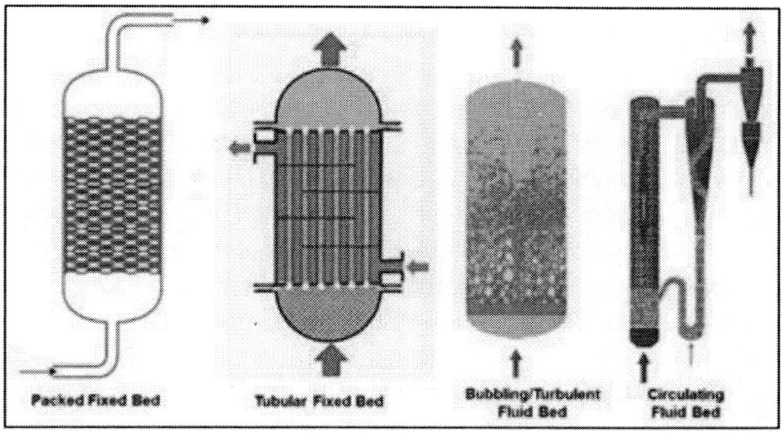


**Figure 1: Basic Commercial solid-gas reactor types**

**Packed/Fixed Bed Reactors**, which employ larger particles in a static bed, are traditionally viewed as the most obvious choices in industry due to easy design and scale-up. For a multi-tubular FXB reactor (TFXB) development and testing, testing in a full-size single tube is ordinarily sufficient for the design of an industrial size reactor of virtually any capacity – the industrial size simply having multiple tubes (hundreds or thousands of the single tube). An FXB reactor also eliminates many potential issues typically associated with a fluidized bed reactor, like solid attrition, entrainment loss and dust emission. However, a multi-tubular FXB reactor suffers from many serious drawbacks that include the following:

1. High heat and mass transfer resistances leading to a steep axial temperature profile in the reactor.
2. Poor catalyst utilization or low catalyst effectiveness factor, due to its use of relatively large particles.
3. Possible maldistribution of gas between all tubes (including impact of catalyst not loaded equally).
4. No continuous solid (catalyst or material) regeneration and replacement capability.
5. No load-following capability.
6. Hot spot formations possibly leading to catalyst activity/selectivity loss and even catalyst/material vitrification.
7. Higher pressure drop.
8. Restricted window of allowable feed composition, temperature and/or pressure to avoid possible temperature runaway for highly exothermic reactions.

**Fluidized Bed Reactors, both Bubbling/Turbulent (BFB) and Circulating (CFB),** do not suffer from the drawbacks of PKB/FXB reactors indicated above. The smaller particle sizes in these particulate systems lead to better heat and mass transfer rates, allowing operation at uniform bed temperature, which is normally preferred. Close to 100% catalyst utilization can be achieved, with no gas maldistribution. These reactors also provide continuous catalyst/material regeneration and replacement capability due to fluidizability of the chosen particulates that can be continuously discharged and recharged after regeneration and/or feeding a make-up stream. These reactors also have load-following capability, allowing operation across a broad window of production capacity as desired. They can be operated with lower pressure drops and are usually also free of hot spots. The windows of allowable feed composition, pressure and temperature are much wider compared with a fixed bed reactor, with less chance of reactor temperature runaway. On the other hand, the key drawbacks of a fluidized bed reactor include the following:

1. Potential dust emission with possible environmental concern.
2. Loss of expensive catalyst by entrainment, to the extent that fines are not fully captured and returned.
3. Particulate attrition causing excessive fines buildup, leading to excessive solid loss or depletion of fines that are needed for good fluidization, and other potential problems like line plugging and agglomeration.
4. Potential erosion of lines or internal heat transfer tubes.
5. Poorer solid-gas mass and heat transfer rates for BFB through the interface films of gas bubbles or voids.
6. Potential neglect of safety concerns (even unacceptable bypassing of safety instrumentation and procedures) if operators do not fully understand, and become complacent with respect to the high flame-arresting safety benefits of a BFB, e.g., for partial oxidation (POX) reactions, including knowing when or where they do not apply.

**Circulating Fluidized Bed Reactors (CFBs)** are usually designed for fast reactions and for reactor-regenerator dual-bed systems. In the dual-bed system, the solid particulates, catalyst or non-catalytic material, can be regenerated in the regenerator bed, or sometimes in a transfer line, and be returned back to the reactor in a continuous circulation of the solid. (This is also done with some BFBs.) The primary section of a CFB system operates without gas bubbles or voids, and has minimal solid-gas heat and mass transfer resistances. However, it suffers from the following drawbacks compared with the other systems:

1. The bed voidage throughout these reactors is usually very high (in the range of 88-98%), thus, solid concentration per unit bed volume very low, resulting in much lower reaction rate per unit bed volume.
2. These reactors operate at relatively much higher gas velocity and therefore are applicable to fast reactions only.
3. At the high gas velocities, which includes the catalyst movement, they demand much tougher or more attrition-resistant catalyst or material, unless the material is desired to be broken down (or reacted away) and to pass once-through as in gasifiers or combustors.
4. The reactors need to be much taller (in the order of 30 meters or even more) for adequate reaction residence time to achieve satisfactory conversions at their high gas velocities.

## DIFREX Reactor Solutions with Particle Technology

DIFREX software, services and solutions integrate chemical reaction engineering with particle technology to provide the right solutions for design, development, scale-up and successful commercialization of gas-solid reactors, enabling one to promptly establish the design and performance of virtually any particle or solid-gas reactor type for reaction systems of different complexities. Since it has fundamental basis in the two engineering sciences (reaction and particle) and avoids much of the empiricism of traditional process development, especially in the HC processing industry, it is better suited for new designs and also revamp or upgrade of an existing technology/ reactor/ process, particularly for a new feed or product stream. This refers to both catalytic and non-catalytic solid-gas reactions, including the use of a catalyst (active components with or without carrier) as the solid for catalytic processes or a non-catalytic solid particle like coal, biomass or a mineral (FeO or NiO, for example) for combustors, gasifiers and metal extraction/roasting processes, and also solid particles as adsorbents to transfer atoms and other molecular species from one stream or vessel to another.

**Figure 2** illustrates the extensive and convenient framework and easy use of the DIFREX software as represented as a ***dial-a-reactor***. It incorporates detailed flow and hydrodynamic behavior for many different reactor types, allowing prompt and easy calculation switching (dialing) from one reactor type to another, for the same chemistry and kinetics of the reaction system. Better, faster initial determination of the best conceptual/ preliminary design for a reaction system, catalyst or process can therefore ***now*** be achieved by comparing the design and performance details projected by the software at the earliest project stages. The best choice could be a packed bed (PKB), a tubular fixed bed (TFXB), a bubbling/turbulent fluid bed (BFB), a circulating fluid bed (CFB), all with or without quench and/or recycle, or a hybrid reactor (CSTR-CFB-BFB) as in a typical FCC reactor or other system. The best **Reactor Of Choice (ROC)** can now be determined along with the **Catalyst or Material Of Choice (COC/MOC)** and much earlier – even while the solid catalyst or material is being developed. Final determination of the best ROC, however, would be determined by **consideration of other criteria** that include cost, particulate handling/recycle, strength and attrition resistance of solid catalyst, and miscellaneous other characteristics like particulate fluidizability and agglomeration problems for fluidized bed systems. However, the software allows one to reach each **GO/NO-GO decision gate** much sooner, leading to substantial savings in both cost and time for catalyst, process or technology development.

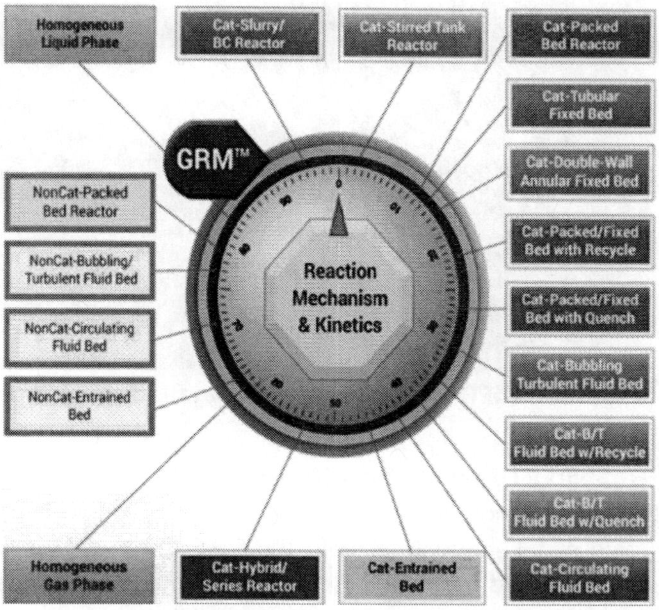

**Figure 2: DIFREX: Selection for Reactor of Choice**

Current DIFREX team members were instrumental in successful commercialization of several new process technologies, involving all four basic solid-gas reactor types: PKB, TFXB, BFB and CFB reactors and their variations. A prime example is the ALMA process, a fluidized bed POX process for production of maleic anhydride from n-butane and air. These designs were products of seamless integration of particle technology (including particulate flow and hydrodynamic behaviors) with chemical reaction engineering principles. Results of some of these works are available in a series of publications by Dutta, Arnold and other authors[1-20]. Many of the theories and procedures used in these works can be found in the two classical books on Fluidization Engineering and Chemical Reaction Engineering by Kunii and Levenspiel[21-22]. The following discussion provides information on some of these reactor designs and the results achieved.

**Packed/Fixed Bed Reactors**

The Fischer-Tropsch process is of great industrial significance. Projected axial profiles for temperature, conversions and pressure drop of a typical Fischer-Tropsch reaction system are shown in **Figure 3**, for syngas conversion to paraffins, olefins and alcohols, carried out in a steam-cooled Tubular Fixed Bed (TFXB) reactor system. PR and SR refer to primary reactant CO and secondary reactant $H_2$. There is a steep temperature rise near the reactor entrance, due to high reaction rate at the concentrated inlet composition, plus relatively high heat transfer resistance typical of this solid-gas fixed-bed reactor system. Temperature runaway with inadequate temperature control, hot spot formation, catalyst selectivity loss, deactivation and possibly even vitrification are potential problems of this system. **Figure 3** also shows the profiles of primary, secondary and undesirable product (PP, SP, UDP) selectivity, per-tube productivity, catalyst productivity (kg/day of PP per kg catalyst), and catalyst effectiveness factor (% utilization) for the primary reaction.

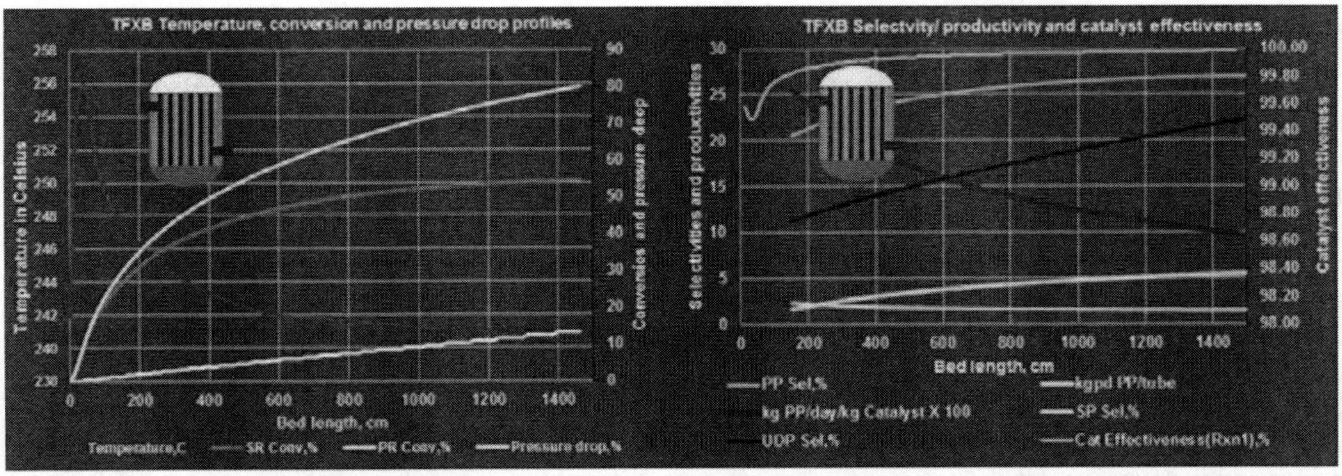


**Figure 3: DIFREX capabilities in detailed reactor design studies: Case of FT synthesis in a TFXB reactor**

**Bubbling/Turbulent Fluidized Bed Reactors**

The conversion of n-butane to maleic anhydride (B→M) presents a unique example where extensive performance tests simultaneously in both tubular fixed bed (TFXB) and bubbling or turbulent fluidized bed (BFB) reactors, accompanied by modeling and scale-up, led to successful commercialization of the world's leading fluidized bed process for this reaction system, called the ALMA Process. Unification of particulate technology principles, flow and hydrodynamic behavior with chemical reaction engineering formed the core of the reactor model that was developed. The model was also applied to simulation and scale-up analysis of two other catalytic reaction systems (N→P and P→A) that vary considerably in mechanism and kinetics. The three reaction systems are demonstrated in **Figure 4**. The data for the N→P system were obtained from published data of a commercial reactor.

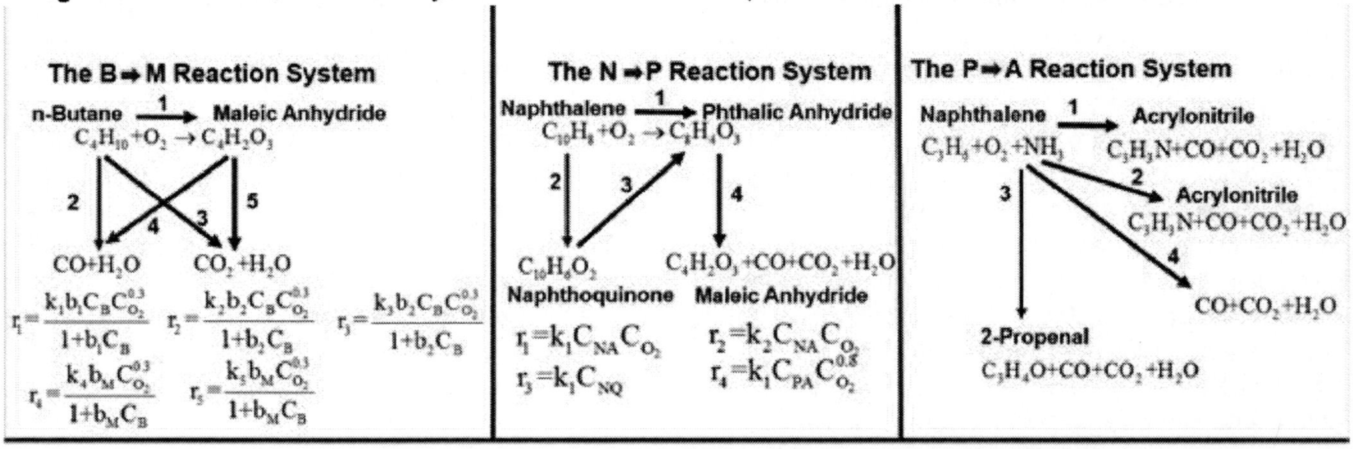


**Figure 4: Catalytic reaction systems modelled and studied in TFXB and BFB reactors**

In the unified model, the entire height of the BFB is viewed to consist of 5 possible zones: 1) ***grid zone*** where bubbles form and grow continuously until they become slugs or reach a constant average size, 2) ***bubble zone*** where bubbles of a constant average size (maximum stable size) rise up through surrounding solid-gas emulsion phase, 3) ***slug zone*** where slugs of constant size rise, filling up the vessel diameter (mostly in smaller-diameter lab or bench-scale reactors), 4) ***splash zone*** where the bubbles burst and splash the particles above the dense bed below and then the catalyst density decreases with height in the dilute phase, and 5) ***post-TDH zone*** where the density remains constant in the dilute phase ("freeboard") above the Transport Disengagement Height. The DIFREX model accounts for particulate flow and dynamic behavior in each zone, and how it affects the reaction system and reactor performance, including conversions and selectivities of each reaction. Details of these treatments are available in the publications of Dutta et al.[14-15].

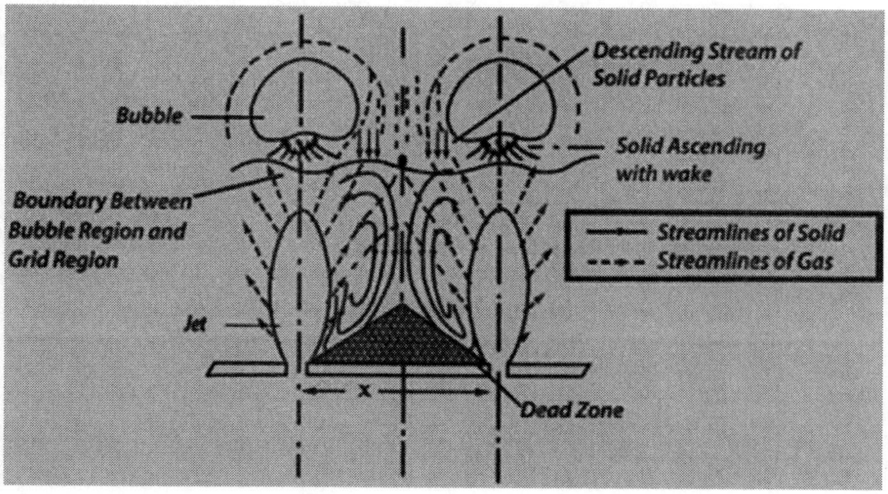


**Figure 5: Schematic of gas and solids flow patterns and dead zone formation in the grid region**

Unlike fixed bed (FXB), the particulate system behavior is quite complex in BFB and may change dramatically from reactor bottom to the top. Proper account of these changes and their effect on the reaction system is necessary for developing a successful BFB scale-up model. A schematic representation is shown in **Figure 5** of gas and solid flow patterns immediately above the grid at the very bottom of a commercial BFB. Due to intense solid-gas mixing, turbulence and smaller bubbles in this zone, plus the higher feed concentration, as much as 60% of the total conversion of a commercial BFB can happen in the ***grid zone***.

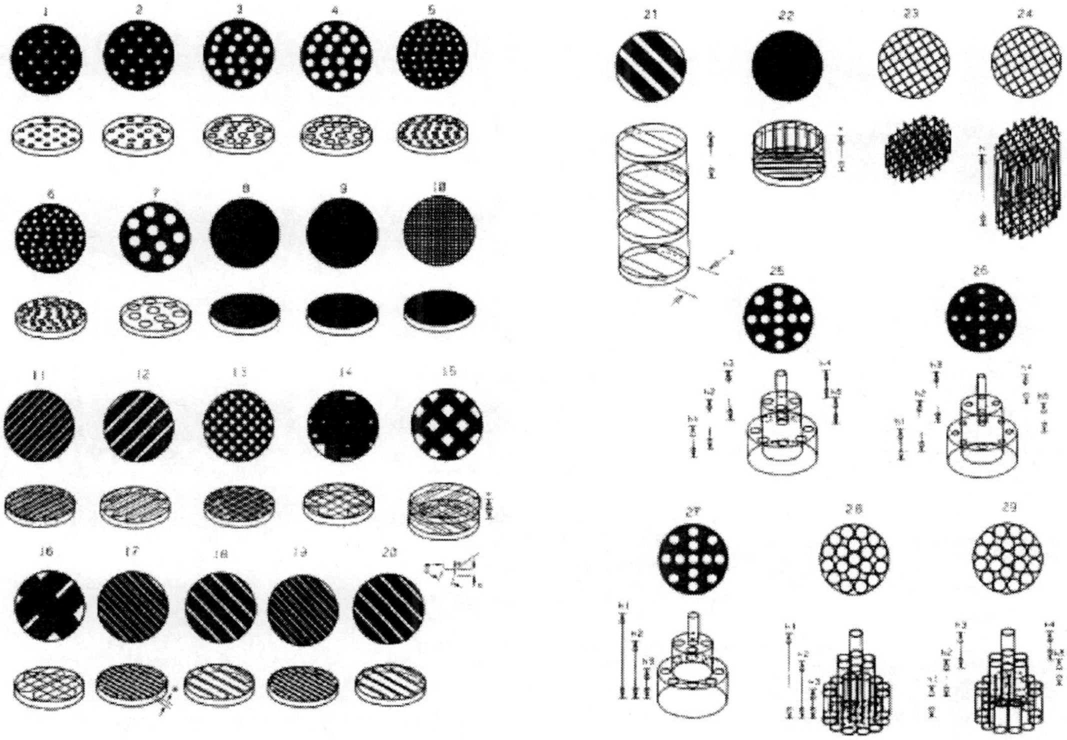

**Figure 6: Top views of experimental baffles [Adapted from Dutta and Suciu[10]]**

Grid design of a BFB is important and a challenging task since several objectives need to be achieved in this design. These include support strength, optimal pressure drop, minimum dead zone, minimum jet attrition of particles, and avoidance of particle slippage through the grid openings. Similar considerations affect the design of internal baffles, etc. A study of designs for the grids and other internals, with different open areas, layers, slot angles, egg-crate shapes, etc., was conducted for the ALMA process as exemplified in **Figure 6**. DIFREX team members have extensive experience in design and commissioning of such internals.

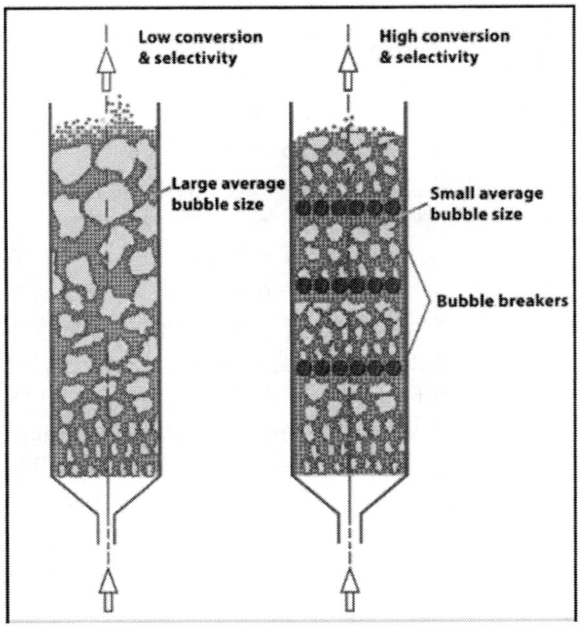


**Figure 7: Bubble size control in BFB**

Bubble size control above the grid zone is also an important consideration for higher reactor conversion, because smaller bubbles mean both higher and more selective conversion, while bigger bubbles mean lower conversion and poor selectivity. This is due to higher mass transfer resistance for the latter case, which slows the exchange (or mass transfer) rate between the rising bubbles with the bulk of the gas and the surrounding emulsion phase containing most of the catalyst. Proper design of bubble breakers therefore can also be an important part of a BFB design. DIFREX can provide help in this area as well. The concept of bubble size control is shown schematically in **Figure 7**.

Applicability of three BFB models were examined and compared during the ALMA process development, as described in a publication by Dutta, Arnold, Suciu and Verde[16]. Schematics of these models[21, 23, 24] are shown in **Figure 8**. Modification of the Kunii-Levenspiel (K-L) model[21] to account for growing bubbles and also the ***splash zone*** as proposed by Hatano et al.[25] are three important features of the DIFREX BFB reactor model. More details are available in Dutta and Suciu[14].

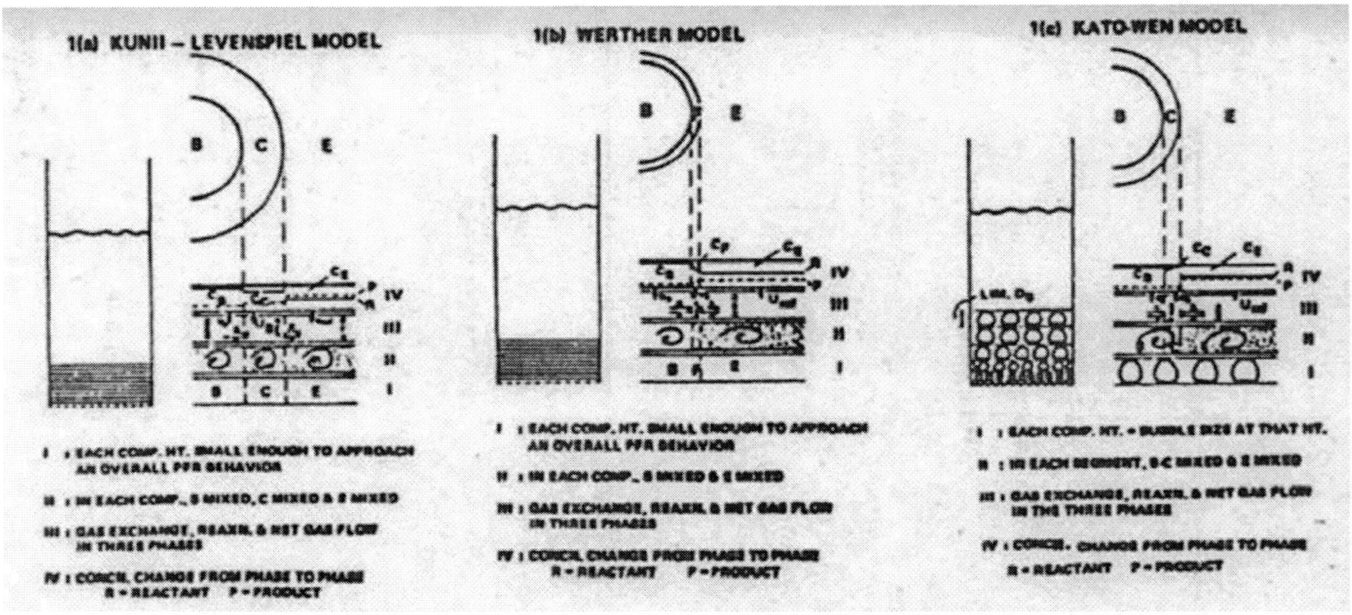


**Figure 8: Schematic representation of three BFB models [Adapted from Dutta et.al [16]]**

Projected BFB reactor performances and the key particulate hydrodynamic behaviors affecting these for the same FT reaction system used in the FXB reactor (those projections shown in Figure 3) are shown in **Figure 9**. The figure shows bubble size growth and expanded ***dense phase*** - the two critical properties of the particulate system that primarily influence the performance of a BFB reactor. The case shown is for a reactor with bubble size control to a maximum size of 10 cm. This figure also shows solid loadings and dilute phase voidage up to the TDH in the dilute phase region, where the declining solid loading (shown by the dotted curve) characterizes the ***splash zone***. The model estimates the TDH, the information necessary to decide the minimum height necessary for the dilute phase of the reactor before the exit gas nozzle location (or entry into internal cyclones, if used). The BFB temperature is ordinarily constant within the bed, and so the temperature profile is not shown in the figure. No significant reaction is noticed in the ***dilute phase*** for this particular case and reaction system used. However, depending on the design and operating conditions, reaction system, relative volume and temperature of the two zones (dilute phase vs. dense phase), significant reaction may occur in the dilute phase, too.

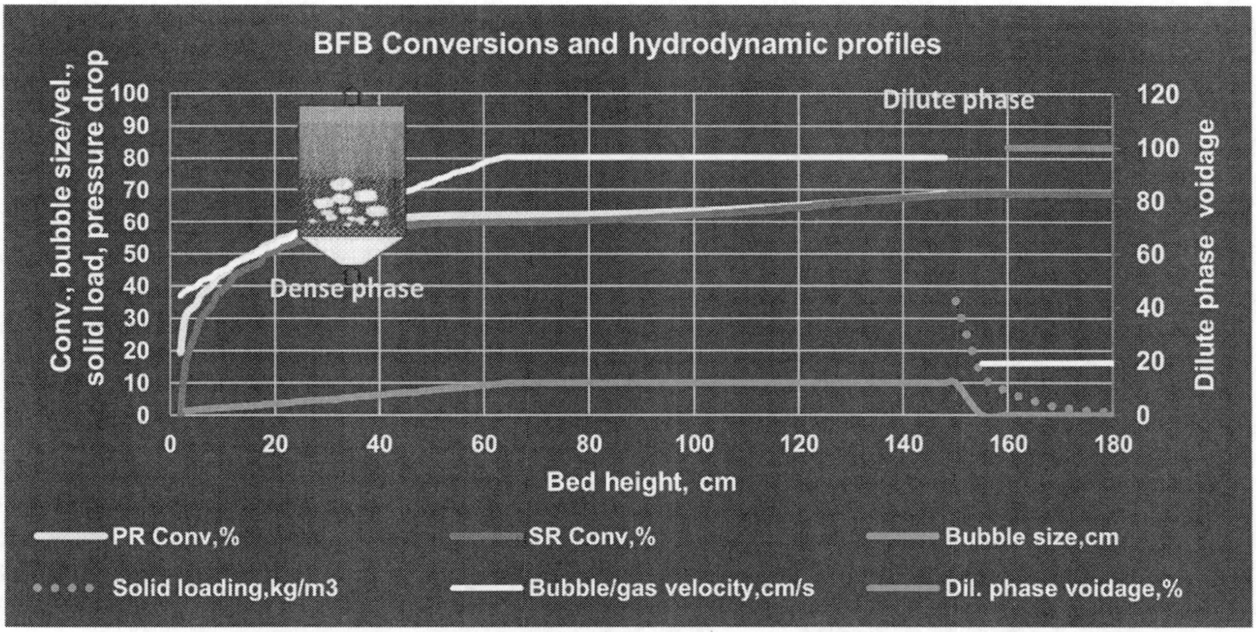


**Figure 9: Projected BFB reactor performances and key particulate hydrodynamic effects**

The DIFREX reactor design and model package GRM™ enables quick generation of the complete design and operating conditions and performance comparison of alternative reactors (FXB, BFB or CFB) if the information required for complete definition of the reaction system involved can be provided or is generated with DIFREX help. **Figure 10** shows one such comparison of complete design and operating conditions of three different solid-gas reactor types - a TFXB (Tubular Fixed Bed), a BFB and a QTFXB (Quenched Tubular Fixed Bed), for 100 TPD production of the primary product (PPX) or product group (like alcohols, paraffins or olefins) from syngas for an FT (Fischer-Tropsch) reaction system (as in Figures 3 and 9). **Figure 11** shows the performance comparison of the three reactors. The best Reactor Of Choice (ROC) can be quickly determined based on the criteria of the user that usually include cost, mechanical and other considerations. In **Figure 11**, the best choice shown is based solely on the highest yield.

Uncertainty/ fear of scale-up of a BFB reactor from bench to commercial scale can now be drastically reduced if proper account of advances made in particulate technology is incorporated in the design and scale-up of this reactor system.

**Circulating Fluidized Bed Reactors**

Successful design and operation of a circulating fluidized bed (CFB) reactor also require proper interplay of particulate flow and hydrodynamics behavior with reaction chemistry and kinetics and heat and mass transfers. The two key advantages of this reactor are: 1) It offers an elegant means of continuous regeneration of spent solid (catalyst or non-catalytic/ reactive particles) in a recirculation loop that removes the solid after only a short exposure (residence) time in the reactor, and subsequently returns the regenerated material back to near the reactor bottom. It is therefore particularly suited for fast-deactivating catalyst and for regenerable catalysts as in an FCC reactor. It is also a prime candidate for applications in **coal/ biomass/ tar sand combustion and gasification and minerals processing**, fast reactions where particle ***attrition*** is not an issue of concern. 2) CFB is also a ***bubble-less*** reactor above the bottom mixing zone, unlike BFB, thereby enjoying minimum heat and mass transfer resistances when compared with both FXB and BFB reactors.

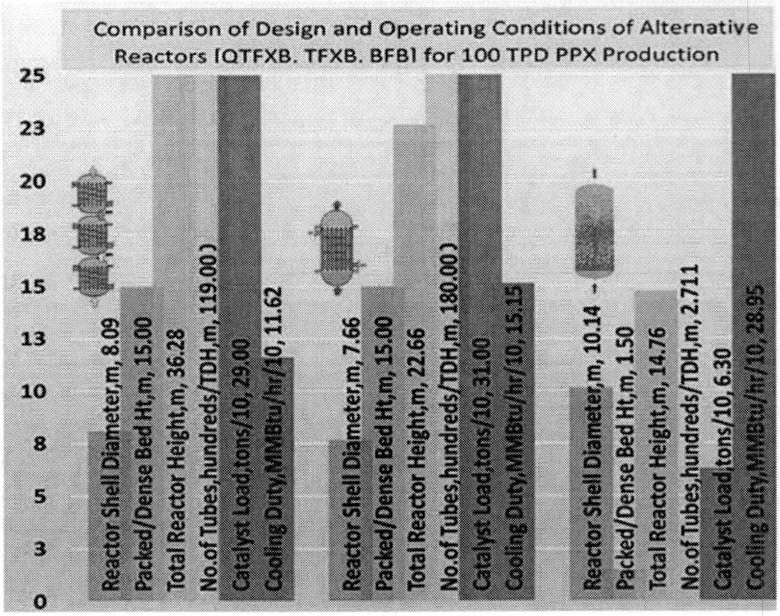


**Figure 10: Reactor design and operating conditions for 3 different solid-gas reactor types**

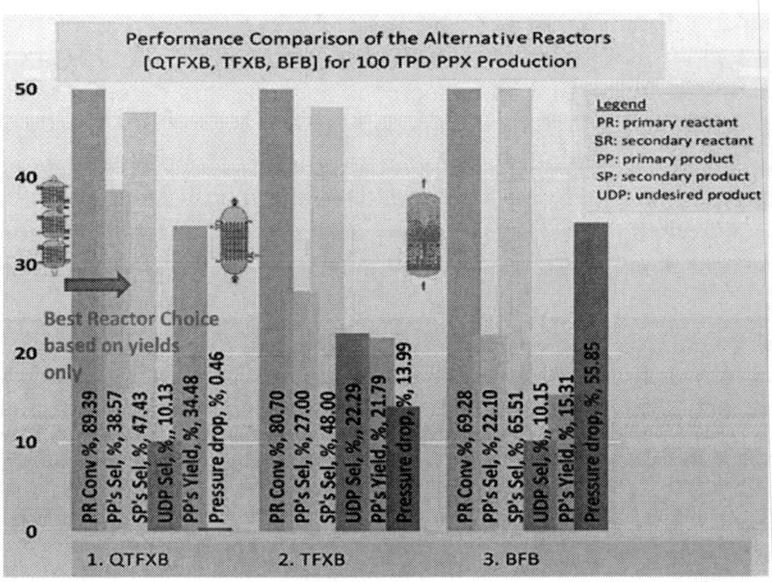


**Figure 11: Performance Comparison for 3 solid-gas reactors**

The two most important particulate system parameters used in the design of a CFB are reactor voidage (**ε**) profile and solid loading or "flux" of the up-flow solid-gas stream. Voidage profile correlations proposed by Kwauk et al.[26] used in the CFB reactor design and model have been first tested for valid simulation of voidage profiles of common solids of commercial interest and found to be satisfactory, as shown in **Figure 12.** For the CFB, reactor pressure drop was calculated by using four major pressure drop components: bed hydrostatic pressure, solids acceleration, solids friction and gas friction, as proposed by Yang and Keairns[27]. More details of these treatments are available in Dutta et al[13]. The DIFREX CFB design and modeling tool has been applied to both catalytic and non-catalytic reactor designs that include ***acrylonitrile production from propylene, FCC catalyst regeneration***[13], ***high-grade crude extraction from oil shale, iron ore reduction*** **and** ***hot gas desulfurization by zinc titanate***[6]. Application of particulate technology and principles is integral to the design of these reactors.

Temperature, conversions and bed/ particulate hydrodynamics for a propylene-to-acrylonitrile CFB system is shown in **Figure 13**, where PR and LR denote primary reactant propylene ($C_3H_6$) and limiting reactant ammonia ($NH_3$). In this reactor design, the bed bottom gas-solid mixing zone has a smaller diameter than the top/main bed. Profiles of temperature, reactor voidage, local gas velocity and solid loading of the up-flowing gas are shown.

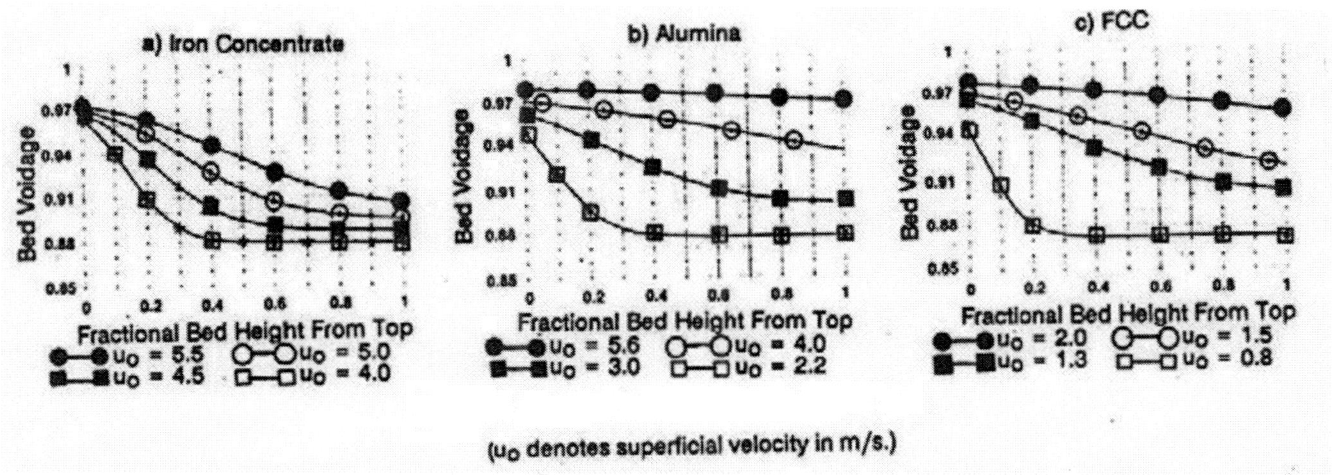


**Figure 12: Simulation of CFB model based on voidage profile correlations proposed by Kwauk et al.[26]**

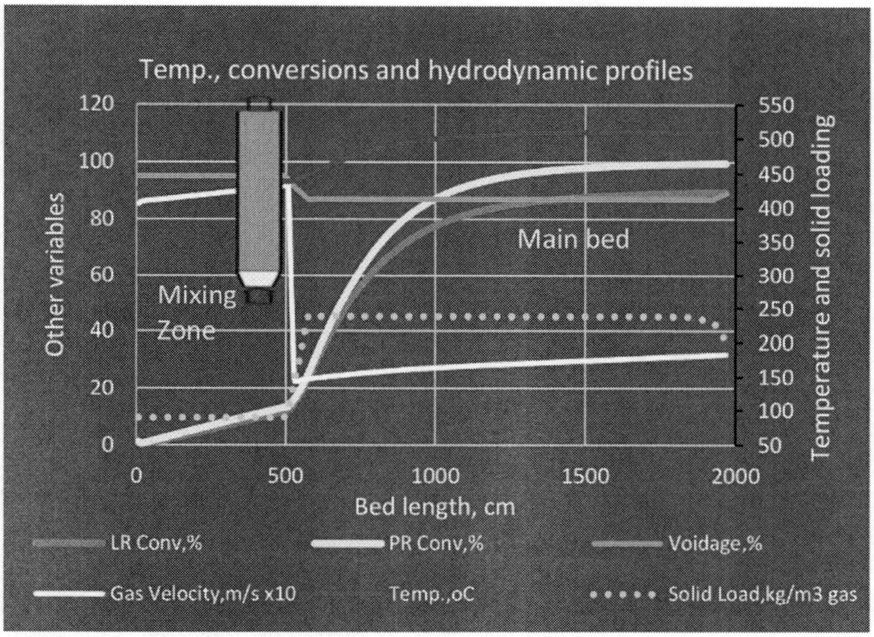


**Figure 13: Profiles (Temperature, conversion and hydrodynamics) for a CFB system**

## An Interesting Observation

An interesting observation while studying the effect of CFB hydrodynamics on a reaction system is worth mentioning. A *CFB* can be viewed as a suspended and moving fixed bed (*FXB*) of particulates (fine particles) but with a very high bed voidage and with significant solids slippage (falling backward in the annulus along the wall). However, due to the intimate best heat and mass transfer rates possible, the bed temperature (rise) profile (for an exothermic reaction) in an adiabatic CFB reactor does not show as steep a temperature rise or spike as can be problematic in an FXB. Also, the bed temperature rise can be controlled by both gas flow and solid circulation rate.

Due the various impacts, the upper portion of a CFB that has a higher average bed voidage (a CFB "suspended bed" with lower solid/catalyst concentration per unit volume) may achieve higher conversion than a CFB "suspended bed" with lower bed voidage. Lower solid loading in this more diluted suspended bed allows a higher temperature rise due to lower volumetric heat capacity because of its lower solid content. The effect of rising temperature in the CFB may then more than compensate for the effect of lower solid/catalyst concentration. This is illustrated in **Figure 14**. Higher temperature and higher conversions are observed for the case with higher voidage and lower solid loading (kg/m$^3$ gas) in spite of the low solid concentration. On the other hand, since the inlet temperature is the same for both cases and the average and upper zone temperatures are higher for the case with higher bed voidage (due to extra temperature rise), the catalyst deactivation is likely to be faster and the selectivity is also likely to be different (for several reasons).

It is important to understand that, unlike the dense bed of a BFB, a CFB is not isothermal even though it is more isothermal than an FXB.

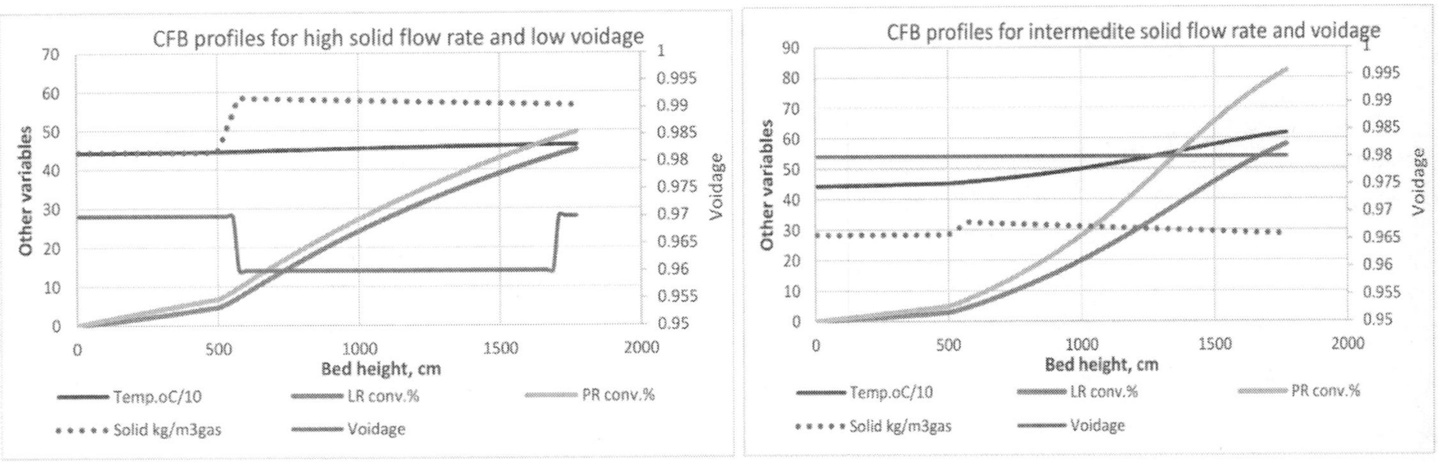


**Figure 14: Effect of solid flow rate and bed voidage in a CFB case study**

## *Hybrid Particulate Systems*

A prime application of particulate technology is in FCC reactors. There are a variety of configurations of these reactors. In many installations, the FCC reactor is a "riser" type, with high gas velocity carrying all the catalyst upward in a dilute phase, with minimal residence time, etc. For this riser kind of FCC reactor, it can be represented as a hybrid/ combination of three particulate-gas contact types: a bottom regenerated solid-feed gas mixing (CSTR or BFB) zone, a middle which is the main CFB (riser) section, and a gas-solid disengaging zone at the top (which may create a further bed that can be simulated as a BFB). This representation is shown in **Figure 15.**

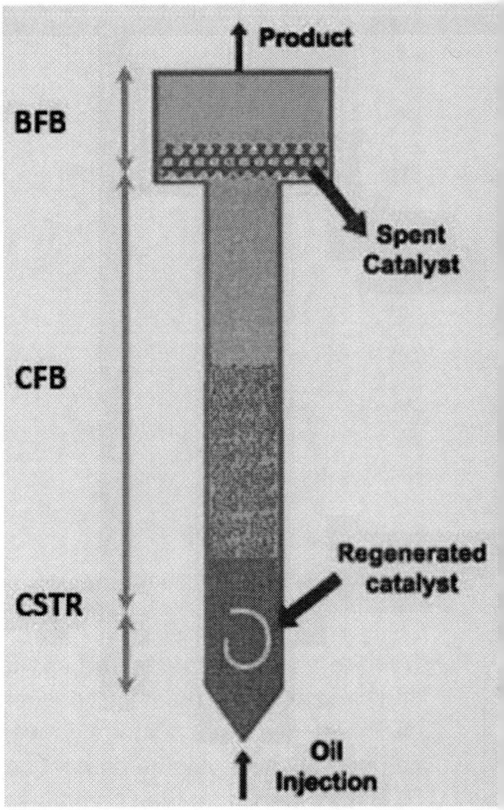


**Figure 15: 3-zone simulation of an FCC performance**

Using 1) a 5-lump reaction scheme with gasoil feed as shown in **Figure 16**, 2) surrogate compounds to represent the components involved, and 3) particulate system behavior in the three zones, a simple FCC reactor model was developed by Dutta and Gualy[8] that simulated the observed performance data reasonably well. The simulation results are shown in **Figure 17(a-d)**.

**Reaction system: FCC 5-Lump reactions with GASOIL feed**

| | | |
|---|---|---|
| GASOIL | GASOLINE | Rxn 1 |
| GASOIL | DRYGAS | Rxn 2 |
| GASOIL | LPG | Rxn 3 |
| GASOIL | COKE | Rxn 4 |
| GASOLINE | DRYGAS | Rxn 5 |
| GASOLINE | LPG | Rxn 6 |
| GASOLINE | COKE | Rxn 7 |
| LPG | DRYGAS | Rxn 8 |

**Figure 16: FCC 5-lump reactions with GASOIL feed**

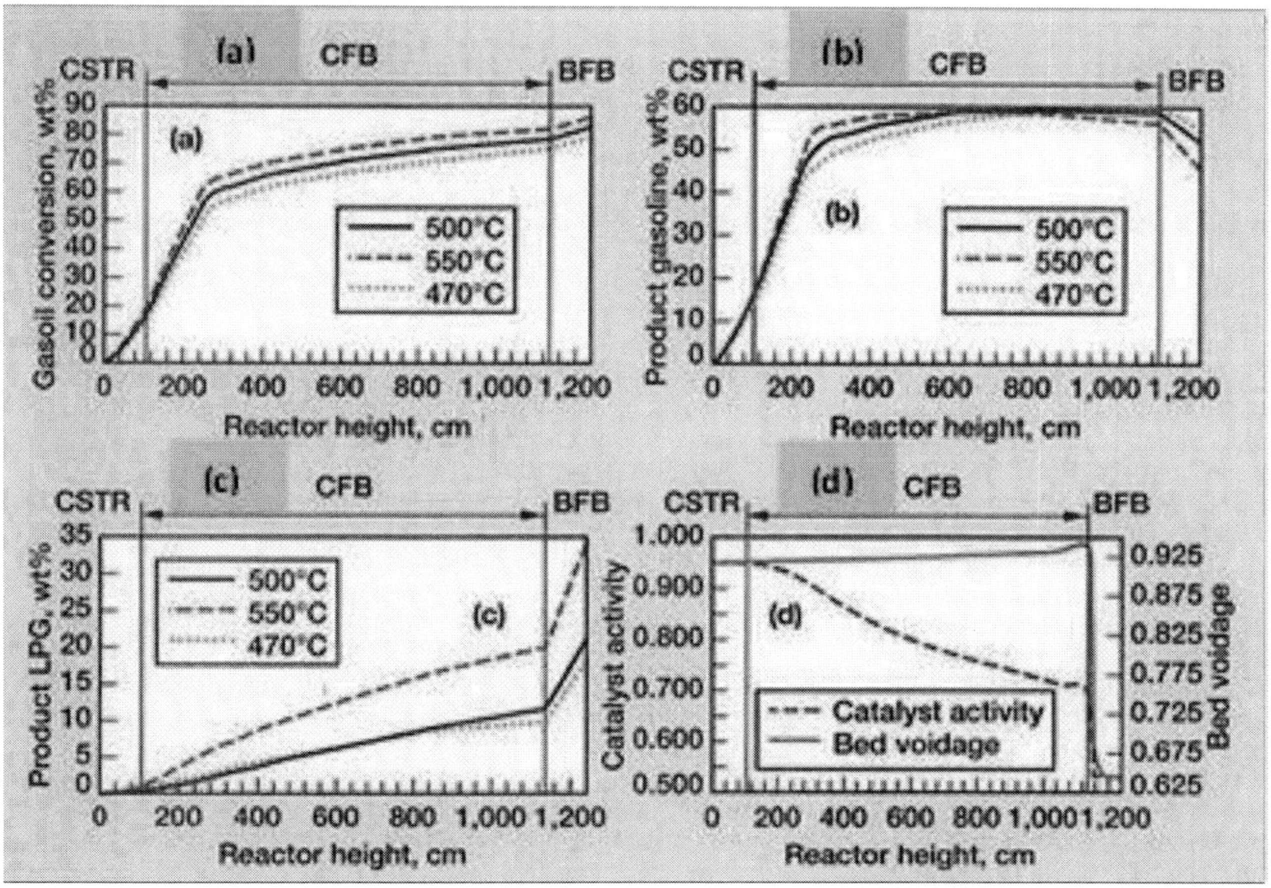


**Figure 17: FCC simulation using a 5-lump reaction scheme**

## Emerging Technologies and Particulate Systems

Is ***Methanol*** the fuel of the future? There is a worldwide push to capture $CO_2$ from atmosphere and power plant exhausts and convert this and $H_2O$ into methanol. One of the emerging technologies is ***PTM*** [(solar) Power to Methanol] that may lead to zero carbon footprint – as solar energy seems to be becoming increasingly more competitive and economical. The particulate technology can play a vital role in the PTM process, where the $CO_2$ could be captured and concentrated ***economically*** by using a regenerable particulate sorbent in a fluidized bed process. The other is the conventional route of ***syngas-to-methanol*** through syngas production from coal or biomass – the ***renewable energy route***. In both of these technologies, the particulate technology plays a vital role. Syngas production by gasification of coal or biomass in a fluidized bed reactor is one of the most attractive means for the second route. Due to a rather unique and perhaps the most diverse expertise and experience in both fluidization and reaction engineering, and also in coal combustion and gasification technology[28-31], DIFREX can play an instrumental role in implementing a rapid and successful move to ***the fuel of the future***. An integrated CTSM/ BTSM (Coal-to-Syngas-Methanol/ Biomass-to-Syngas-Methanol) plant can be designed and built promptly with help from DIFREX. This will be further facilitated by DIFREX already having an advanced and proprietary syngas to methanol (S2M) reactor design available.

## References


1. Dutta, S. and R. Gualy, "Build Robust Reactor Models", Chemical Engg. Progress, 96(10), October 2000, p. 37.
2. Dutta, S. and R. Gualy, "Super-modeling Reactors for Profitability", Chem. Engg., 107(6), 2000, p. 72.
3. Dutta, S. and R. Gualy, "Modernize Process Reactors", Hydrocarbon Processing, November, 1999, p.91.
4. Dutta, S. and R. Gualy, "Overhaul Process Reactors", Hydrocarbon Processing, September 1999, p. 43.
5. Dutta, S. and R. Gualy, "General Reactor Model Improves HPI Applications", Hydrocarbon Processing, July 1999, p. 45.
6. Dutta, S., "Circulating Versus Bubbling Fluidized Bed for Hot Gas Desulfurization by Zinc Titanate", Proc. 5th Int'l. Circulating Fluidized Bed Conf., Beijing, PRC, May, 1996, p. pr-7.
7. Dutta, S., "Reactor Model: Usefulness of a Lesser Known Tool in Process Plants", paper presented at SABIC Tech. Meeting, May, 1996.
8. Dutta, S., "Fluidized and Moving Bed Desulfurization by Zinc Titanate", AIChE Symp. Ser., 301(90), 1994, p. 157.
9. Dutta, S. and B. Jazayeri, "Alternate Reactor Concepts for Oxidative Coupling of Methane", Proc. 7th Int'l. Fluidization Conf., ed. Potter and Nicklin, Brisbane, Australia, May, 1992, p. 445.
10. Dutta, S. and G. D. Suciu., "An Experimental Investigation on the Effectiveness of Baffles in Breaking Bubbles in Fluid Beds", J. Chem. Engg. Japan, 25(3), 1992, p. 345.
11. Dutta, S., "A Novel CFB Application: A Chemical Compressor", Proc. 3rd Int'l. Circulating Fluidized Bed Conf., ed. Basu, Horio and Hasatani, Nagoya, Japan, Oct., 1990, p. 465.
12. Dutta, S., "A Novel Circulating Bed Metal Hydride Compressor", paper presented at the 8th World Hydrogen Conf., Honolulu, Hawaii, 1989.
13.Dutta, S., B. R. Christian and M. F. Ratterman, "Circulating Fluid Bed Reactor Model Developed for FCC/ART Catalyst Regenerator", Proc. 6th Int'l. Fluidization Conf., ed. Grace, Shemilt and Bergougnou, Banff, Canada, May, 1989, p. 9.
14. Dutta, S. and G. D. Suciu, "Unified Model Applied to the Scale-up of Catalytic Fluid Bed Reactors of Commercial Importance", Proc. 6th Int'l. Fluidization Conf., ed. Grace, Shemilt and Bergougnou, Banff,Canada,May,1989, p.311.
15. Dutta, S., "A Unified Approach to the Modeling of Catalytic Fluidized Bed Reactors", Proc. World Cong. III Chem. Reaction Engg., Tokyo, 1986, p. 287.
16. Dutta, S., S. C. Arnold, G. D. Suciu and L. Verde, "Scale-Up of a Commercial Catalytic Fluid Bed Reactor Involving Complex Kinetics", Inst. Chem. Engg. Symp. Ser. 87, 1984, p. 517.
17. Dutta, S., "Measurement of Bubble Characteristics by Optical Probe", paper presented at the AIChE Annual Meeting, Chicago, Nov., 1980.
18. Dutta, S., "A Simple Probe for Fluidized Bed Measurements", Cana. J. Chem. Engg., 57, 1979, p. 115.
19. Dutta, S., "Dead-zone Heights near the Grids of Fluidized Beds", Fluidization, Cambridge University Press, 1978, p. 32.
20. Wen, C. Y. and S. Dutta, "Research Needs for Analysis, Design and Scale-up of Fluidized Bed Reactors", AIChE Symp. Ser., 73(161), 1977, p.1.
21. Fluidization Engineering, 2nd edition, by Daizo Kunii and Octave Levenspiel, Butterworth-Heinemann, 1991.
22. Chemical Reaction Engineering, 3rd edition, by Octave Levenspiel, John Wiley & Sons, 1999.
23. Werther, J., Chem. Engg. Sci., 35, 1980, p. 372.
24. Kato, K., and C. Y. Wen, Chem. Engg. Sci., 24, 1969, p. 1351.
25. Hatano, H., T. Okuma and M. Ishida, J. Chem. Engg. Japan, 17, 1984, p. 491.
26. Kwauk, M. N., N. Wang, Y. Li, B. Chen and Z. Shen, Circulating Fluidized Bed Technology, ed. P. Basu, Pergamon Press, 1986, p. 33.
27. Yang, W. C. and D. L. Keairns, AIChE Symp. Ser., 141(70), 1974, p.27.
28 Wen, C.Y. and S. Dutta, Book Chapter "Kinetics of Coal Pyrolysis and Gasification Reactions", Ch.2 Coal Conversion Technology, Addison-Wesley Publishing Company, 1979.
29. Wen, C.Y. and S. Dutta, "Solid-Gas Reactions in Coal Conversion Processes", Coal Processing Technology, CEP Tech. Manual, IV, 1978, p. 40.
30. Wen, C.Y. and S. Dutta, "Reactivity of Coal and Char: Part I – In Carbon Dioxide Atmosphere", I&EC Process Design & Development, 16, 1977, p.20.
31. Wen, C.Y. and S. Dutta, "Reactivity of Coal and Char: Part II – In Oxygen-Nitrogen Atmosphere", I&EC Process Design & Development, 16, 1977, p.31.
32. www.difrex.com
33. http://difrex.com/pdf/BrochureContents.pdf.
34 Dutta, Subhash, Stephen Arnold, Aashish Gaurav and Jim Brenner, "DIFREX Nexgen Expands and Advances Reactor & Technology Solutions", paper presented at AIChE Spring Meeting, Orlando, April, 2018.

# 508152 Erosion and Mobilisation of Highly ACTIVE Simulant Suspensions with Impinging Vertical Jets

**Abstract:**

Based on the nuclear industry in the U.K., storage of nuclear waste poses a particularly complex risk. Part of the waste during storage before being disposed of begins to form colloidal particles. These particles begin to settle out, and can compromise the integrity of their storage container. Using an Impinging vertical liquid jet is the means to prevent this. However the operational effectiveness of these jets is unknown due to the nature of the containers. Reduced scale modelling is required to assess their efficacy, and understand the dimensionless results as they allow systems of scale to be compared. This is of particular interest due to the wide reaching applications to all industries that may require utilisation of impinging jets, or require mass transfer, or even heat transfer.

An experimental rig of 0.5 $m^3$ has been used to model the central jet of the storage tank system. A Honite simulant ($SiO_2$, glass spherical particles) of varying size distributions has been used for the initial non-cohesive particle testing, and modelling. Each experiment uses 10 different jet diameters. Data is collected with an Ultrasonic Velocity Profiler, which scans the simulant bed in real time under impingement utilising an automated traverse. Acoustic backscatter data was recorded to yield a dynamic scour depth, and dynamic crater radius to allow dimensionless analysis of the erosion for each size distribution. Dimensionless analysis takes the form of the Erosion Parameter, $E_c$, which allows comparison for systems of scale and a means of correlating multiple simulants against one another, so patterns are easily identifiable. Part of the dimensionless analysis was also to deduce the Shieldâ€™s Parameter from vane viscometry to allow for a comparison of erosion behaviours taking into account particle properties, flow properties and the overall influence these have on the erosion profile.

# 505818 REAL WORLD Examples of Influences of Particles and Cake Formation on SOLID-Liquid Separation Technology Operation

**Abstract:**

The paper begins with a general discussion of the bench-top laboratory tests that are conducted for solid-liquid separation technology problem analysis, process development, technology selection and scale-up. The tests include pressure, filter media, filter aids, and similar process parameters.

Solid-liquid separation technologies are explained for continuous pressure filtration and batch pressure filtration. This provides an overview for the testing and troubleshooting.

Three different case histories are presented illustrating different scenarios of the impact of particles, particle size distribution (PSD) and cake formation on the separation technology operations. The first is a specialty chemical application for a continuous rotary pressure filter where the particles changed from lab scale to production scale. The second case is a clarification application where the system was running successfully for over one-year but inexplicitly, the performance changed drastically and the filter media began plugging very quickly during the cycle. The final application looks at lab testing and pilot scale testing where the cake formation changed resulting is a unique operating technique.

Sherlock Holmes and Dr. John Watson are fictional characters of Sir Arthur Conan Doyle. Process engineers who live in the real-world can learn many things from the two of them for solving process filtration problems. One example that Holmes proves time and again is that there is no benefit to â€œjumping to conclusions.â€ As discussed in the illustrations, there are many choices of filtration technologies to achieve the necessary level of quality with a cost-effective and a reliable process. Also, in terms of troubleshooting, there are many causes of filtration symptoms. With careful analysis and taking a creative and "non-conventional approach" by looking at what is behind the process data and the operating parameters, the optimum solution will be realized.

# 508019 Scale up Practices for Today's Breakthrough Technologies

**Abstract:**

Fluidized bed reactors have distinct advantages over other reactor unit operations such as superior heat transfer and solids mobility. While the superior heat transfer has sparked breakthrough technologies in processes such as gasification, acrylonitrile and oxychlorination, it is the solids mobility benefits that are being exploited today. This mobility allows solids or sets of solids to move from reduction to oxidation zones in a continuous fashion thereby reducing operational complexity and capital costs. As a result, processes once considered impossible are being implemented where particulate feeds stocks and media are combined and separated in fluidized bed unit operations to provide superior yields unattainable on a commercial-scale with other reactor designs.

Encina Chemicals has capitalized on this solids mobility benefit for a coal upgrading process. Along with coal upgrading, the process provides the continuous regeneration of heat transfer media together with the separation of ash byproduct. The effect of complex economic drivers and resulting design criteria on required fluidization hydrodynamics are managed by combining cold-flow hydrodynamic testing with CFD modeling insights. This enables rapid hypothesis testing and scale-up practices with faster time to commercialization of an optimum design. This paper discusses the approach used to combine cold-flow hydrodynamics research with CFD simulation for rapid process commercialization considering relevant economic and technological drivers.

# 509649 Study of FCC Riser Hydrodynamics Using CFD Tools

**Abstract:**

Past experiences of commercial FCC units indicate that hydrodynamics have a pronounced effect on the riser reactor performance. A study of FCC riser hydrodynamics was undertaken to understand the phenomena observed in the field. It encompassed computational fluid dynamics (CFD), cold flow modeling, radioactive tracer studies, kinetic modeling and yield estimation. The ultimate goal of the study was to isolate the effects of riser diameter, lift steam velocity, and solids flux on the hydrodynamic behavior and yield performance of FCC risers.

Results of the cold flow studies were used to select and validate the best CFD modeling approach. Predictions of several models developed within the Eulerian-Eulerian and Eulerian-Lagrangian frameworks were compared with the experimental solids flux profiles in the cold riser operating at commercially-relevant velocities. Both model classes showed good agreement with experiment in the "normal" riser operation. The Eulerian-Lagrangian approach showed better predictions of experiment when the riser was "pushed" in terms of velocity and solids flux.

The general trends of riser hydrodynamics, per CFD predictions, were identified as follows. Lower vapor velocity, larger riser diameter and higher solids flux through the riser all contribute to hydrodynamic non-idealities, such as catalyst downflow at the wall and vapor channeling through the core of the riser. In practice, these "non-idealities" lead to undesirable shifts in conversion (under-conversion in the core) and selectivity (overcracking in the annulus) of the riser.

# 508584 Investigation of Particle Sheeting of Monodisperse Particles with an Euler-Euler Two-Fluid Model Incorporating Electrostatic Charging Effects

**Abstract:**

Electrostatic charging of particles is a disruptive phenomenon commonly encountered in gas-solid fluidization. In commercial applications such as catalytic polymerization of polyethylene, such events lead to particle aggregation as well as "sheeting" on the reactor wall. The sheets lead to halting the production process for reactor clean up and in turn results in large revenue losses.

Computational Fluid Dynamic (CFD) modeling has received interest over the past few years to predict the occurrence and magnitude of charging in gas-solid fluidized beds. Rokkam et al. (2013) had carried out CFD simulations using fixed particle charge densities to demonstrate the effect of the electrostatic forces in influencing the hydrodynamics of fluidization and wall sheeting. The model used three types of particles, namely wall, dropped and fines. The size and charge density of each of these types was based on the average values of these variables as observed in the fluidization experiments carried out by Sowinski (2012). It was shown that at the end of fluidization, a small layer of highly charged particles sticks to the wall in a similar way to that observed in experiments. The goal of the present work is the extension of Rokkam's work to include a model for charge generation and transport. To this purpose, two phenomena need to be described by the model: charging due to particle-wall collisions, and charging due to particle-particle collisions. The model for particle-wall charging hypothesized by Matsusaka et al. (2000) and Matsusaka and Masuda (2003) was used to obtain a boundary condition describing particle charging due to contact with the wall. The methodology based on the kinetic theory of the granular flow described by Johnson and Jackson (1987) was used to derive this boundary condition.

The charging model of Matsusaka et al. (2000) was used to describe charging due to particle-particle collisions by determining the rate of change of charge due to collisions through the solution of the appropriate collision integral, as defined in accordance to the kinetic theory for monodisperse granular spheres (Jenkins and Savage, 1983). Consequently, expressions for the charge transport coefficient due to particle-particle collisions and charge flux due to particle-wall collisions are found. The model was implemented into an Euler-Euler two-fluid model with kinetic theory closures (Jenkins and Savage, 1983) to describe the properties of the particulate flow. The resulting model was applied to the simulation of gas-solid fluidized beds.

The domain size and initial conditions used to perform the numerical simulations reproduce the experimental conditions of Sowinski (2012). The fluidized bed was 0.1 m in diameter and the particles had a diameter of 300 μm. The fluidization velocity was varied between 1.5 and 4 times the minimum fluidization velocity, encompassing both bubbling and slug flow regimes. For the electrostatic properties, work function values of 4.4 eV and 5.6 eV (Masuda et al., 2006) were used for the reactor wall (carbon steel) and polyethylene particles respectively. A saturation charge density of 117 μC/kg, based on experimental observations was used. The bed was fluidized until a steady state of electrification was observed. The gas flow was then interrupted, and particles were allowed to settle at the bottom of the bed. Results of the simulations showed that the profile of charge density had its highest value near the wall, and decreased towards the center of the fluidized bed. A force balance was used to determine regions where particles would stick to the wall if the distributor plate was removed and the particles be allowed to fall. The radial component of the electrostatic force predicted by the model was significantly larger than its axial counterpart, and its magnitude also declined rapidly with distance from the wall. It was determined that the particles within a 2 mm thickness layer on the wall would not drop, in agreement with experimental observation in Mehrani's group. The average charge density within this layer was 47 μC/kg while it was 70 μC/kg for a thickness layer of 0.5 mm on the wall. The results are in agreement with the experimental observations obtained by Sowinski (2012).

**References**


Jenkins, J.T., Savage, S.B., 1983. A theory for the rapid flow of identical, smooth, nearly elastic, spherical particles. Journal of Fluid Mechanics 130, 187. doi:10.1017/S0022112083001044

Johnson, P.C., Jackson, R., 1987. Frictional–collisional constitutive relations for granular materials, with application to plane shearing. Journal of Fluid Mechanics 176, 67. doi:10.1017/S0022112087000570

Masuda, H., Higashitani, K., Yoshida, H. (Eds.), 2006. Powder technology handbook, 3rd ed. Taylor & Francis

Matsusaka, S., Ghadiri, M., Masuda, H., 2000. Electrification of an elastic sphere by repeated impacts on a metal plate. Journal of Physics D: Applied Physics 33, 2311–2319

Matsusaka, S., Masuda, H., 2003. Electrostatics of particles. Advanced Powder Technology 14, 143–166. doi:10.1163/156855203763593958

Rokkam, R.G., Sowinski, A., Fox, R.O., Mehrani, P., Muhle, M.E., 2013. Computational and experimental study of electrostatics in gas–solid polymerization fluidized beds. Chemical Engineering Science 92, 146–156. doi:10.1016/j.ces.2013.01.023

Sowinski, A., 2012. The study of reactor wall fouling in gas-solid fluidized beds caused by electrostatic charge generation (Thesis dissertation). University of Ottawa

# 506383 Large-Scale DEM-CFD Method for an Industrial Fluidized Bed

**Abstract:**

Numerical simulations of fluidized bed are often performed for the better understanding of complex phenomena. In these numerical simulations, Eulerian-Lagrangian approach [1] is often employed, where solid and fluid phases are modeled by discrete element method (DEM) and computational fluid dynamics (CFD), respectively. Indeed, the DEM-CFD method is applied to various solid-fluid coupling problems. Adequacy of the DEM-CFD method is proven not only in a gas-solid flow system such as a fluidized bed [2,3] but also a wet ball mill [4,5]. Although the DEM-CFD method seems to be established, there is a couple of critical problems such as substantial limit of number of calculated particles and difficulty for stable computation for a gas-solid flow in a complex-shaped boundary. In order to solve these problems, we develop a new methodology that is referred to as DEM-CGM-CDF-SDF-IBM method. In the DEM-CGM-CFD-SDF-IBM method, signed distance function (SDF) [6] and immersed boundary model [7] are used for boundary wall of solid and fluid phases. The coarse grain model (CGM) [8] of the DEM is employed as the scaling law model, where a model particle represents a group of original particles. To show adequacy of the DEM-CGM-CDF-SDF-IBM method, a validation test is performed in a fluidized bed. In the simulation, 200,000 spherical particles whose density, original particle diameter and coarse grain ratio are, respectively, 2500 kg/m$^3$, 0.1 mm and 5. Experiments are performed under fair conditions of the numerical simulations. It is shown that the DEM-CGM-CDF-SDF-IBM method can simulate the bed height and pressure drop obtained from the oroginal system accurately through the validation test. Consequently, we conclude that the DEM-CGM-CDF-SDF-IBM method can simulate industrial scale fluidized bed, and hope that our method becomes standard in numerical simulation method in industrial systems.

References

[1] Y. Tsuji, T. Kawaguchi, T. Tanaka, Powder Technol., 77 79â€"87 (1993)

[2] M. Sakai, H. Takahashi, C.C. Pain, J-P Latham, J. Xiang, Adv. Powder Technol., 23, 673-681 (2012)

[3] M. Sakai, M. Abe, Y. Shigeto, S. Mizutani, H. Takahashi, A. Vire, J.R. Percival, J. Xiang, C.C. Pain, Chem. Eng. J., 244, 33-43 (2014)

[4] X. Sun, M. Sakai, Y. Yamada, J. Comput. Phys., 248, 147-176 (2013)

[5] X. Sun, M. Sakai, M-T. Sakai, Y. Yamada, Chem. Eng. J., 246, 122-141 (2014)

[6] Y. Shigeto, M. Sakai, Chem. Eng. J., 231, 464-476 (2013)

[7] X. Sun, M. Sakai, Chem. Eng. Sci., 139, 221-240 (2016)

[8] M. Sakai and S. Koshizuka, Chem. Eng. Sci., 64, 533-539 (2009)

# 513870 Predicting Performance of Industrial FCC Regenerator Using Computational Fluid Dynamics


**Abstract:**

In the Fluid Catalytic Cracking (FCC) process, uniform regeneration of spent catalyst is key to the successful operation of the FCC units. The degree and uniformity of catalyst regeneration has a substantial impact on the product yields. The regenerator combustion efficiency strongly depends on good distribution of both air and spent catalyst, and air catalyst mixing and flow patterns within the regenerator are greatly influenced by both process conditions as well as regenerator hardware configuration. Therefore, understanding the regenerator hydrodynamics and kinetics, and the impact of hardware modifications on regenerator performance is essential for efficient regenerator design, optimization and troubleshooting.

In this work, a CFD based kinetic tool has been developed to simultaneously model both hydrodynamics and coke combustion characteristics in the FCC regenerator. The tool enables us to evaluate the influence of the fluidized bed hydrodynamic on the catalyst regeneration in an industrial FCC regenerator, and serves as a useful tool for future technology development. TechnipFMC employed Barracuda CFD software and coupled the code with proprietary coke combustion kinetic equations and rate constants, developed experimentally at IFP Energyies nouvelles (IFPEN). TechnipFMC and IFPEN have tested and compared the model against several commercial regenerator designs. This study also provides an example of how the CFD based kinetic tool is used in the modeling of an industrial FCC regenerator. The results provide detailed qualitative mapping of the regenerator behavior in terms of flue gas composition, temperature distribution including afterburning, and regenerator bed density in both axial and radial planes. The results proved to be instrumental to recommend hardware and operational modifications.

# 507532 The Effect of Drying on the Metal Distribution ofÂ Supported Catalysts with High Metal Loadings


**Abstract:**

The Effect of Drying on the Metal Distribution of Supported Catalysts with High Metal Loadings



Cody Bishop, Anusha V. Noorithaya, Prateek Sarkar, Johannes G. Khinast and Benjamin J. Glasser

Department of Chemical and Biochemical Engineering,

Rutgers University, Piscataway, NJ, 08854



Supported catalysts are essential components in a variety of industrial processes, ranging from catalytic converters to production of new drugs. The demand for these supports is due to their high surface area, high mechanical and thermal stabilities. The performance of a catalytic process is intimately related to the catalyst design which results in various metal profiles - uniform, egg-yolk, egg-shell and egg-white profiles. Although catalyst preparation and catalytic processing have been investigated for many years, many aspects of catalyst manufacturing are still not fully understood. In the industry, catalyst design is usually based on trial and error, which is expensive, time-consuming, and does not offer assurances on the final results.

It is generally believed that the metal profile is controlled by the conditions that are applied during impregnation, however, experiments have shown that drying can also significantly impact the metal distribution within the support. Therefore, to achieve a desired metal profile we need to understand both impregnation and drying. Controlling the impregnation and drying conditions can enhance catalyst performance, and minimize the production of useless batches that have to be disposed, or recycled.

In this work we have investigated the metal distribution during the high concentration impregnation and subsequent drying of supported catalysts. Previous work found that high concentration impregnation (3-4 molar) and drying of Ni/Alumina catalysts yield a uniform metal distribution. At intermediate concentrations (~1 molar), the Ni\Alumina catalysts yielded egg-shell profiles. Theoretical models to simulate the impregnation and drying processes of Ni/Alumina had been previously developed. The high concentration drying model was developed for the higher concentrations of nickel nitrate (above 0.1 M). This included the effects of the metal concentration on the solution density, viscosity, surface tension, vapor pressure and the volume ratio of metal. Good agreement was found between experimental and simulation post-drying metal distributions for this model using nickel nitrate. To make the above model widely applicable to many more metals, we designed a simplified model, the reduced parameter model. A detailed analysis on all the solution properties and adsorption parameters was done using experimentally determined properties of nickel nitrate as a reference. The reduced parameter model was tested using properties and parameters of copper nitrate.

Our experiments have allowed us to better understand the fundamental mechanisms that occur during drying, and to develop a strategy that can generate desired metal profiles. Although the results presented are based on a particular metal/support system, they serve to provide physical insight into the fundamentals of the drying processes.

# 507809 Online Pellet Drying for Inline Polymer Analysis

**Abstract:**

Some polymers extrude with high levels of moisture-holding cracks. This phenomena, often called melt fracture, makes drying of these particles particularly difficult. At-line or in-line analyses of polymer pellets using film extrusion, for example, require moisture-free streams of pellets on a continuously feeding basis. The low feed rate of most sampling systems, however, means that industrial scale dryers are not appropriate. Similarly, such a dryer should be able to give close to a first-in/first-out or plug flow residence time characteristic with as little backmixing as possible. A prototype-sampling device has been developed based upon first principles. The novel drying device utilizes a cyclone separator combined with a fluidized bed to rapidly dry the pellets prior to cast film analysis. Initial results from the novel drying system applied to polyethylene applications suggest that it may be superior to pellet drying using currently practiced technology. Details of the design, testing and evaluation of the dryer will be presented.

# 508434 Short-Time Hydrothermal Sterilization of Thermally-Labile Powders

**Abstract:**

Powdered solid is one of the most important raw materials that attributes for tailoring the functionality and prolonging shelf-life for the food and pharmaceutical industries. With the increasing consumer demand in incorporating added-value functional ingredients to the final dried products, researches have been focusing on the development of effective processes that preserve these heat- and moist-sensitive materials while keeping microbial load within the legal-acceptable range [1, 2]. Suggested technologies cover both thermal and non-thermal processes of bulk powders, however, a technically feasible industrial solution for continuous minimal processing is yet challenging.

Among different food ingredients, starch is the fundamental natural polymer that has a broad application only after modification enzymatically, chemically through cross-linking and/or substitution or by physical means such as high-pressure treatment, annealing (ANN) and heat-moisture treatment (HMT) [3, 4, 5]. In particular, the two hydrothermal treatments, ANN and HMT, have received significant interest for alternating the starch structure and properties but preserving the granular structure that improves heat-, shear-stability and digestibility of starches for further processing. Currently, most heat-moisture treatments are based on lab-scale production which are time- and energy-intensive with treatment time ranges from 15 min to 1 day at temperature range from 50°C to 120°C [6].

Therefore, this investigation focuses on the short-time hydrothermal sterilization in the novel Falling Particle Cloud tubular reactor prototype that was developed in the laboratory. Our previous study validated the decontamination process for model particles based on the promising Vacuum-Steam-Vacuum (VSV) concept which targets efficient heat transfer to the surface of food powders [7]. The application through the reactor prototype is further expanded to hydrothermal treatment of starch and its thermally-labile derivatives. Within 1 s steam treatment time, surface microbial decontamination and hydrothermal modification of the starches can be achieved in the reactor while maintaining powder quality. This research aims to give better understanding on the dynamic process conditions during the short-time hydrothermal sterilization and its effects on granule morphologies and gelatinization behaviors of starch, as well as on the microbial inactivation for powder products. A detailed processing mechanism and device description will be presented together with results from decontamination experiments.

1. Ohlsson, T. (2000) Minimal Processing of Foods with Thermal Methods, Innovations in Food Processing. CRC Press LLC.

2. Morris, C. Brody, A. L., Wicker, L. (2007) Non-Thermal Food Processing/Preservation Technologies: A Review with Packaging Implications. Packaging Technology and Science, 20, 275 – 286.

3. Jayakody, L., Hoover, R. (2008) Effect of annealing on the molecular structure and physicochemical properties of starches from different botanical origins – A review. Carbohydrate Polymers, 74, 691 – 703.

4. BeMiller, J. N. (1997) Starch modification: Challenges and prospects. Starch-Stärke, 49, 127 – 131.

5. Hoover, R. (2010) The Impact of Heat-Moisture Treatment on Molecular Structures and Properties of Starches Isolated from Different Botanical Sources. Critical Reviews in Food Science and Nutrition, 50, 835 – 847.

6. Chung, H.-J., Liu, Q., Hoover, R. (2009) Impact of annealing and heat-moisture treatment on rapidly digestible, slowly digestible and resistant starch levels in native and gelatinized corn, pea and lentil starches. Carbohydrate Polymers, 75(3), 436 - 447.

7. Hörmansperger, J., Buchmann, L., Merz, S., Schmitt, R., Beyrer, M., Windhab, E. (2016) Microbial decontamination of porous model food powders by Vacuum-Steam-Vacuum treatment. Innovative Food Science & Emerging Technologies, 34, 367 - 375.

# AUTHOR INDEX

# AUTHOR INDEX

# AUTHOR INDEX

# AUTHOR INDEX

# AUTHOR INDEX

**AIChE**
120 Wall Street, FL 23
New York, NY 10005-4020

ISBN 978-1-5108-6969-1